D0590276

Patrick Moore's
HISTORY OF
ASTRONOMY

Patrick Moore's
HISTORY OF
ASTRONOMY

MACDONALD & CO
LONDON & SYDNEY

A MACDONALD BOOK

First published in 1961 by
Oldbourne Book Company Limited
(under the title *Astronomy*)

Second revised edition, 1964

Third revised edition, 1967

Fourth revised edition, 1972
(under the title *The Story of Astronomy*)

Second impression, 1973

Third impression, 1974

Fifth revised edition, 1977

Sixth revised edition published in Great Britain in 1983
(under the title *Patrick Moore's History of Astronomy*)
by Macdonald & Co (Publishers) Ltd
London & Sydney

A member of BPCC plc

ISBN 0 356 08607 0

Printed and bound in Belgium
by Henri Proost.

Macdonald & Co (Publishers) Ltd
Maxwell House
74 Worship Street
London EC2A 2EN

FRONTISPIECE

Colour photograph of the Great Spiral in Andromeda;
102-cm Schmidt telescope, Palomar.

(California Institute of Technology, Pasadena, Cal.)

Contents

Foreword

ASTRONOMY IS THE OLDEST science in the world. Today, it is also one of the sciences which is undergoing the most rapid development. Our ideas today are very different from those of 1900, 1950 or even 1970. Each year brings its quota of new discoveries and new surprises.

What I have tried to do in this book, is to tell the story of astronomy, beginning in the remote past and taking it through to March 1983. Inevitably, much has been left out and much more has been glossed over; but I hope that what I have written may be of interest to those who read it.

My thanks are due to Jock Curle for all his help.

PATRICK MOORE
Selsey, March 1983

1

Introduction: The Sky Above Us

THE ANDROMEDA SPIRAL. *One of the nearest of the external galaxies, photographed with the 508-cm Hale reflector at Palomar Observatory. The two satellite galaxies are clearly shown.*

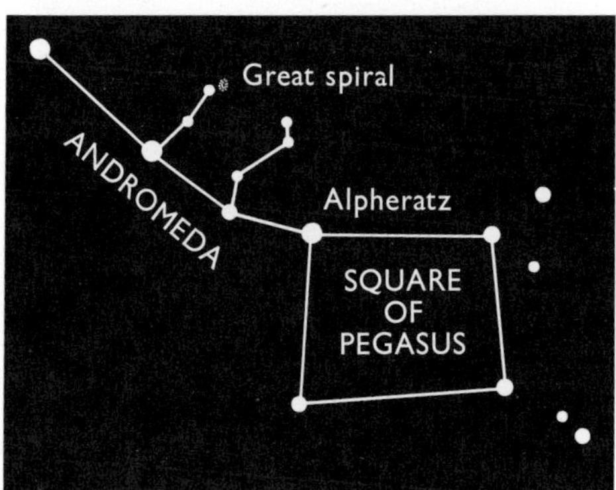

POSITION OF THE ANDROMEDA SPIRAL. *The Spiral is just visible to the unaided eye on a clear night.*

GROUP OF GALAXIES IN LEO, *photographed with the 508-cm Hale reflector at Palomar Observatory.*

FAR away in space, so remote that it looks like nothing more than a misty patch of light, lies an object which astronomers know as the Great Nebula in Andromeda. It is just visible with the naked eye when the sky is really dark; binoculars show it clearly, and when it is photographed with large telescopes the vague blur becomes a magnificent spiral – a whole system of stars. It has been known for centuries, but until less than seventy years ago we had no real idea of how big or how distant it really was.

Astronomy is a fast-moving science. It has developed more quickly during the past few decades than at any period in the history of mankind. We know more about the Sun, the Moon, the stars and planets today than would have seemed possible at the end of the last war, and a book written as recently as – say – 1965 is hopelessly out of date now. Of course, the basic facts have been known for a long time; it had been realized that the stars are huge suns, many of them far larger and hotter than our own Sun, and that the planets are members of our own particular part of the universe which we call the Solar System. But new facts and new discoveries are coming to light every year, and there is no suggestion that the rate of progress will slow down in the foreseeable future. We live in exciting times.

Tracing the story of astronomy is not so straightforward as might be thought, because there are so many different lines of investigation. We have to deal not only with observations made with telescopes, but also with what may be termed 'invisible astronomy', and nowadays with space research also. Astronomy is no longer a subject isolated from other sciences. But before going any further it may be as well to clear the air, so to speak, and outline some of the fundamentals.

The Earth is a planet, moving round the Sun at a mean distance of 150,000,000 kilometres in a period of 365¼ days. It is not the only planet; there are eight others, two of them (Mercury and Venus) closer-in to the Sun than we are, and the rest (Jupiter, Saturn, Uranus, Pluto and Neptune) further out. They have no light of their own; they shine only by reflecting the rays of the Sun, in the manner of large though rather inefficient mirrors. The same is true of the Moon, which is our companion in space, and stays with us as we travel round the Sun. At its mean distance of a mere 384,400 kilometres, the Moon is much the closest natural body in the sky; cosmically speaking it is on our doorstep, and it is not nearly so important as it looks. Other planets have moons or satellites of their own. Jupiter, the giant of the Solar System, has as many as sixteen though admittedly all but four are very small, while Saturn has over twenty.

The Sun itself is a star, made up of intensely hot gas. Even at its surface the temperature is of the order of 6000 degrees Centigrade, and near its core the temperature rises to the unbelievable value of at least 14,000,000 degrees. It is not burning in the conventional sense; for one thing, it is too hot! Its energy is being created by nuclear

PLANETS	DIAMETER IN KM	MEAN DISTANCE FROM THE SUN IN MILLIONS OF KM
MERCURY	4,880	58
VENUS	12,103	108
EARTH	12,756	150
MARS	6,787	228
JUPITER	142,800 (Equatorial)	778
SATURN	120,000 (Equatorial)	1,427
URANUS	51,800	2,870
NEPTUNE	49,500	4,497
PLUTO	3,000	5,900

SEGMENT OF THE SUN
1,390,470 KM IN DIAMETER

SCALE OF THE SOLAR SYSTEM. *The distances of the various planets from the Sun are shown on this scale drawing.*

COMPARATIVE SIZES OF THE SUN AND PLANETS. *(Opposite) The sizes of the Sun and planets, drawn to the same scale. Drawing by D. A. Hardy.*

reactions, and as it grows older it loses mass. The rate of loss is 4,000,000 tonnes every second, so that it 'weighs' much less now than it did when you picked up this book. I hasten to add that there is no immediate cause for alarm. The Sun is at least 5000 million years old, and it will go on shining for at least another 5000 million years yet before anything drastic happens to it. On the other hand, it will not last for ever. Nothing in the universe is eternal – perhaps not even the universe itself.

The stars appear much less splendid than the Sun because they are so much further away. Their distances are so great that to measure them in kilometres would be hopelessly clumsy, just as it would be awkward to give the distance between London and New York in millimetres. A better unit is the light-year, which is the distance covered by a ray of light in one year. Since light flashes along at 300,000 kilometres per second, a light-year is equivalent to 9,460,700,000,000 km. It takes only eight and a half minutes for light to reach us from the Sun, but even the nearest star is more than four light-years away.

It follows from this that when we look into what we may call deep space, we are also looking back in time. We see the Sun as it used to be eight and a half minutes ago; the nearest star as it used to be over four years ago, and so on. Modern instruments can show us objects so distant that their light takes thousands of millions of years to reach us. A scale of this kind is impossible to visualize, and we simply have to accept it – something which our ancestors were quite unable to do.

Because the stars are so remote, they seem to move very little compared with each other. The patterns or constellations which we see today are almost exactly the same as those which must have been seen by Julius Cæsar or even the builders of the Pyramids. The stars do have individual or 'proper' motions, but the shifts are too slight to be noticed by the naked-eye observer even over periods of many lifetimes. The planets, which are much closer, do seem to wander about from one constellation to another, and the word *planet* really means 'wanderer'. True, they keep to certain well-defined regions, so their movements are easy to follow. Our star-system or Galaxy contains about 100,000 million stars, together with many other types of objects; star-clusters, gas-clouds or nebulæ and so on, together with a tremendous amount of thinly-spread material. Yet it is by no means the only galaxy. Many millions of others are known, and there are even a few which can be seen without optical aid. Such is the Great Spiral in Andromeda, which is over 2,000,000 light-years from us. Even more conspicuous are the two Clouds of Magellan in the far south of the sky, which are somewhat closer to us (less than 200,000 light-years).

We are still being parochial. The Great Spiral and the Clouds belong to what is called the Local Group of galaxies, and when we look further out we find that there are other groups, so remote that we can see no details in them. At the moment we have probed out to well over 10,000 million light-years, and we have established that the whole universe is expanding. Beyond the Local Group all the galaxies are racing away from us, and the further away they are the faster they go.

Whether this expansion will continue indefinitely we cannot tell. Neither do we know just how the galaxies came into existence. Our best hope of solving these basic problems is to look out into the far reaches of the universe, and see the galaxies as they used to be thousands of millions of years ago.

STAR-TRAILS. *A time-exposure was made with the camera pointing at the North Celestial Pole; the stars seemed to move slowly across the sky, so producing trails. This apparent movement of the sky is due to the real rotation of the Earth on its axis. Photograph by Allan Lanham.*

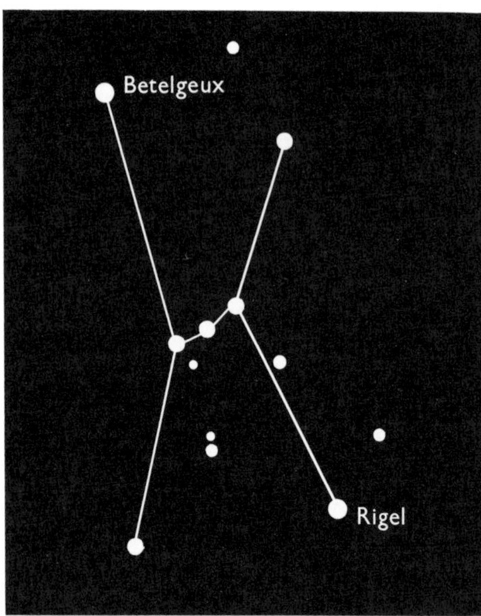

ORION. *One of the most brilliant constellations in the sky, with two first-magnitude stars, Betelgeux and Rigel.*

This is being done by means of equipment which has been developed only during the past few decades. Originally, astronomers had to depend upon their eyes alone. Telescopes were invented early in the seventeenth century, and in 1610 the great Italian scientist Galileo made the first important telescopic observations of the sky. Photography dates from the nineteenth century, and today electronic devices are fast superseding the photographic plate. Moreover, we are no longer restricted to studying the visible light coming to us from beyond the Earth.

Light may be regarded as a wave-motion; the colour depends upon the wavelength (that is to say, the distance between one wave-crest and the next). Red light has the longest wavelength, violet the shortest. If the wavelength is longer than that of red light, our eyes cannot perceive it. First comes infra-red, and then radio waves, while to the short-wavelength end of the visible range we have ultra-violet, X-rays, and the very penetrating gamma-rays.

Radio astronomy began in the early 1930s, and by now it has become a vital part of astronomical science. Most people know of the great 'dish' at Jodrell Bank in Cheshire, which collects and focuses radio waves just as an ordinary telescope collects and focuses light-rays (even though a radio telescope does not give an actual picture, and one certainly cannot look through it). Instruments for studying very short radiations are different again, and have to be taken above the Earth's atmosphere. This means using rockets.

Ideas of space-research go back for many years, but it was only in 1957 that the first artificial satellite or man-made moon was launched: Russia's Sputnik 1, which sped round the world transmitting the 'bleep! bleep!' signals which will never be forgotten by anybody who heard them (as I did). Astronomy had entered the Space Age. The first manned space flight took place in 1961; the Moon was reached in 1969, and by 1983 unmanned probes had travelled out beyond Saturn, the outermost of the planets known in ancient times.

We cannot send rockets out to the planets of other stars. The journeys would take far too long. We cannot even be sure that other planetary systems exist, though most astronomers are convinced that they do. Certainly we have no proof that there is advanced life on other worlds, and there can be none on the familiar planets of the Solar System, but we must never forget that the Sun is an ordinary star, in no way different from many millions of others in our Galaxy alone. Whether we will ever manage to contact 'other men' remains to be seen, but at the moment we have little idea of how to set about it.

Meanwhile, new types of objects have been found. Quasars, pulsars, black holes... none of these were even suspected twenty years ago. Discovery follows discovery, and it is becoming hard to keep track of them all. But I am running ahead of my story; so let us turn back to the distant past, when the nature of the universe and even of the Earth itself was completely unknown.

THE HALE REFLECTOR. *The 508-cm reflector at Palomar Observatory, for many years the largest telescope in the world.*

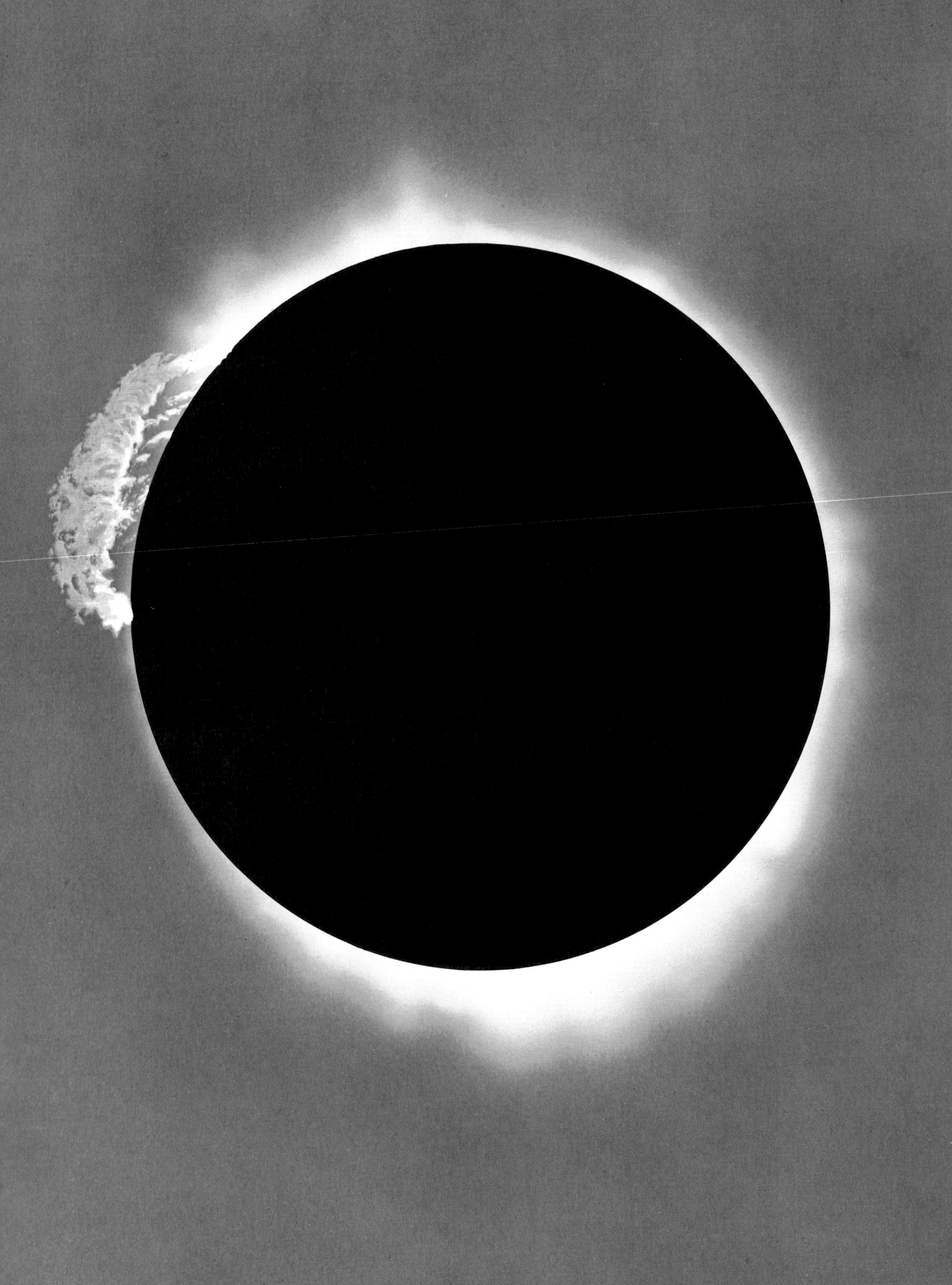

2

Watchers of the Stars

TOTAL ECLIPSE OF THE SUN, 1919. *(Opposite) The drawing shows an extensive prominence which was nicknamed 'the Anteater Prominence'. It is very seldom that a prominence of this size is seen during a total eclipse. Drawing by D. A. Hardy.*

PARTIAL ECLIPSE OF THE SUN, *June 19, 1936, 16h 20m. From an observation made by Patrick Moore, using a 7·5-cm refractor, × 100. Three sunspots are shown. Drawing by D. A. Hardy.*

TO the casual observer, the world looks flat. Allowing for local irregularities such as hills and valleys, there is nothing obvious to show that the Earth is a globe. Naturally, then, ancient men believed that the Earth really was flat, and it was equally natural to regard it as motionless, with the sky turning round it once a day. The idea that we could be whirling round the Sun at a rate of some thirty kilometres per second would have seemed far-fetched.

There were some peculiar theories, not all of which died out before astronomy became a true science. The Vedic priests in India believed the world to be supported upon twelve massive pillars; during the hours of darkness the Sun passed underneath, somehow managing to thread its way between the pillars without hitting them. According to the Hindus, the Earth stood on the back of four elephants; the elephants in turn rested upon the back of a huge tortoise, while the tortoise itself was supported by a serpent floating in a limitless ocean. One cannot help feeling rather sorry for the serpent!

For early astronomical records, we must go back to the great civilizations of China, Mesopotamia and Egypt. The Chinese were careful sky-watchers, and they certainly recorded a grouping of the bright planets which took place around 2500 B.C. Even earlier they

TOTAL ECLIPSE OF THE SUN, *February 15, 1961. The line of totality extended across Southern Europe. The eclipse was shown on B.B.C. television from three stations in Europe in succession; St. Michel (France), Florence (Italy) and Mount Jastrebac (Jugoslavia). Pictures obtained from France and Italy were good; conditions in Jugoslavia, where the author was commenting, were affected by cloud. This photograph was taken off the television screen during transmission from France.*

seem to have adopted a year of 365 days which made it possible for them to draw up a calendar. It mattered little to them whether the Sun went round the Earth, or the Earth went round the Sun; the 365-day year was the same in either case. Obviously the Chinese had no idea of the true make-up of the universe, and they believed that the celestial objects had been created for the benefit of mankind in general and the Chinese Emperor in particular. One story, dating from 2136 B.C., has been re-told many times, and is worth relating again here, because it concerns a total eclipse of the Sun.

As we now know, the Moon moves round the Earth, taking less than a month to complete one journey. As it has no light of its own, half of it is sunlit while the other is dark. If the Moon happens to pass directly between the Sun and the Earth, it will block out the brilliant solar disk for a few seconds or a few minutes, producing an eclipse. When the Sun is completely hidden, during a total solar eclipse, the sight is magnificent. The solar atmosphere flashes into view, and we can see the glorious pearly gas known as the corona, together with the 'red flames' or prominences rising from the Sun's surface. No total eclipse lasts for as long as eight minutes; and because the Moon appears only just large enough to cover the Sun, total eclipses are rare as seen from any particular point on the Earth's surface. One has to be in just the right place at just the right time. Thus the last total eclipse to be seen from anywhere in England was in 1927; the next is not due until 1999.

An eclipse begins as a 'bite' at the edge of the Sun's disk as the dark, invisible body of the Moon moves forward. The Chinese had different ideas. They believed that an eclipse was due to the onslaught of an unfriendly dragon which was doing its best to gobble up the Sun, and that the only remedy was to scare the beast away by shouting, screaming, banging gongs and making as much noise as possible. Not surprisingly, this remedy always worked. But it was clearly important to know when an eclipse was imminent, and the Chinese astronomers were able to calculate the danger-times by making use of a cycle known as the Saros. It so happens that one eclipse is apt to be followed by a similar eclipse eighteen years 11.3 days later. The period is not exact, but it is better than nothing at all.

One day in 2136 B.C., the Imperial Court was alarmed to find that the dragon was in action unexpectedly. The chief astronomers of the realm, who had the picturesque names of Hsi and Ho, had given no warning – and yet the Sun was changing from a brilliant disk into a crescent. There was widespread consternation, followed by relief when the crisis passed and things returned to normal. The sad result was that Hsi and Ho were summarily executed for neglecting their duties.

THEORY OF AN ECLIPSE OF THE SUN. *The Moon's shadow just reaches as far as the Earth; a partial eclipse is seen to either side of the belt of totality. [The drawing is not to scale]*

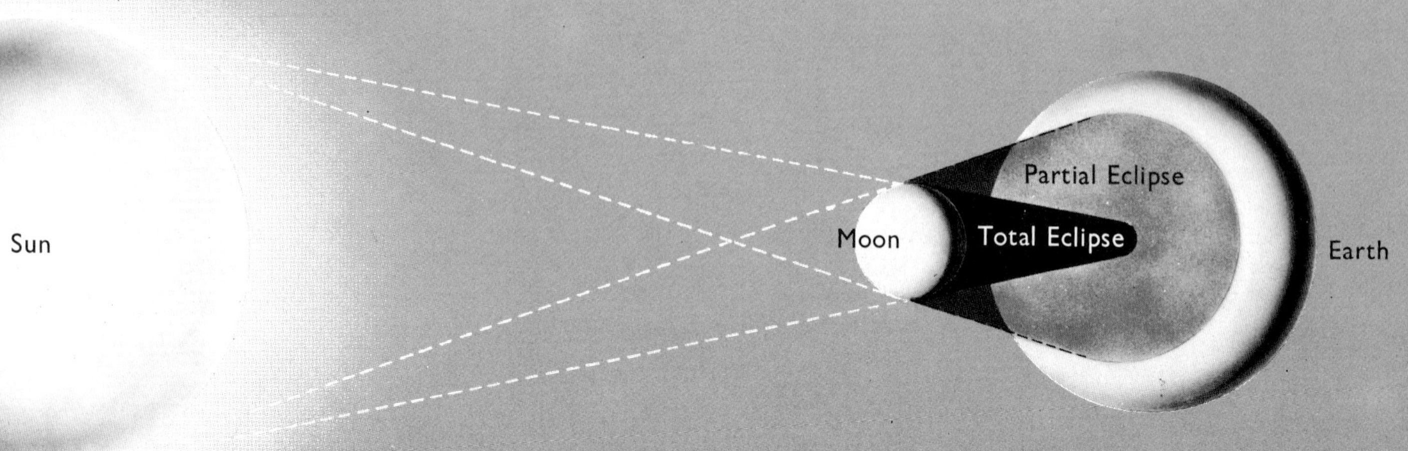

I have no idea whether the story is true. I have an inner feeling that it is not, but at least we can learn something from it! There is no mystery about the Saros. If the Moon's path or *orbit* were in the same plane as that of the Earth, an eclipse would happen every month; but the lunar orbit is tilted at an angle of five degrees, so that in most cases the Moon passes undetected either above or below the Sun in the sky. Conditions have to be exactly right, and it so happens that the Sun, Moon and Earth come back to almost the same relative positions every eighteen years 11.3 days.

The Chinese were not the only sky-watchers of ancient times. There were, for instance, the Mesopotamians. The Middle East was as turbulent in those far-off days as it is now, and this is no place to go into details about the wars between the Babylonians, the Assyrians and the various other races, but there is no doubt that regular observations were made. We still have the famous Venus Tablet, which is thought to date from the reign of Hammurabi, between 1792 B.C. and 1750. It says, for example, that when Venus is brilliant 'rains will be in the heavens'. Obviously this is not true, but it does show that the Babylonians had recognized Venus as being something exceptional, and the

COMET AREND-ROLAND, 1957. *This was one of the two naked-eye comets seen in 1957; it was a fairly conspicuous object for some weeks during May, when it was in the northern part of the sky. Photograph by F. J. Acfield, Forest Hall Observatory, Northumberland.*

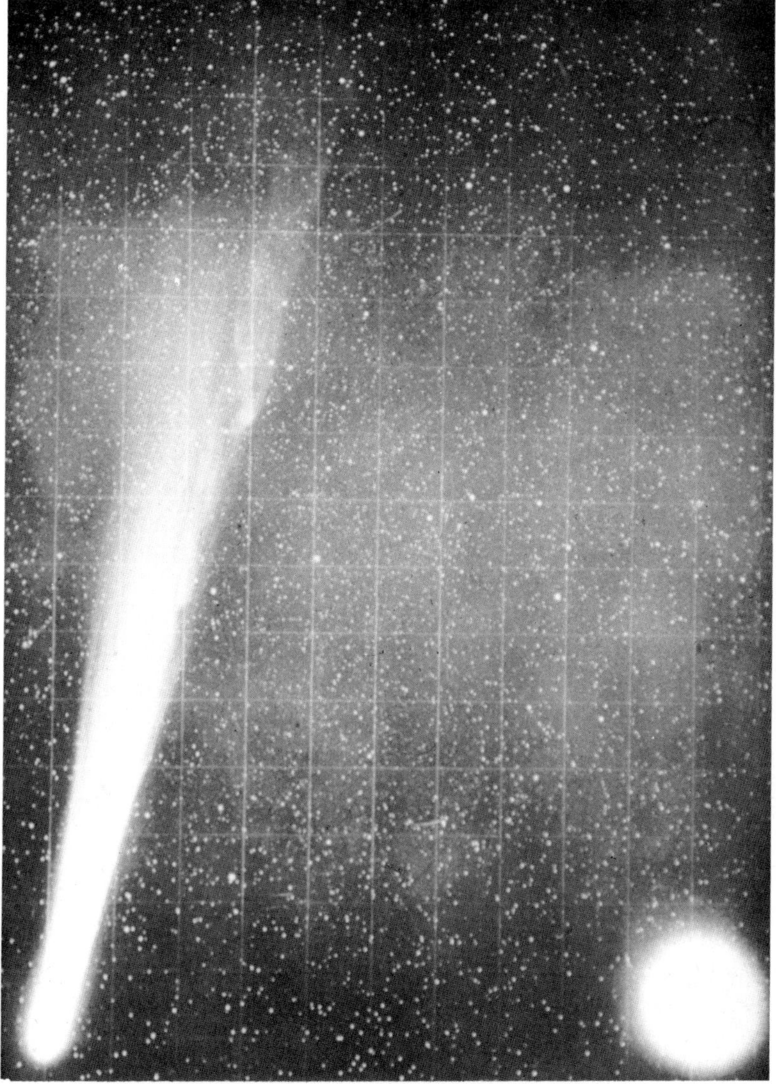

VENUS AND HALLEY'S COMET, 1910. *When this photograph was taken, the comet was almost at its brightest, but Venus is necessarily over-exposed. Halley's Comet will next be brilliant in 1986. Photograph by H. E. Wood, Union Observatory, Johannesburg.*

library of King Ashurbanipal, who reigned from 668 B.C. to 626, is another mine of information. Then there were the Chaldæans, who fought against their neighbours, and in 625 B.C. captured Babylon and destroyed it. It may even be that the Chaldæans drew up the original constellation patterns.

Of special importance were the Egyptians. (Of course, these various ancient civilizations overlapped each other in time, but we have to take them in some sort of order.) The Egyptians were remarkable people, capable of carrying out very accurate measurements, but they made the initial mistake of supposing that the universe takes the form of a rectangular box, with the longer sides running north-south. There is a flat ceiling, supported by four pillars at the cardinal points, and the pillars are connected by a mountain chain, below which runs a ledge carrying the celestial river Ur-nes. Along this river sail the boats carrying the Sun, Moon and other gods; when a corner is reached, the boats make right-angle turns in a manner which would defeat any modern sailor. The people of the cities in the eastern part of the Nile Delta went so far as to claim that the sky was formed by the body of a goddess with the appropriate names of Nut, who was permanently suspended in what must have been an uncomfortable as well as an inelegant position. Egypt lay in the centre of the flat Earth, and was surrounded on all sides by a vast ocean. The Egyptians were ruled, in fact if not in theory, by the priests, and there have been suggestions that these priests knew much more about astronomy than is generally believed. Some of the early Greek philosophers certainly visited Egypt, and learned a great deal. But, of course, astronomy and religion were merged into one. In 1367 B.C. the young Pharaoh Akhenaten even founded a new city as the centre of his new religion, which was based upon sun-worship. The experiment failed; but Akhenaten's *Hymn to the Sun* is worth reading, and it can teach us much.

Egyptian constellations were different from those which we now use. Some of the patterns, such as Leo (the Lion) are much the same; Orion is shown as a man in the famous Dendereh Zodiac, an Egyptian star-map which was discovered in 1798 by Napoleon's General Desaix. The Zodiac is 1.55 metres in diameter, and is remarkably well-preserved. On the other hand, our Cassiopeia is the Egyptian Hippopotamus, while our Lyra (the Lyre) is their Hawk. The two systems differ widely.

The Egyptians needed a good calendar. The whole economy of their country depended upon the annual flooding of the Nile, and it was essential to know when this was due to happen. In particular, the priests observed the heliacal rising of Sirius, the brightest star in the sky, which they called Sothis. When Sirius could first be seen rising in the dawn sky, they had an accurate time-check. They finally worked out a year of 365¼ days.

Of all their achievements, the Ancient Egyptians are probably best remembered today for their Pyramids, which have stood in the desert for so many centuries, and which are still regarded as among the world's greatest wonders. Many theories have been put forward to explain them, but it is certainly true that the Great Pyramid of Kheops is astronomically aligned, because it is oriented with respect to what was then the north pole of the sky.

The Earth's axis of rotation is not perpendicular to the plane of its orbit round the Sun. It is inclined at an angle of 23½ degrees, which accounts for the seasons; the northern hemisphere is tilted toward the Sun around June, while the southern hemisphere has the full advantage of the Sun's rays around December. Northward, the axis points to a position within one degree of the bright star Polaris, in our Ursa Minor, the Little Bear (or, if you prefer it, part of the Egyptian Hippopotamus). Therefore Polaris appears to stay almost motionless in the sky, with everything else moving around it once in approximately twenty-four hours. When the Pyramid was built, however, the pole of the sky was in a different place, close to a fainter star, Thuban in Draco (the Dragon). The Earth is not a perfect sphere; it is slightly flattened, so that the equatorial zone bulges out, and the diameter measured through the poles is forty kilometres shorter than the diameter as measured through the equator. The Sun, Moon and planets pull upon this bulge, and the result is that the Earth's axis wobbles very slowly, rather in the manner of a gyroscope which is running down. This causes an effect known as precession. The polar point describes a circle in the sky, taking 26,000 years to complete a full turn. The effect may be slight, but over the centuries it adds up. In 12,000 years from now the north pole star will be the brilliant blue Vega, in Lyra.

We must admire the Egyptians for their observations, their measurements and above all for their buildings; the Pyramids remain as testimony to their skill. Yet all in all, they were content to 'look' without making much effort to 'explain', and the idea that the Earth could be an unimportant world in a vast universe was quite outside their comprehension.

Ancient astronomy was not confined to the great civilizations. There can be little doubt that Stonehenge, the famous stone circle in Wiltshire, is astronomically aligned, and the same is true of many other stone circles found in other parts of the world. The monument we now see is actually Stonehenge III, built around 2100 B.C. by a somewhat shadowy race known to us as the Beaker People (the earlier Stonehenges date back considerably earlier). All sorts of theories have been proposed, ranging from the sane to the absurd – the latter being typified by the suggestion that the Druids might have been involved; in fact the Druidic cult did not reach England much before 1300 B.C., so that in time the Druids were as far removed from Stonehenge as we are from the Crusades.

Gerald Hawkins, who has carried out a tremendous amount of research, believes that the monument was used as a primitive computer, with special emphasis upon the prediction of eclipses. Others disagree, but in any case the astronomical nature of Stonehenge is beyond doubt. In North America we have the 'medicine wheels' (nothing to do with medicine!) which also are astronomically aligned; that at Moose Mountain, in South Saskatchewan, seems to date back for some 2,000 years, though others are younger. And the Inca and Maya civilizations certainly paid great attention to the sky.

To delve more deeply into all these ancient astronomies would mean too much of a digression, particularly as there is again a wide overlap in time; the Inca civilization was not destroyed by the invading Spaniards until a few centuries ago. So let us next look at the Greeks, who were the first to change astronomy from a mystical cult into a true science.

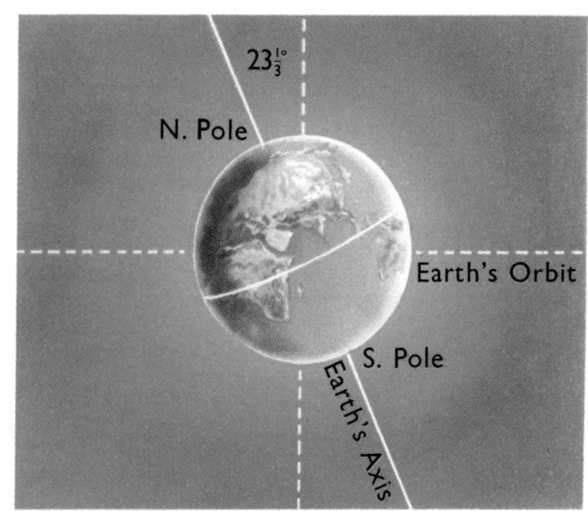

INCLINATION OF THE EARTH'S AXIS. *The equator is inclined by 23⅓ degrees to the plane of the Earth's orbit.*

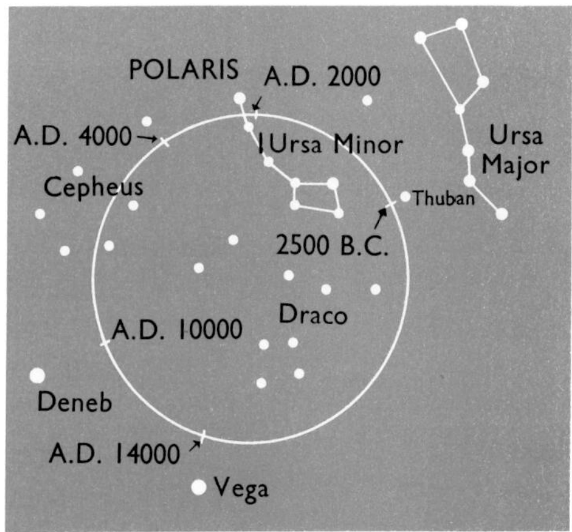

APPARENT MOVEMENT OF THE NORTH CELESTIAL POLE. *The shifting is due to precession effects, or the slow movement in the direction of the Earth's axis. In ancient times the Pole Star was Thuban in Draco; at present it is Polaris in Ursa Minor; by the year* A.D. *14000 the northern polar star will be the brilliant Vega, in Lyra.*

3

The Greek Astronomers

THALES, first of the great Greek philosophers, was born about the year 624 B.C.. Ptolemy, last of the famous astronomers of the Greek school, died around A.D. 180. This means that the whole story was spread over eight centuries – the same period as that which separates Sputnik 1 from Henry II. It was not a period of quick, spectacular progress, but in the end it changed the whole outlook of mankind. On the whole, it is quite remarkable how much the Greeks managed to find out.

Thales of Miletus seems to have owed much to the teachings of the Babylonians and the Egyptians, and in some ways his ideas were no more advanced than theirs; he believed the Earth to be flat, and to float on water in the manner of a log or cork. (He was also one of the first of all absent-minded scientists. The story of how he once walked into a deep well while gazing upward at the stars is quite possibly true.) He is said to have predicted the eclipse of 585 B.C., which put a prompt end to a battle going on between the armies of the Lydians and the Medes, in which case he was well aware of the Saros cycle.

Thales' younger contemporary, Anaximander, is credited with the suggestion that the Earth is suspended freely in space, remaining fixed because it is at an equal distance from all other celestial bodies. He may even have thought that the Sun is as large as the Earth. Yet he believed the Earth to be shaped like a short cylinder; to him the stars were fiery jets, and the Sun 'like a chariot-wheel, the rim of which is hollow and full of fire'. Another contemporary, Anaximenes of Miletus, was of the opinion that the Sun, Moon and stars were made of fire, with the stars fixed on to a crystal sphere.

Then, somewhat later, came Pythagoras, who was born around 572 B.C.. Nowadays we remember him chiefly because of his famous geometrical theorem, but he seems to have known that the Earth is a globe, and he was certainly well aware that the planets move against the starry background (though in fact this must have been discovered much earlier). Pythagoras accepted the idea of a solid sphere with the stars fixed on to it, but he was well in advance of his contemporary Xenophanes of Colophon, who taught that the world had been born from a mixture of earth and water, and will come to an end when it is absorbed back into the sea and changed into mud. According to Xenophanes, there is a new sun every day, while the Earth's upper side touches the air and its lower side extends without limit. And Heraclitus of Ephesus, born about 544 B.C., maintained firmly that the diameter of the Sun was about thirty centimetres!

Anaxagoras of Clazomenæ, born about 500 B.C., is the next major character in the story. He was a flat-earth supporter, but at least he was bold enough to suggest that the Sun is a red-hot body larger than the peninsula upon which Athens stands. This led him into trouble with religious leaders, and he was accused of impiety; he was banished from Athens, and but for his friendship with Pericles, the most powerful man in the city, he might have fared even worse. We have a hint here of the religious persecution which lay far ahead.

Anaxagoras made two major advances. First, he believed the Moon to be 'earthy', with mountains and ravines; he also knew the causes of eclipses. There was a strange aftermath. In 413 B.C. the Peloponnesian War between the city-states of Athens and Sparta was at its height, and the Athenians had unwisely dispatched an army to attempt the conquest of Sicily. Things went wrong almost from the start, and the only course was prompt withdrawal. Unfortunately the Athenian commander, Nicias, delayed evacuation because of an eclipse, which he interpreted as being divine advice to stay where he was. By the time he did make up his mind to embark, it was too late; the whole Athenian expedition was wiped out, and it was this disaster which led to the surrender of Athens itself. If Nicias had believed in Anaxagoras' theory of eclipses, the whole course of European history might have been different.

One of the most influential of all the Greek philosophers was Aristotle, who lived from 384 B.C. to 322. His authority was so great that for many centuries afterwards few people dared to question it, and it was lucky that he came to the conclusion that the flat-earth theory was wrong. Moreover, he gave three experimental proofs. First, he reasoned that a sphere is 'the shape that a body naturally assumes when all parts of it tend toward the centre' – a first glimmer of the idea of gravitation. Secondly, he pointed out that the stars appear to change in height above or below the horizon according to the observer's position on the Earth; for instance the brilliant southern star Canopus can be seen from Alexandria, but never rises over the more northerly city of Athens – something which is only to be expected if the Earth is a globe. Thirdly, he drew attention to eclipses of the Moon. As the Earth's shadow on the Moon is curved, it follows that the surface of the Earth must also be curved.

All this was reasonable enough. Aristotle said that the world was round; so round it must be. The next step was to measure its size, and this was done by Eratosthenes of Cyrene, who lived from 276 B.C. to 196. His method was simple and sound. It was based on the fact that when the Sun is directly overhead from one observing site, it is some way from the zenith or overhead point as seen from elsewhere.

Eratosthenes lived in Alexandria, where he was in charge of an extensive library which contained the works of Aristotle and many older writers. Eratosthenes learned that at the time of the summer solstice, when the Sun reaches its northernmost point in the sky (in mid-June each year), the noonday Sun is directly overhead as seen from Syene, known today as Assouan. At this moment, the Sun is seven and a half degrees from the zenith as seen from Alexandria, which lies due north of Syene. A full circle contains 360 degrees, and seven and a half is about one-fiftieth of 360, so that if the Earth is spherical its circumference must be fifty times the distance between Alexandria and Syene. This distance was known to be 5000 stadia, and it followed that the Earth's circumference must be 250,000 stadia.

The stadion is an ancient measure of distance, but unfortunately there were several varieties of it, and we are not sure which one Eratosthenes used. One value for it is 158 metres, in which case the Earth's circumference works out at 39,984 kilometres and the diameter 12,631 kilometres. The true figures are 40,064 kilometres and

THREE PHOTOGRAPHS OF AN ECLIPSE OF THE MOON. *It is easy to see that the Earth's shadow on the lunar surface is curved.*

1
2
3
4

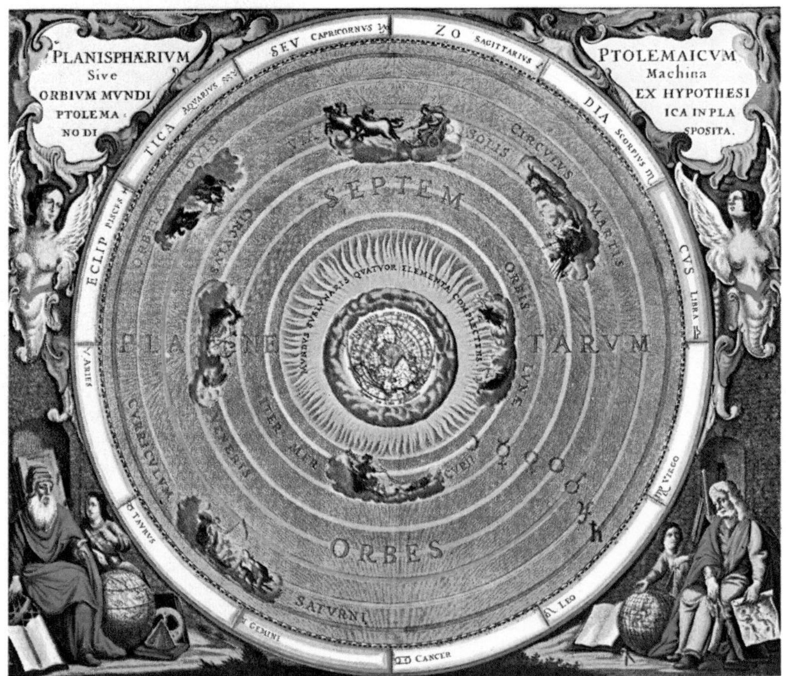

THE UNIVERSE, ACCORDING TO PTOLEMY. *From an old print, 1600. The arrangement of the celestial bodies according to the Ptolemaic theory is clearly shown, though no attempt has been made to make the distances even approximately correct. However, Ptolemy realized that the actual motions of the planets must be complex; as we have seen, he was compelled to introduce numerous epicycles in an attempt to reach agreement with the observational data.*

12,756 kilometres. In any case, his estimate was quite amazingly accurate, and was much nearer the truth than the value used by Christopher Columbus during his voyage to the New World in 1492. Columbus took the circumference to be less than 30,000 kilometres, which explains why he came home without having had much idea of where he had been.

To use a hackneyed modern term, the discovery that the Earth is a globe with a diameter of around 12,000 kilometres was the first great breakthrough in astronomy. Note that it was achieved not by guess-work, but by sound scientific reasoning, which is why the Greeks – at any rate, the later ones – were so much in advance of the Chinese, the Egyptians, the Babylonians and the rest. Aristotle did not simply assume the Earth to be spherical; he demonstrated it by observation. And nobody can fault Eratosthenes, either for his method or for the way in which he applied it.

The Greeks very nearly managed the second great breakthrough: the discovery that instead of being the centre of the universe, the Earth is an ordinary planet moving round the Sun. Aristarchus of Samos, who lived from 310 B.C. to 230, put forward precisely this idea, and also gave the Sun's distance as 8,000,000 kilometres, which was a gallant attempt even though it was much too low. Aristarchus' way of measuring the Earth-Sun distance was perfectly sound in theory, and it was not his fault that he was unable to make the essential measurements accurately enough.

It was a pity that Aristarchus' views about the movement of the Earth found so few supporters. Much more credence was given to the ideas of his contemporary Apollonius of Perga, who agreed that the planets move round the Sun, but thought that the Sun itself was in orbit round the Earth.

Two more great figures pass across the stage before the age of classical astronomy comes to an end. Hipparchus of Nicæa is the first. We know nothing about his career, but he lived around 150 B.C.. He

'REVOLVING TABLE' FOR ESTIMATING THE POSITIONS OF THE ZODIACAL CONSTELLATIONS BETWEEN 7000 B.C. AND A.D. 7000. *Published in the famous book* Astronomicum Cæsareum *by P. Apian (Ingolstadt, 1540). The apparent movements of the stars are of course affected by the shifting of the celestial pole due to precession, so that conditions were not precisely the same in Apian's time as they had been in Ptolemy's. Yet even in 1540, the Ptolemaic theory of the universe was still generally accepted, and few scientists even considered questioning it.*

AN ECLIPSE OF THE MOON, *in four stages. (Opposite) Drawings by D. A. Hardy.*

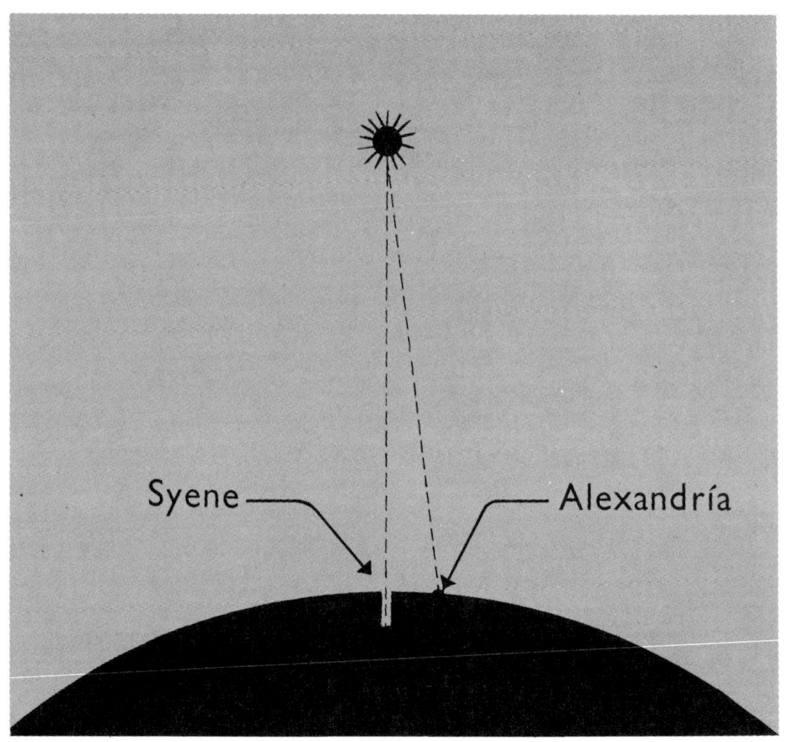

ERATOSTHENES' METHOD OF MEASURING THE SIZE OF THE EARTH. *At noon at the time of the summer solstice, the Sun was vertical at Syene, but not at Alexandria. Eratosthenes measured the Sun's altitude at this time, as seen from Alexandria, as 7 degrees away from the zenith, and was thus able to measure the circumference of the Earth with remarkable accuracy. It is significant that the value which he gave was more correct than that used many centuries later by Christopher Columbus on his voyage of discovery to the New World, which is a remarkable tribute to Eratosthenes' theoretical and practical skill.*

drew up an important star catalogue; he discovered precession; he made systematic use of trigonometry in his calculations, and he improved Aristarchus' estimates of the distances of the Sun and the Moon. Finally there was Claudius Ptolemæus of Alexandria, who takes us into the Christian era; he was born in or about A.D. 120, and died in 180.

Ptolemy, as he is always called, was an excellent observer and mathematician. He was a philosopher, an astronomer and a geographer; he made the first tolerable map of the world, and many features on it are recognizable (even though Scotland was joined to England in a back to front position). He wrote a great book, the *Syntaxis*, which has came down to us by way of its Arab translation, and is usually called the *Almagest*; it is nothing more nor less than an encyclopædia of ancient science, and our debt to it is immeasurable. Occasional attempts to discredit Ptolemy, and to claim that he was a copyist at best, have been singularly unsuccessful, and he well merits his nickname of 'the Prince of Astronomers'.

Ptolemy undertook a thorough revision of Hipparchus' star catalogue. (But for this, the catalogue itself would have been lost.) He laid down his 'system of the world', which is always known as the Ptolemaic System; Ptolemy himself did not invent it, but he brought it to its highest pitch of development. The Earth lay at rest in the centre of the universe, and round it revolved, in order, the Moon, Mercury, Venus, the Sun and then the more remote planets Mars, Jupiter and Saturn, beyond which lay the sphere of the fixed stars. Since the circle was the 'perfect' form, and nothing short of perfection could be allowed in the heavens, all orbits had to be strictly circular.

This led to immediate difficulties, because Ptolemy was much too good an observer to believe that the planets move across the sky in a regular manner. For instance, Mars, Jupiter and Saturn periodically stop their westward motion against the stars, pause, and then 'retrograde' or move eastward for a while before stopping again and resuming their westward march. Ptolemy's solution was to assume that each planet moved in a small circle or epicycle, the centre of which – the deferent – itself moved round the Earth in a perfect circle. Another device was to use an 'eccentric circle', whose centre did not coincide with that of the Earth. To account for all the irregularities, more and more epicycles had to be introduced. Finally the whole system became hopelessly clumsy and artificial, but – and this is the vital point – it did account for the movements of the planets, and since Ptolemy knew nothing about the nature of gravitation he can hardly be blamed for his failure to find the right answer.

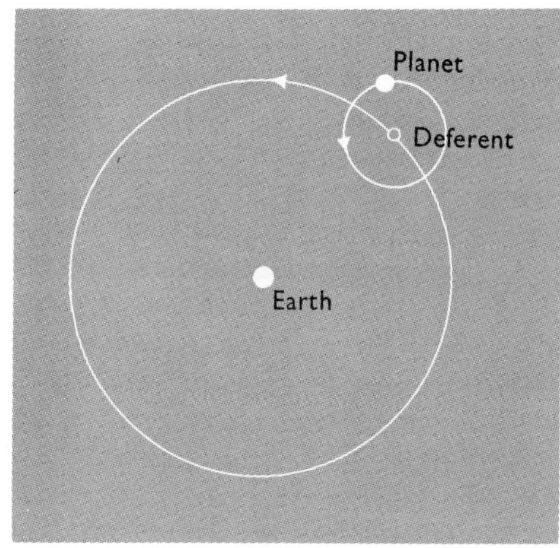

THE PTOLEMAIC THEORY. *According to Ptolemy, a planet moved in a small circle or* epicycle, *while the centre of this circle (the* deferent*) itself moved round the Earth in a perfect circle. This system was not invented by Ptolemy, but its greatest development was due to him. Ptolemy – who was an excellent mathematician – realized that this comparatively simple arrangement would not explain the actual movements of the planets in the sky, and he was compelled to introduce extra epicycles, thus making the whole system clumsy and unwieldy. However, the Ptolemaic theory was almost universally accepted by scientists up to the time of Copernicus.*

From our point of view, Ptolemy lived at the best possible time, because the year of his death also marked the end of Roman greatness. Marcus Aurelius, the philosopher-emperor, died at about the same time (actually, we know much more about his career than we do about Ptolemy's). Had Ptolemy lived a few centuries earlier, he would have been able to do no more than leave us an incomplete account; had he lived later, at least part of the old knowledge would have been forgotten. One particularly disastrous loss was that of the Alexandrian Library. It is usually said that it was destroyed in A.D. 640 by order of the Arab caliph Omar, who pronounced that either the books opposed the teachings of the Koran and were heretical, or else supported the Koran and were therefore unnecessary. On the whole it is more likely that much of the Library had been dispersed by the time of Omar, but the salient point is that the Library ceased to exist, to the everlasting regret of scholars all over the world.

The Romans were not scientifically-minded, and did not share the Greek love of learning. However, they were essentially practical, and they took the trouble to reform the calendar. The true 'year', the time taken for the Earth to go once round the Sun, is not exactly 365 days, but more nearly 365¼, so that to draw up a calendar which will not become out of step with the seasons is not so easy as it might appear. Julius Cæsar realized this, and instructed a Greek astronomer, Sosigenes, to form a more accurate calendar. Sosigenes did his work well; for instance he invented the leap-year, which took care of the extra quarter-day in the Earth's revolution period. The Julian calendar was not perfect, and later tampering with it made matters no better, but it was a considerable improvement, and it was quite good enough to satisfy the Romans.

But all in all, it is fair to say that astronomy of the classical era came to an end with Ptolemy. The Dark Ages lay ahead.

4

The Rebirth of Astronomy

IT is not entirely true to say that astronomy stagnated for over 600 years after Ptolemy's death. Nothing came out of the collapsing Roman Empire, and little from China; but a certain amount of observation was carried on elsewhere, and in India an astronomer named Aryabhātta, who was born around the year 476, taught that the Earth is a rotating globe, as well as giving the correct causes of eclipses and building primitive stone 'observatories'. In Central America – the modern Belize, Honduras and parts of El Salvador and Mexico, including Yucatán – the Maya civilization reached its peak from the third to the ninth centuries; astronomers there prepared planetary tables, and paid special attention to Venus, upon whose movements their system of timekeeping was based.

Perhaps the most significant announcement of the period was made by a bishop, Isidorus of Seville, born in 570. He presided over many Church councils, and wrote a book in which he dealt with elementary astronomy, but his main importance is that he was the first man to draw a definite distinction between astronomy and astrology.

I must digress here, because even today there are still many people who confuse the two – not only in India and the Middle East, but also in Europe and the United States. (I have lost count of the number of times that I have been asked whether I can cast horoscopes.) Let it be said at once that there is no connection whatsoever between astronomy and astrology. Astronomy is an exact science, while astrology is a relic of the past – irrational and totally without foundation.

Astrology is the superstition of the stars. It teaches that the character and destiny of a man (or woman) depends upon the positions of the Sun, Moon and planets in the sky at the moment of birth. Thus if you were born at, say, ten o'clock in the morning of March 1, in London, you will be quite different from someone who was born at midnight on November 5 in Manchester.

It does not take much intelligence to show how absurd this is. To begin with, a 'constellation' is not a genuine grouping, because the stars are at very different distances from us. As an example, look at the familiar Great Bear or Plough (known in America as the Big Dipper). The seven main stars make up a pattern which cannot be mistaken, but this pattern depends solely upon our own position in space. Mizar, the second star in the Bear's tail, is less than 60 light-years from us; Alkaid, the end star, is over 100 light-years away. Alkaid is therefore many light-years away from Mizar. There is no connection between the two. We are dealing with nothing more than a line-of-sight effect.

Also, the constellation names are purely of our own making. Astrologically Leo, the Lion, is a 'masculine' sign, while Pisces, the Fishes, is 'watery' – but what exactly does this mean? Neither group looks anything like an animal or a fish, and if we had followed the Chinese or Egyptian system instead of Ptolemy's our whole sky-map would have been different.

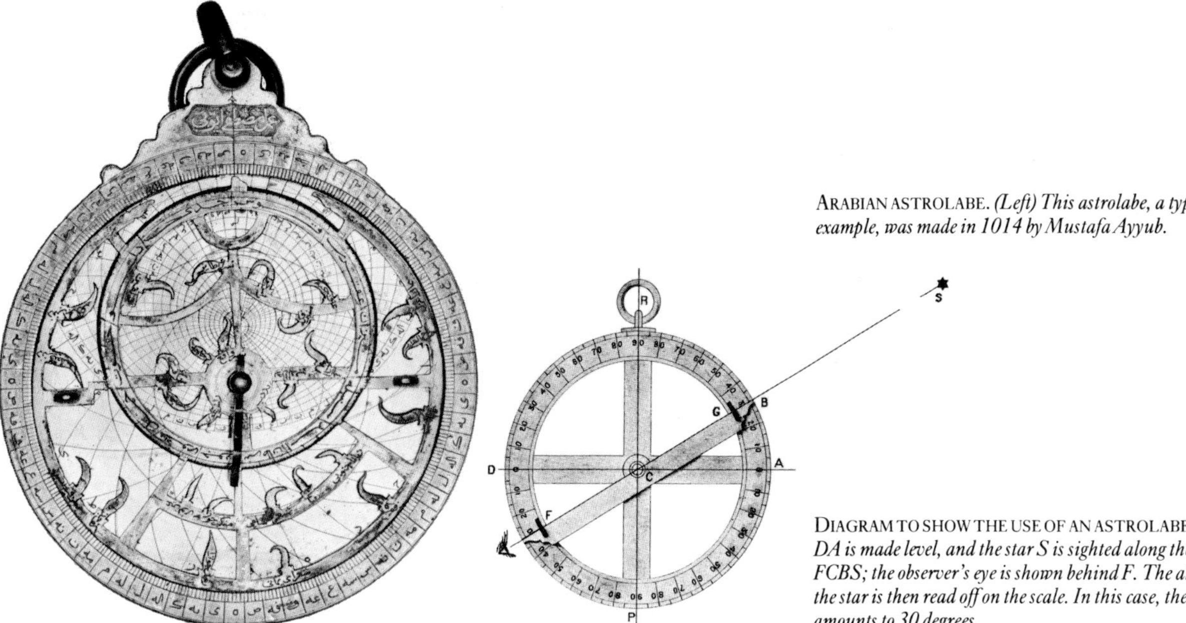

ARABIAN ASTROLABE. *(Left) This astrolabe, a typical example, was made in 1014 by Mustafa Ayyub.*

DIAGRAM TO SHOW THE USE OF AN ASTROLABE. *The axis DA is made level, and the star S is sighted along the direction FCBS; the observer's eye is shown behind F. The altitude of the star is then read off on the scale. In this case, the altitude amounts to 30 degrees.*

Quite apart from this, the planets are much closer to us than the stars. If the Earth-Sun distance is scaled down to thirty centimetres, even the nearest star will be over six kilometres away. To say that a planet is 'in' a constellation is no more logical than holding up a finger, aligning it with a cloud, and claiming that your finger is 'in' the cloud.

Astrologers blithely disregard all this; they draw up elaborate horoscopes, make predictions, confer degrees upon themselves, and do their best to set themselves up as genuine scientists. Of course many of them are quite sincere, but the same may be said of the Flat Earth Society (which, believe it or not, still exists). The best that can be said of astrology is that it is fairly harmless so long as it is confined to circus tents and seaside piers.

Yet astrology did have some use in the past, because the old astronomers were also astrologers – Bishop Isidorus was very much the exception – and in order to draw up horoscopes, it was necessary to know not only the exact positions of the stars, but also the movements of the planets. When the revival began, it came by way of astrology, and those responsible were the Arabs.

By a stroke of luck, Ptolemy's great book reached the Arab world, and in 820 it was translated into Arabic. Other books followed. Astronomical centres were set up at Damascus and Baghdad, and the Arabs proved themselves to be skilful observers. One caliph, Al-Mamun – son of Harun al Rashid of *Arabian Nights* fame – went so far as to build a fine observatory. It was quite unlike a modern observatory, since telescopes still lay far in the future, but it contained measuring instruments as well as an excellent library. By the time that Al-Mamun died, in 833, Baghdad had become a true astronomical centre.

Probably the most famous astronomer of the Baghdad school was Al-Battani, who was born around the middle of the ninth century. He was an excellent mathematician, and made observations which compared favourably with those of Hipparchus and Ptolemy. He also wrote an important book, the English title of which may be given as *The Movements of the Stars*.

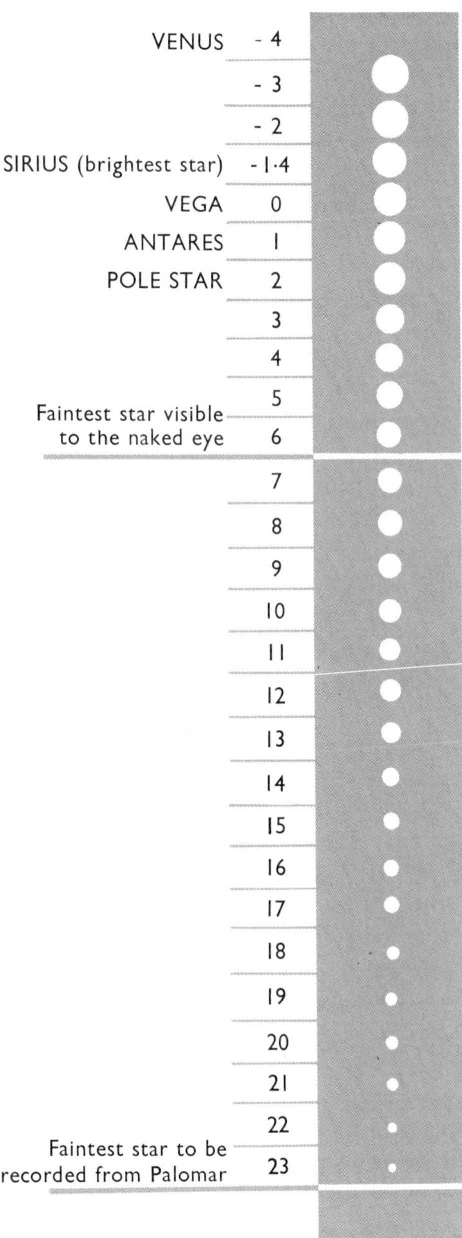

STAR MAGNITUDES. *The magnitudes are shown here as disks, though the stars themselves appear only as points of light.*

Another skilful Arab astronomer was Al-Sûfi, whose life extended between 903 and 986. He, too, wrote a book – *Uranographia*, in which he dealt with the apparent brilliancies of the stars. This is an important matter, and is worth describing in a little more detail.

The stars are graded into classes or 'magnitudes' of apparent brightness. The scale seems confusing at first, because the brightest stars have the lowest magnitudes; 1 is brighter than 2, 2 is brighter than 3, and so on. The faintest stars visible with the naked eye under good conditions are of magnitude 6. The scale is rather like that of a golfer's handicap; the most brilliant performers have the lowest numerical values.

Rigel in Orion is very brilliant, and on the modern scale its magnitude is 0.1, or only one-tenth of a magnitude below zero. The four brightest stars in the sky (Sirius, Canopus, Alpha Centauri and Arcturus) have negative magnitudes; that of Sirius is −1.4. On the same scale Venus, the brightest planet, may reach −4.4, while theSun is −26.8.

Nowadays it is possible to measure star magnitudes very accurately. Electronic devices can reach a precision of at least one-thousandth of a magnitude, and the world's most powerful telescopes can reach down to about magnitude +26. Al-Sûfi, of course, had to make his estimates simply by using his eyes, but most of his values agree quite well with those of today. There are a few interesting discrepancies; for instance, he graded Denebola in Leo as of the first magnitude, whereas it is now below the second. It is difficult to be sure whether or not these long-term changes are real, but we do know that some stars are variable in light, so that we must not be too quick to accuse Al-Sûfi of inaccuracy.

Many of the star-names now in use are Arabic; Rigel (the foot of Orion), Deneb in Cygnus (the 'Hen's Tail'), and so on – though a few are Greek, notably Sirius, which really means 'the Scorching One'.* Astronomical terms such as 'zenith' for the overhead point of the sky are also Arabic.

Yet meticulous though the Arabs were, they remained strangely silent about two spectacular events of the eleventh century. In 1006 a brilliant new star flared up in the southern constellation of Lupus (the Wolf), and according to Chinese and Japanese sources it became as bright as the quarter-moon for a brief period before fading away. In 1054 there was another stellar outburst, this time in Taurus (the Bull); it was not the equal of the first, but it rivalled Venus, and became visible in broad daylight for a few weeks. Our knowledge of both these stars depends upon the Far East, though it is possible that the second was recorded in 'cave drawings' by the American Indians. From the Arab world – nothing, which is truly remarkable.

We know what these outbursts were; they were supernovæ, colossal stellar explosions violent enough to blow a star to pieces. The 1054 supernova has left the superb gas-cloud we call the Crab Nebula, and I will have much more to say about it later.

(*En passant*, it has often been suggested that the Star of Bethlehem might have been a supernova. Unfortunately our knowledge is far too fragmentary for us to come to any definite conclusion, but I have no

*The correct pronunciation is 'Sȳ-rius', though most people call it 'Sirrius'.

ULUGH BEIGH MEMORIAL. *This memorial now stands in Samarkand, the site of Ulugh Beigh's old observatory.*

faith in the supernova theory, because anything of the sort would have been widely reported elsewhere – which it was not. Neither can the Star have been due to the close conjunction of two planets, though this idea is trotted out at regular intervals. We have to admit that there is no scientific answer, and it may be that the whole story *is* a story and nothing more.)

The Arabs did not concern themselves only with star-catalogues. They carried out observations of all kinds, and they were careful students of the movements of the Moon and planets, admittedly for astrological purposes. Mention should be made of King Alphonso X of Castile, who called a number of Arab and Jewish astronomers to Toledo, and was responsible for improved tables of planetary positions which were known as the Alphonsine Tables. For their time they were excellent, and they remained in use for the next three hundred years.

The last of the great Arab astronomers was Ulugh Beigh, who established an observatory at Samarkand in 1433. He was a powerful ruler – his grandfather was the Oriental conqueror Timur, or Tamerlane – and he equipped his observatory with the best instruments which could be made, so that he could draw up new star catalogues and prepare planetary tables. He even established an Academy of Science. Unluckily for him he was also a firm believer in astrology. He cast the horoscope of his eldest son, Abdallatif, and found to his alarm that the boy was destined to kill him. He therefore dismissed his son from Court and sent him into exile. Abdallatif had no wish to be set aside. He rebelled, invaded his father's kingdom, and had Ulugh Beigh murdered. That was one astrological prediction which came true, and it also marked the virtual end of Arab astronomy.

Much later – in the early eighteenth century – Jai Singh set up elaborate stone observatories in India, and these, notably the great buildings at Delhi, still exist. But meanwhile the old wish to learn had been revived in Europe, and observatories were established.

The first of these was at Nürnberg in Germany, and was due to Johann Müller, better remembered by his Latinized name of Regiomontanus. With his tutor, Georg von Peuerbach (Purbach) he revised the Alphonsine Tables, and after Purbach's death he joined his own pupil, Bernard Walther, to introduce new and better methods of observation. Printing had been invented, and Regiomontanus set up his own press, so that he could publish astronomical information for

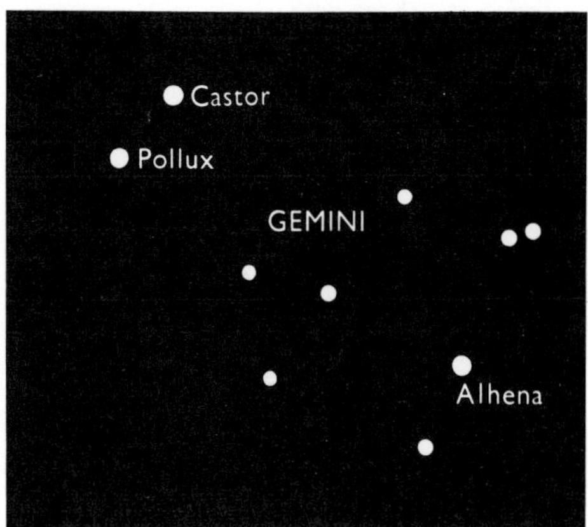

GEMINI, THE TWINS. *According to Al-Sûfi and others of his time, Castor used to be brighter than Pollux, but it is now half a magnitude fainter. Alhena, on the other hand, is now a magnitude brighter than as given by Al-Sûfi.*

These photographs show one of the greatest observatories of pre-telescopic times, erected at Delhi in India. The upper picture shows part of the building, and a general view of the observatory is given below. An ancient observatory of this kind was, of course, very different from a modern astronomical observatory; all work had to be carried out with the naked eye only, and was therefore limited largely to positional measurements. Some of the measures made were nevertheless, of surprising accuracy. The Delhi Observatory contained instruments which were capable of yielding very valuable results. Photographs by W. T. O'Dea.

the use of others. He continued to produce these almanacs up to the time of his death in 1476.

Then there was the 'forerunner', Leonardo da Vinci, whose life extended from 1452 to 1519. Leonardo was not an astronomer, but he made one discovery which cleared up an old mystery. This concerned the Moon.

The cause of the Moon's phases, or apparent changes of shape each month from new to full, was not in the least mysterious. More puzzling was the fact that when the Moon is a crescent, the unlit portion can often be seen shining faintly, giving the appearance often called 'the Old Moon in the Young Moon's arms'. Leonardo realized that this must be due to light reflected on to the Moon from the Earth. In fact, earthshine is nearly always seen when the crescent moon shines down from a dark sky.

All this work showed that astronomy had really 'woken up'. Not only had star catalogues been improved, but careful observations had enabled astronomers to predict the movements of the planets for years in advance. Yet the official outlook was still purely Ptolemaic. The second major breakthrough had yet to come.

The first signs of it came with Nikolaus Krebs, son of a German winegrower, who was born in 1401 and died in 1464. His boyhood was unhappy; he ran away from home, studying first at Heidelberg and then in Italy. He entered the Church, and became a cardinal; he is often known as Nicholas of Cusa, since he was born at Cues on the Moselle. Krebs proposed alterations in the calendar, and in his book *De Docta Ignorantia* he suggested that the Earth might not, after all, be the centre of the universe.

Not many astronomers paid any attention to him, and for half a century after his death little more was heard of the theory of a moving Earth. The next step – the realization that our world is a mere planet – was one which mankind found very hard to take.

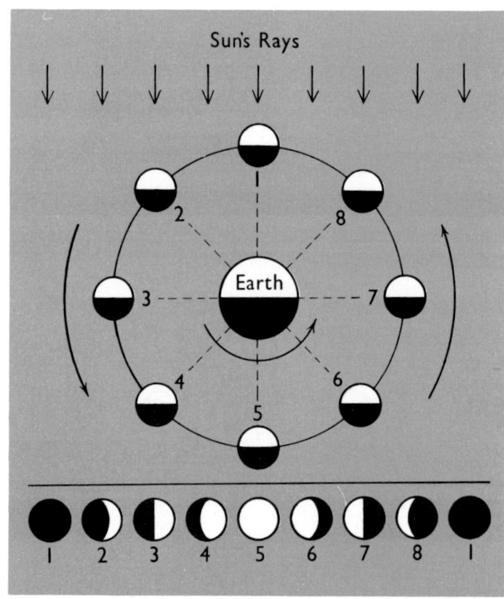

PHASES OF THE MOON. *The Moon has no light of its own, and depends on reflecting the rays of the Sun, so that half of it is luminous while the other half is dark. When the Moon has its dark face turned towards us, so that it is almost between the Earth and the Sun, we cannot see it at all, and this is what the astronomer terms 'New Moon'. (If the three bodies are exactly lined up, the result is an eclipse of the Sun, but owing to the tilt of the lunar orbit this does not happen at every New Moon.) In the diagram, New Moon is shown in position 1. At other times the Moon shows as a half (positions 3 and 7), three-quarter shape (4, 6), and full (5).*

5

The Design of the Universe

SINCE Greek times, when Aristotle and Ptolemy had taught that the Earth is the centre of the universe, the whole development of astronomy had been held up. Aristarchus, who had hit upon the truth, met with little support, and officialdom remained content with the Ptolemaic theory, which certainly did fit the facts as they were then known. The man who altered all this, and finally changed Man's whole conception of the universe, was a Polish churchman named Mikolaj Kopernik, better known to us as Copernicus.

Copernicus was born at Toruń, on the river Vistula, in 1473. He studied at Cracow University, and then in Italy. Later he went back to Poland, and became Canon of Frombork in Ermland, which he described as 'the remotest corner of the Earth'. He had a varied career; he practised medicine, and became a statesman as well as organizing resistance to the attacks of the Order of Teutonic Knights, who made persistent efforts to subdue the country. Yet Copernicus' main interest was in astronomy. He was not an observer; he was purely a theorist, and he was concerned mainly with the design of the Solar System.

Early in his life he became very doubtful about the correctness of Ptolemy's system. The main trouble, as he saw it, was that the theory was so complicated. To account for all the observed movements of the planets in the sky it had become necessary to introduce an alarmingly large number of small circles or epicycles. In science, a simple and straightforward theory is generally more accurate than a cumbersome one, and Copernicus looked for some way of avoiding the complications which Ptolemy had been forced to introduce.

In one way Copernicus was better off than Ptolemy; he could make use of more accurate measurements of the planetary motions, and he could be sure that these measurements were not greatly in error. Finally hecame to the conclusion that there was only one solution. The Earth must be dethroned from its proud position in the centre of the universe.

Today, we are so used to thinking of the Sun as the most important body in the Solar System that we find it hard to consider any other idea. Yet Copernicus knew that he was taking a bold step. He was claiming that astronomy then as being taught in all universities and schools was utterly wrong. Worse still, he faced opposition from the Church, which would certainly object to the idea that our world is a mere planet. Copernicus was a Church official, and though his theory was probably more or less complete by 1533 he did not feel inclined to publish it.

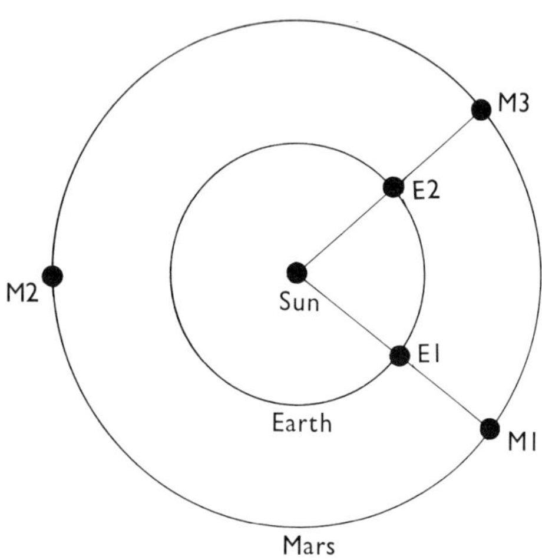

OPPOSITIONS OF MARS. *When the Earth is at E1 and Mars M1, the Sun, the Earth and Mars are in almost a straight line; Mars is opposite to the Sun in the sky, and is at* opposition. *A year later the Earth has returned to E1; but Mars, moving more slowly in a larger orbit, has not completed a full revolution, and lies at M2, so that it is unfavourably placed for observation. Another opposition does not occur until the Earth has 'caught Mars up', when the Earth will be at E2 and Mars at M3. This is why oppositions of Mars occur only at intervals of about 780 days (the* synodic period *of Mars).*

The crux of the whole problem was that the planets do not move regularly and steadily against the background of stars. As we have noted, there are times when the planets move in a westward or retrograde direction. Mercury and Venus have their own way of behaving, because they are closer to the Sun than we are, and even on Ptolemy's system they were nearer to us than any other bodies apart from the Moon. The remaining bright planets, Mars, Jupiter and

Saturn, are easier to follow, because they can be seen against a really dark background. Mars is shown in the diagram, but the same arguments apply to Jupiter and Saturn (as well as to Uranus, Neptune and Pluto, which were of course unknown in Copernicus' time).

For most of the time Mars moves from west to east. We know now that its orbit round the Sun lies beyond that of the Earth, and also that it is travelling more slowly: only 24 kilometres per second, as against nearly 30 kilometres per second for the Earth. In the diagram, we begin with the Earth and Mars in the positions marked **1**. By the time that the Earth has reached **2**, Mars also has reached **2**, and so on. It is obvious that in positions **4** and **5**, Mars is apparently moving across the sky in a retrograde direction, though its real motion around the Sun is unaltered; the Earth is overtaking it, so to speak. In positions **6** and **7** things have reverted to normal.

A planet which performs a slow 'loop' in the sky was very hard to explain on Ptolemy's theory, and this was one of the main reasons why Copernicus rejected it.

Mars is best seen at intervals which are separated, on average, by 780 days (the interval is not quite constant). At such times the Earth is almost directly between Mars and the Sun. Mars is therefore opposite to the Sun in the sky, and is said to be at *opposition*, as shown in the next diagram.

Again we begin with the Earth at **E1** and Mars at **M1**, so that Mars is at opposition. A year later the Earth has completed one revolution, and is back at **E1**, but Mars has not; its 'year' is 687 Earth-days, so that it is travelling more slowly in a larger orbit. The Earth has to catch it up, and does so when the two planets are at **E2** and **M3** respectively, so that there is another opposition. The interval between successive oppositions is known as the *synodic period*. When the planet is on the far side of the Sun it is at *conjunction*, and is to all intents and purposes out of view.

Jupiter and Saturn are much further away than Mars, and move more slowly, so that it takes the Earth less time to catch them up. The synodic period of Jupiter is 399 days, and for Saturn 378 days. This means that Jupiter and Saturn are always well placed for several months in each year, while Mars is not. Thus there were oppositions of Mars in 1978, 1980 and 1982, but none in 1979 or 1981.

It would be wrong to suggest that Copernicus solved all the problems at one stroke. He did nothing of the kind. He still believed that the orbits of all the planets must be perfectly circular, and this led to fresh difficulties. Eventually he was reduced to bringing back epicycles, thereby falling right into the trap which he had been trying so hard to avoid. Yet he had taken the essential step, even if for the moment he kept it more or less to himself.

News of the new theory leaked out, and Copernicus was urged to publish it. His book, *On the Revolutions* (in Latin, *De Revolutionibus Orbium Cœlestium*) was complete, but for many years Copernicus held back. Not all the Church leaders were hostile; the Archbishop of Capua, Cardinal von Schönberg, urged publication, but the lead was eventually taken by Georg Rhæticus, Professor of Mathematics at Wittenberg. Rhæticus visited Copernicus at Frombork, and became highly enthusiastic; he stayed for over two years, and at last Copernicus gave way. He dedicated the book to the Pope, Paul III, and Rhæticus took the manuscript to Nürnberg to have it printed.

In 1543 *De Revolutionibus* appeared, though the cautious publisher,

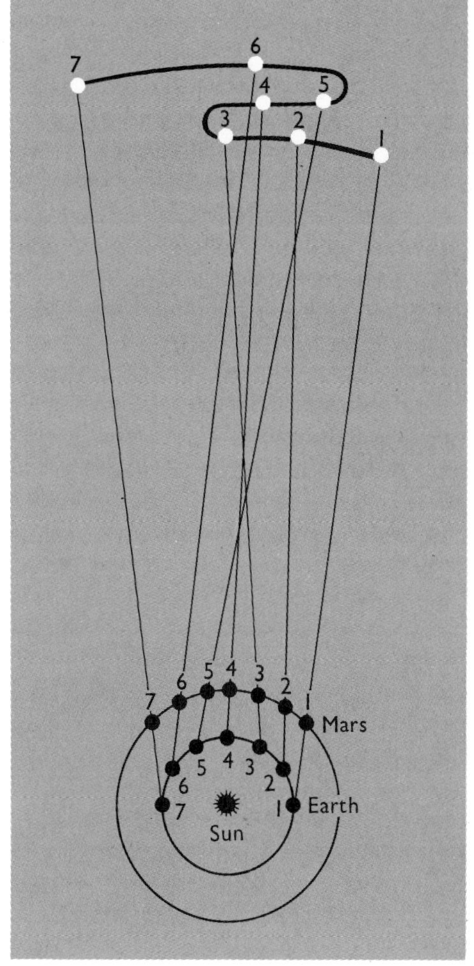

RETROGRADE MOVEMENT OF MARS. *The apparent path of Mars in the sky is given at the top of the diagram, and the actual relative positions of the Earth and Mars at the bottom. It will be seen that between positions 3 and 6 the Earth catches up Mars and passes it, so that for this period Mars seems to move in a retrograde or backward direction among the stars. Behaviour of this sort was very difficult to explain on the old theory according to which the Sun moved round the Earth, and was one of the reasons why Ptolemy was forced to add further epicycles. Of the planets known in ancient times, Jupiter and Saturn behave in similar fashion, but the effects are less obvious because both these planets are so much farther away from the Sun, and their apparent movements in the sky are slower.*

Osiander, had added a note to the effect that the theory was merely a convenient fiction helpful in working out tables for the planets. Copernicus had not authorized any such thing, but by now he was an old man, and was seriously ill. It is said, perhaps with truth, that the first printed copies of the book reached him a few hours before he died.

His fears had been well-founded, and the Church in general showed itself bitterly hostile. The famous religious leader Martin Luther commented that 'This fool seeks to overturn the whole art of astronomy. But as the Holy Scriptures show, Jehovah ordered the Sun, not the Earth, to stand still'. At first only Rhæticus and his friend Erasmus Reinhold, also of Wittenberg, dared to teach the new theory openly. It is also important to note that the first tables of the planets prepared on the basis of the Copernican system – Reinhold's 'Prutenic Tables' of 1551 – were very little better than the Ptolemaic ones had been, mainly because even Copernicus, with all his insight, could not bring himself to break free from the concept of perfectly circular orbits.

We have to admit that Copernicus realized only part of the truth. He made many mistakes, and parts of his book were as unsound as the *Almagest* had been. But the breakthough had been made, and history will never forget the quiet man from 'the remotest corner of the Earth'.

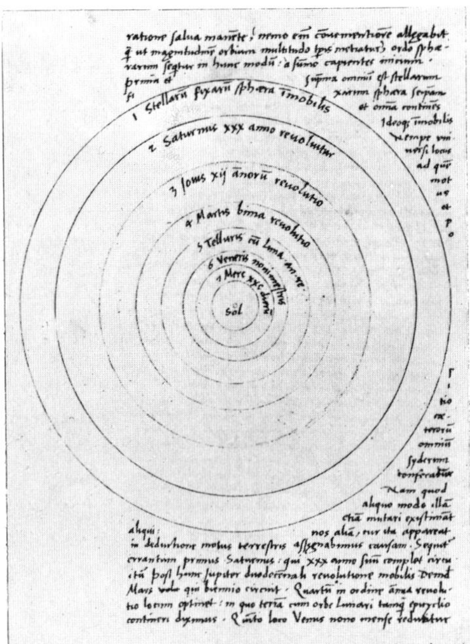

PAGE FROM COPERNICUS' GREAT BOOK. *A reprint of a typical page from the* De Revolutionibus Orbium Cœlestium *of Copernicus, published in 1543.*

ARMILLARY SPHERE, *containing twenty circles; the Zodiac is clearly shown. From an old print.*

6

The Story of Tycho Brahe

THREE years after the death of Copernicus and the publication of *De Revolutionibus*, a boy of very different character was born in Denmark. His name was Tycho Brahe, and in every way he was unlike Copernicus – except that he had the same love of astronomy, and the same urge to find out as much as he could.

He was a Dane of noble ancestry; his father Otto was Governor of Helsingborg Castle, and a man of very considerable influence. Tyge or Tycho, his eldest son and the second of his five children, first saw the light of day on December 14, 1546, at the family seat at Knudstrup in Skaane. (To be accurate there were six children, but one of them – Tycho's twin brother – died either at birth or a few hours afterwards.) Early events set the pattern for the tempestuous life which was to follow. For some unknown reason Otto had promised his brother Jørgen, an officer in the Danish Navy, that as soon as he had another son it would be time to hand little Tycho over to be brought up in Jørgen's household. When another baby boy was born, Otto and his wife Beate apparently had second thoughts, with the result that Uncle Jørgen kidnapped Tycho and removed him. We do not know the full story, but it seems that Tycho spent his youthful years at Torstrup, Jørgen's home, rather than with his parents. It also seems that after a while the situation was amicably resolved, and there were no hard feelings on either side.

Actually we do not know much about the boy's upbringing, but probably it was conventional enough, and the first step after school was University. So in 1559 Tycho went off to Copenhagen (it was then, incidentally, that he started to sign himself 'Tycho' instead of the Danish 'Tyge'). During his stay there he saw a partial eclipse of the Sun. It impressed him, and it may be that his love of astronomy began then.

Star-gazing was all very well, but it did not suit his uncle, who was set upon making Tycho an important figure in Danish political circles. Clearly, then, the law should be the chosen study, and Leipzig University had an excellent law school. However, Jørgen did not trust Tycho's dedication to legal studies, and probably cast unfavourable looks at the Latin translation of Ptolemy's *Almagest* which the boy had bought when still at Copenhagen. So a travelling companion, Anders Vedel, was selected to go to Leipzig too, and, metaphorically, keep Tycho's nose to the legal grindstone. It was an impossible task. Vedel did his best, but had to agree to a compromise; the student could look at the stars by night provided that he put in a full day's work first.

In 1563 there was a conjunction of the two bright planets Jupiter and Saturn; that is to say, they appeared side by side in the sky. Tycho realized that the date of the conjunction as given in the old Alphonsine Tables was a full month wrong, and even Reinhold's Prutenic Tables were in error by several days. This was Tycho's first recorded observation, and he made it in a decidedly primitive way. His only

TYCHO BRAHE, *the great Danish observational astronomer*

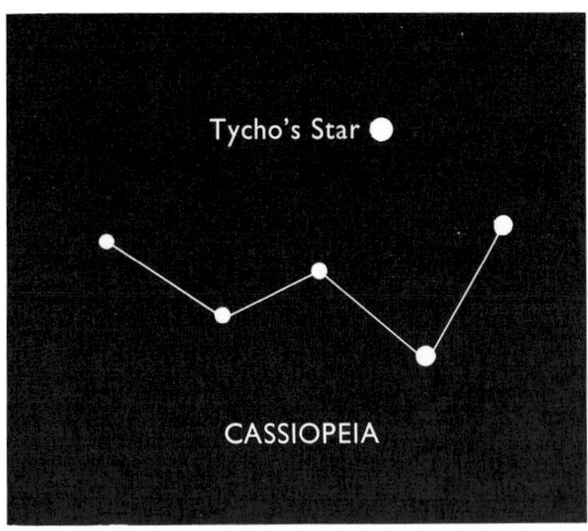

POSITION OF TYCHO'S STAR. *Tycho made careful observations of the brilliant supernova of 1572, which appeared in Cassiopeia not far from the famous 'W' of stars. This was the brightest supernova of the past thousand years. Only three others have been seen in our own Galaxy during this time; the 1054 star was certainly a supernova, and so was Kepler's Star of 1604, while it has recently been established that a star seen in 1006 in the constellation of Lupus (the Wolf) must also have been a supernova.*

equipment consisted of a pair of compasses. By holding the centre close to his eye, and pointing one arm to Jupiter and the other to Saturn, he was able to find the angular distance between them. This did not satisfy him, and he began to make proper measuring instruments.

In 1565 Tycho's uncle died as the result of a curious accident; he caught pneumonia after soaking himself in pulling the King of Denmark out of a moat into which he had fallen. Tycho was genuinely distressed, but there was nothing now to keep him at his legal studies, and he promptly abandoned them. He went to Rostock University in Germany, and it was here that he met with a famous misadventure; he quarrelled with another student, one Parbsjerg, and after a duel fought in the best mediæval tradition he had part of his nose sliced off. Ever resourceful, he repaired the damage with gold, silver and wax, and this was how his nose remained for the rest of his life, apparently without causing him any discomfort!

Another episode of the Rostock period demonstrates Tycho's faith in astrology. He predicted that the eclipse of the Moon in October 1566 foretold the death of the Sultan of Turkey. Before long news came through that the Sultan had indeed departed this earthly life; but it later transpired that he had died before the eclipse, not afterwards, which cast some doubt upon the validity of the prediction.

By 1571 Tycho was back in Denmark, not for pleasure but because his father Otto was dying. After the funeral he stayed for a while, and then came the turning-point of his career: the appearance of a brilliant new star in the constellation of Cassiopeia. It blazed out in November. Let us hear the story in Tycho's own words:

> 'In the evening, after sunset, when, according to my habit, I was contemplating the stars in a clear sky, I noticed that a new and unusual star, surpassing all the other stars in brilliancy, was shining almost directly above my head; and since I had, from boyhood, known all the stars of the heavens perfectly (there is no great difficulty in attaining that knowledge), it was quite evident to me that there had never before been any star in that place in the sky, even the smallest, to say nothing of a star so conspicuously bright as this. I was so astonished at this sight that I was not ashamed to doubt the trustworthiness of my own eyes. But when I observed that others, too, on having the place pointed out to them, could see that there really was a star there, I had no further doubts. A miracle indeed, either the greatest of all that have occurred in the whole range of nature since the beginning of the world, or one certainly that is to be classed with those attested by the Holy Oracles.'

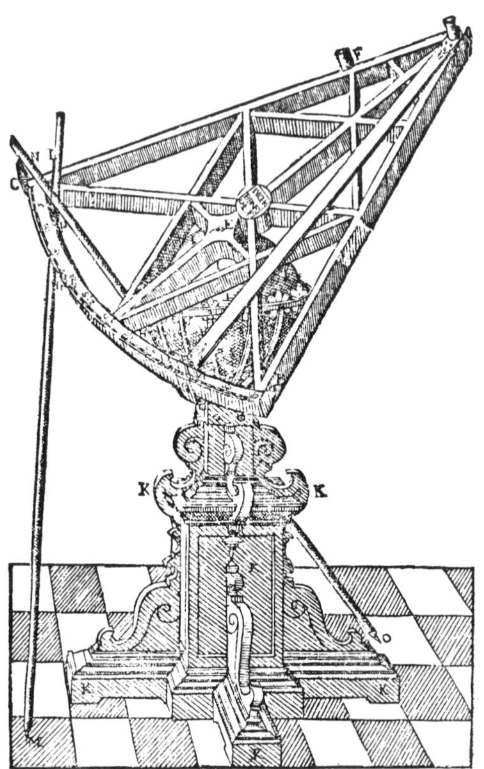

SEXTANT USED BY TYCHO BRAHE. *This sextant was in use by Tycho in 1577. It was one of the elaborate instruments which were set up on the island of Hven, where Tycho worked for so many years.*

What made this star so important? Simply the fact that according to Aristotle, the heavens and the starry sphere were changeless. Yet this new star was more than normally obtrusive; it could be seen even in daylight. Tycho was not the first to observe it. From all accounts it was discovered by Wolfgang Schuler of Wittenberg on 6 November, and Tycho did not see it until the 11th, but because of his careful studies of it we know it as Tycho's Star.

It was nothing more nor less than a supernova, not so brilliant as that in Lupus in 1006, but fully equal to the 1054 outburst in Taurus. Tycho, of course, had no idea of its real nature, but at least he could make precise measurements of it – and he did. He found that there was no detectable change in its position, and concluded that it must be very remote. As the weeks went by, it faded. By December 1572 it was about equal to Jupiter; by March 1573 it had fallen to the first magnitude, and

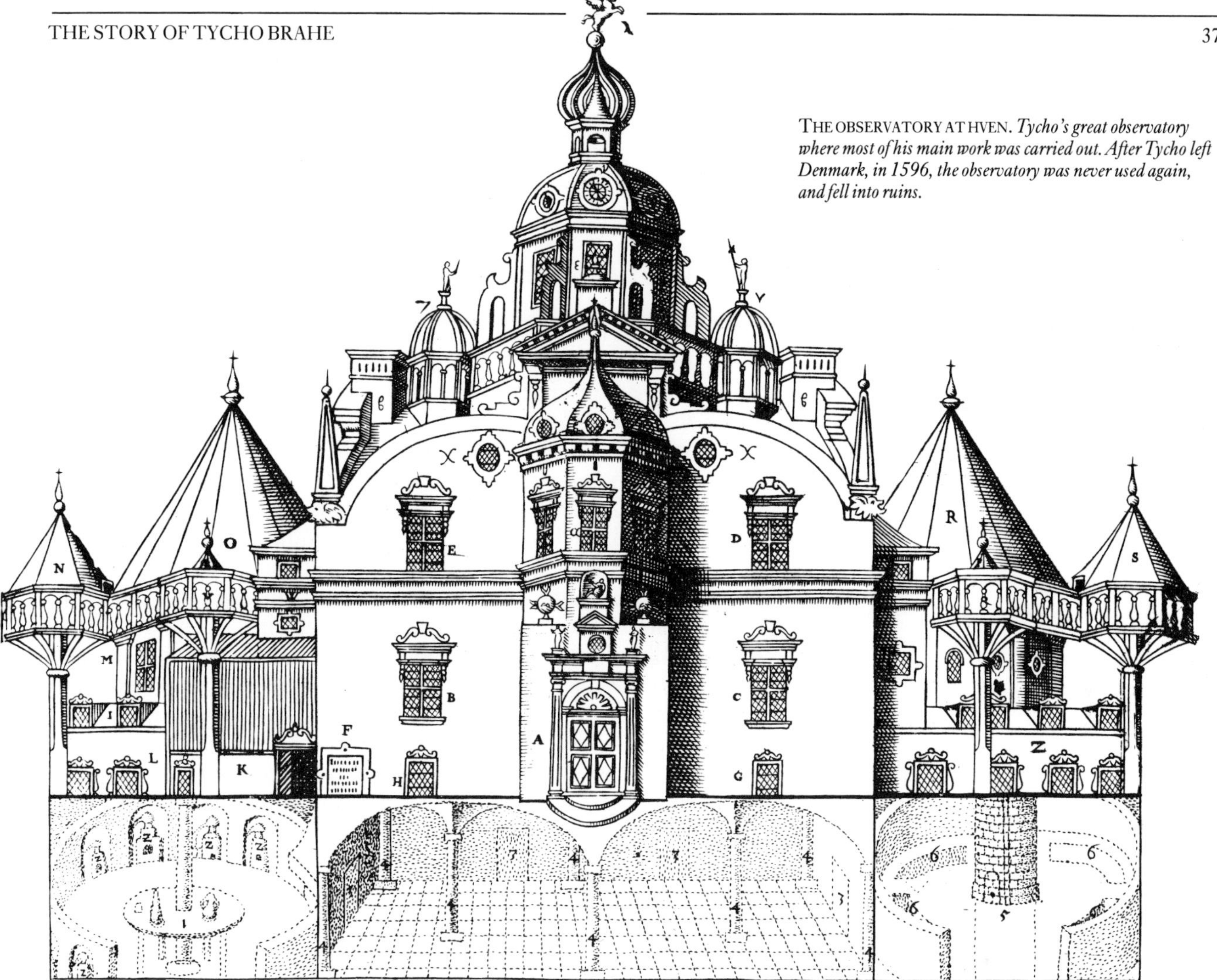

THE OBSERVATORY AT HVEN. *Tycho's great observatory where most of his main work was carried out. After Tycho left Denmark, in 1596, the observatory was never used again, and fell into ruins.*

it disappeared in March of the following year. During the decline its colour changed, and Tycho was not slow to point out the astrological implications:

> 'The star was at first like Venus and Jupiter, giving pleasing effects; but as it then became like Mars, there will next come a period of wars, seditions, captivity and death of princes, and destruction of cities, together with dryness and fiery meteors in the air, pestilence, and venomous snakes. Lastly, the star became like Saturn, and there will finally come a time of want, death, imprisonment and all sorts of sad things.'

This extract, like the first, comes from Tycho's book *De Stella Nova* (On the New Star), which was published in 1573. Naturally enough, Tycho believed that the star would eventually vanish completely, and that by the time it had faded below naked-eye visibility it had shrunk down to the size of the Earth. This may seem very wide of the mark, and indeed it is, but it pales beside the theories of Georg Busch of Erfurt, who wrote two pamphlets about the star. He regarded it as a comet, 'formed by the ascension from Earth of human sins and wickedness... This poisonous stuff falls down upon people's heads, and causes all kinds of unpleasant phenomena, such as diseases, sudden death, bad weather and Frenchmen.' (Comments from French authorities are, perhaps fortunately, not on record.) However, the vital point was that the 1572 supernova made Tycho decide to devote the whole of the rest of his life to astronomy.

STJERNEBORG. *This is a view of Stjerneborg, the second of Tycho's two observatories on Hven. The buildings shown are of course modern, since nothing now remains of the original observatory. Photograph by Gösta Persson, 1958.*

By now Tycho was becoming well-known, and in 1576 the King of Denmark, Frederik II, offered to provide him with a full-scale observatory, together with funds to maintain it. The chosen site was Hven, a low-lying island in the Baltic between Copenhagen and the Swedish town of Malmö; it is today known as Ven. Tycho accepted, and in 1576 he began the construction of his observatory – Uraniborg, the 'Castle of the Heavens'. It was built in the middle of a large square enclosure laid out as a garden, the corners of which pointed north, south, east and west. It contained a library and a chemical laboratory as well as living quarters and the rooms for the instruments themselves. Later, in 1584, he built a second 'Castle of the Stars', Stjerneborg, in which some of the instruments were located below ground level, because when the wind blew strongly the instruments above the ground were shaken to and fro. Tycho also added a printing press and a paper-mill. Hven became a hive of scientific activity, and many distinguished people from all over the world visited it. One guest was James VI of Scotland, afterwards James I of England.

Tycho lived in magnificent style, and those who came to Uraniborg were royally entertained. Banquets, games and hunts were held, and one member of the retinue was a pet dwarf named Jep, who was said to be gifted with second sight. On the other hand, the islanders were not well treated. Tycho was a harsh landlord, and even built a prison to hold those tenants who were reluctant to pay their rents.

The instruments themselves were by far the best of their time, and since Tycho was a most careful and accurate observer he obtained excellent results. He measured the positions of 777 stars, and drew up a catalogue; his star positions were never in error by more than one or two minutes of arc. When we remember that he had no telescopes, we can see how good an astronomer he must have been. Yet he was still devoted to astrology, and it is said that he never began observing without dressing himself in special robes.

As well as compiling his star catalogue Tycho measured the movements of the planets, and it was these observations which proved to be so valuable later on. He observed seven comets during his stay at Hven, and correctly stated that they were much more distant than the Moon, thereby disposing of the old theory that comets were 'atmospheric exhalations' a few miles up. On the other hand he could never accept Copernicus' theory of a moving Earth, partly on religious grounds but mainly because he believed that on the Copernican system the stars would have to be impossibly remote. He preferred a kind of hybrid system, according to which the planets revolved round the Sun while the Sun and Moon were in orbit round the Earth.

Unkind people have referred to this system as Tycho's Folly, but the great Dane was inordinately proud of it, and reacted furiously when he believed that it had been copied and re-presented by an avowed enemy of his, Reymers Bär, who had begun life as a swineherd and had risen to become a professor of mathematics. But more mundane matters had begun to intrude upon his life at Hven, and the long period of fruitful observation was coming to an end.

STATUE OF TYCHO BRAHE

King Frederik II died in 1588, and about the same time Tycho lost another valuable ally, the Danish Chancellor Niels Kaas. His hot temper, his treatment of his tenants, and his neglect of various official duties led to a break with the Court. His funds were cut off; Tycho abandoned Hven, taking the main observing instruments with him, and after a spell in Germany arrived in Prague, capital of Bohemia, to become Imperial Mathematician to the Holy Roman Emperor, Rudolph II. A castle at Benatek, outside Prague, was placed at his disposal, but conditions there were very different from those at Hven. The Emperor himself was a curious character; he was mainly interested in astrology and alchemy (the so-called making of gold out of baser elements), and he was melancholy and incompetent. His reign was one succession of disasters, and finally, in 1611, he was deposed. Even at the time of Tycho's arrival he was short of money, and it was not easy to pay the salary which had been promised. Before long Benatek was given up, and Tycho settled in Prague itself.

By this time he had been joined by a younger man, Johannes Kepler, who was later to make the best possible use of the observations which Tycho had collected. The two men were not always on good terms, and it is very probable that the main fault was Tycho's, but they managed to work together for some time. Then, on November 24, 1601, Tycho died suddenly.

In every way Tycho Brahe was a picturesque figure. Hasty, intolerant, proud and often cruel, he was at the same time brilliantly clever, sincere and hard-working.

There is little more to be said about Hven, which was ceded to Sweden in 1658 and is still Swedish. Uraniborg and Stjerneborg were never used again, and nothing now remains of them. The site of Tycho's castle is occupied by a grassy dip and some trees, with only a few markers to indicate where the building once stood. Yet when I visited Hven a few years ago, and walked up to the old observatory, I found Tycho's statue – vast, commanding and powerful, surveying the scene where so much pioneer work had been done. The spirit of the Master of Hven still lingers on.

TYCHO BRAHE'S QUADRANT. *With instruments of this kind, all built by himself, Tycho drew up his remarkably accurate star catalogue.*

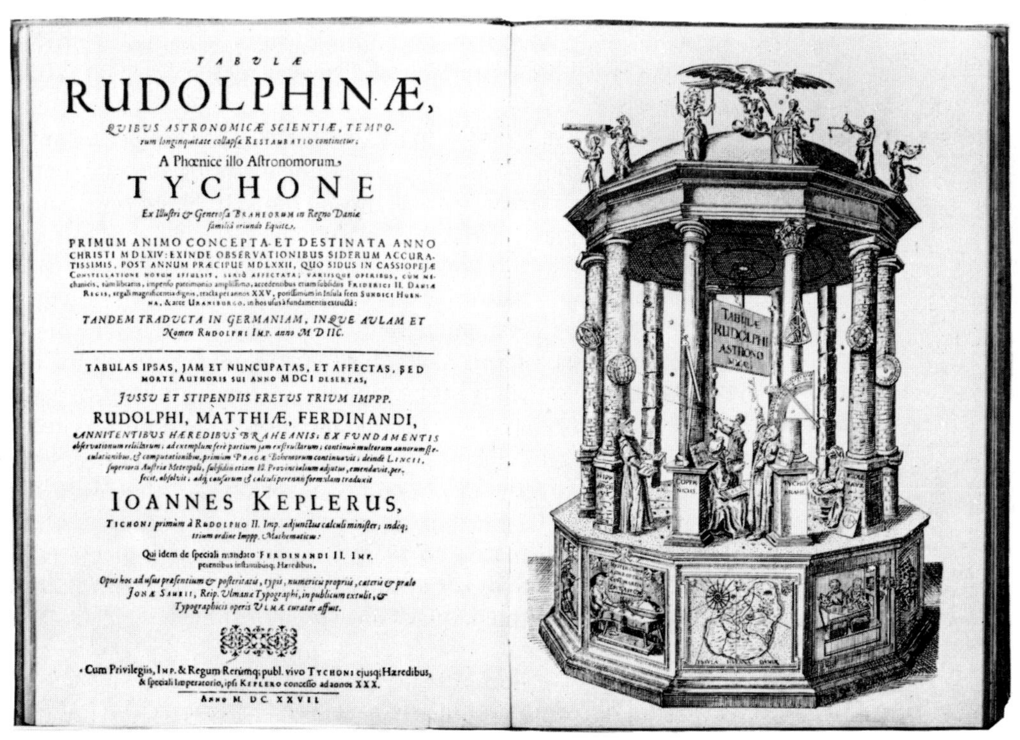

TABULÆ

RUDOLPHINÆ,

QUIBUS ASTRONOMICÆ SCIENTIÆ, TEMPOrum longinquitate collapsæ RESTAURATIO continetur;

A Phœnice illo Astronomorum

TYCHONE

Ex Illustri & Generosa BRAHEORUM in Regno Daniæ familiâ oriundo Equite,

PRIMUM ANIMO CONCEPTA ET DESTINATA ANNO CHRISTI MDLXIV: EXINDE OBSERVATIONIBUS SIDERUM ACCURATISSIMIS, POST ANNUM PRÆCIPUE MDLXXII, QUO SIDUS IN CASSIOPEJÆ CONSTELLATIONE NOVUM EFFULSIT, SERIÒ AFFECTATA; VARIISQUE OPERIBUS, CÙM MEchanicis, tùm librariis, impenso patrimonio amplissimo, accedentibus etiam subsidiis FRIDERICI II. DANIÆ REGIS, regali magnificentia dignis, tracta per annos XXV, potissimùm in Insula freti SUNDICI HUENNA, & arce URANIBURGO, in hos usus à fundamentis extructâ:

TANDEM TRADUCTA IN GERMANIAM, INQUE AULAM ET Nomen RUDOLPHI IMP. anno MDIIC.

TABULAS IPSAS, JAM ET NUNCUPATAS, ET AFFECTAS, SED MORTE AUTHORIS SUI ANNO MDCI DESERTAS,

JUSSU ET STIPENDIIS FRETUS TRIUM IMPPP.

RUDOLPHI, MATTHIÆ, FERDINANDI,

ANNITENTIBUS HÆREDIBUS BRAHEANIS; EX FUNDAMENTIS observationum relictarum; ad exemplum ferè partium jam exstructarum; continuis multorum annorum speculationibus, & computationibus, primùm PRAGÆ Bohemorum continuavit; deinde LINCII, superioris Austriæ Metropoli, subsidiis etiam Il. Provincialium adjutus, emendavit, perfecit, absolvit; adq; causarum & calculi perennis formulam traduxit

IOANNES KEPLERUS,

TYCHONI primùm à RUDOLPHO II. Imp. adjunctus calculi minister; indeq; trium ordine Imppp. Mathematicus:

Qui idem de speciali mandato FERDINANDI II. IMP. petentibus instantibusq; Hæredibus,

Opus hoc ad usus præsentium & posteritatis, typis, numericis propriis, cæteris & prælo JONÆ SAURII, Reip. Ulmanæ Typographi, in publicum extulit, & Typographicis operis ULMÆ curator affuit.

Cum Privilegiis, IMP. & Regum Rerúmq; publ. vivo TYCHONI ejusq; Hæredibus, & speciali Imperatorio, ipsi KEPLERO concesso ad annos XXX.

ANNO M. DC. XXVII.

FRONTISPIECE OF THE RUDOLPHINE TABLES. *These tables represented Kepler's last astronomical work, and were published shortly before his death. Kepler's acknowledgement to Tycho is prominently featured on the title-page. The tables were so named in honour of Kepler's old benefactor, the Holy Roman Emperor Rudolph II, who was interested mainly in mysticism and astrology, but who also encouraged astronomical science. Rudolph himself had of course been dead for many years by the time that the tables appeared.*

7

Maps of the Sky

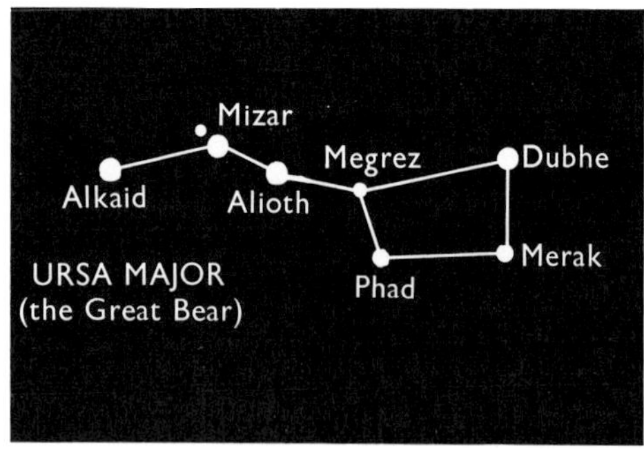

URSA MAJOR, THE GREAT BEAR. *The seven famous stars make up the 'Plough' or 'Big Dipper', but the whole of Ursa Major is a very extensive constellation.*

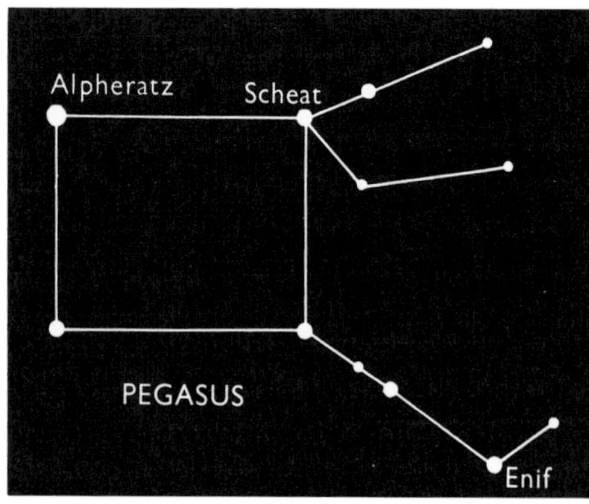

PEGASUS, THE FLYING HORSE. *Pegasus is a prominent constellation; it is marked by four stars arranged in a square. For some reason, however, one of the stars (Alpheratz) is now included in the neighbouring constellation of Andromeda and is officially termed Alpha Andromedæ: it used to be known as Delta Pegasi. In mythology, Pegasus was the horse upon which the hero Bellerophon rode in his expedition against a terrible fire-breathing monster, the Chimæra.*

TYCHO Brahe's star catalogue was the most accurate of its time, but it was not the only one. Johann Bayer, born in Bavaria in 1572 (the year of Tycho's supernova) published his *Uranometria* in 1603; it listed the positions and magnitudes of 500 stars in addition to Tycho's total of 777. It was particularly important, because in it Bayer introduced a new system of stellar nomenclature. Bayer was a lawyer by profession, so that it is not surprising that his system was strictly logical.

What he did was to give Greek letters to the stars in each constellation, beginning with Alpha and working through to Omega. Alpha Arietis is therefore the brightest star in Aries, the Ram; the second brightest is Beta Arietis, the third Gamma and so on. Some stars have individual or proper names; thus Alpha Orionis is Betelgeux and Beta Orionis is Rigel, while the catalogue designation of Sirius is Alpha Canis Majoris, the brightest star in Canis Major (the Great Dog). Sometimes the letters are out of order, so that, for instance, Rigel is brighter than Betelgeux. Still, the system works satisfactorily on the whole, and it is still used.

The constellation patterns which we know have come down to us from Ptolemy, so presumably they are of Greek origin, though it has also been suggested that they were due originally to the Chaldæans. At any rate, Ptolemy listed a total of forty-eight groups, including familiar ones such as Ursa Major, Ursa Minor, Cassiopeia and Orion. Twelve of the constellations were in the Zodiac; of the rest, twenty-one were wholly or mainly in the northern hemisphere of the sky, and the other fifteen in the south. Not all the sky was covered; patches were left out, as well as the far southern regions which never rose above the horizon from Alexandria.

This was unsatisfactory. Later sky-mappers modified Ptolemy's constellation boundaries, and introduced new groups, not only in the far south but also in other regions. Tycho himself added one: Coma Berenices, or Berenice's Hair, not far from the Great Bear. Bayer was responsible for eleven, all southerly. But before long extra groups were added for no obvious reason; it was said that no astronomer seemed comfortable until he had created a constellation of his own. One of the worst offenders was Johann Elert Bode, an eighteenth-century German whom we will meet again later in connection with Bode's Law. His proposed constellations had hideous names such as Sceptrum Brandenburgicum (the Sceptre of Brandenburg), Honores Frederici (the Honours of Frederick), Globus Ærostaticus (the Balloon), Lochium Funis (the Log Line) and even Officina Typographica (the Printing Press). Fortunately these have now passed into merciful oblivion, though one of Bode's groups, Quadrans (the Quadrant) has given its name to the annual Quadrantid meteor shower which is seen in early January each year.

It cannot be said that the constellations are either logical or convenient. They are very unequal in area; the largest (Hydra, the

Watersnake) covers over 1300 square degrees of the sky, while the smallest, Crux Australis (the Southern Cross) has only 68 square degrees. Some are very brilliant; others are dim and shapeless. However, the patterns are now so well established that they will certainly not be altered. The situation was ratified by order of the International Astronomical Union in 1933, when the boundaries were finally laid down. One casualty was Argo Navis, the Ship Argo – one of Ptolemy's originals – which was so large that it was unwieldy. It was brutally chopped up into a Keel, Sails and a Poop!

Many of the newer constellations have modern names; Telescopium (the Telescope), Sextans (the Sextant) and so on. The old ones are mainly mythological, and it has been said that the sky is a cosmical story-book. Who has not heard, for instance, of the legend of Perseus and the sea-monster?

According to this tale, there was once a queen, Cassiopeia, whose daughter Andromeda was exceptionally beautiful. Cassiopeia went so far as to boast that she was lovelier than the sea-nymphs or Nereids, children of the sea-god Neptune. Now Neptune, like most Olympians, was decidely touchy, and to show his displeasure he sent a monster to lay waste the kingdom. Things went from bad to worse, until Cassiopeia and her husband, King Cepheus, were in despair.

What was to be done? Cepheus consulted the Oracle at Delphi, and was told bluntly that the only way to save his country was to chain Andromeda to a rock and leave her by the sea-shore, so that the monster could have a tasty breakfast. Reluctantly, Cepheus agreed. Luckily, it so happened that the hero Perseus had been on an expedition to kill the Gorgon, Medusa – a woman with snakes instead of hair, and whose glance could turn any living creature into stone. Perseus had been helped by the gods; he had been provided with a pair of winged sandals, and given a magic shield which protected him from the Gorgon's baleful stare. He had duly cut off Medusa's head, and was flying home in triumph when he saw Andromeda chained to the rock. With exemplary gallantry he swooped down, turned the monster to stone, and saved the situation, after which he was suitably rewarded with Andromeda's hand in marriage. That was one of the few mythological legends with a happy ending!

THE LEGEND OF PERSEUS AND ANDROMEDA. *This is one of the most famous of the mythological legends associated with the constellation patterns.*

All the main characters are to be found in the sky. Cepheus and Cassiopeia are there; so is Perseus, with the Gorgon's head marked by the famous variable star Algol, the 'Demon'. Even the sea-monster Cetus is among those present, sprawling across the southern sky, though it is true that some versions demote it to the status of a harmless whale.

Tycho, Bayer and others had charted the stars as accurately as was possible in pre-telescopic days, and the right ascensions and declinations of the stars were known with reasonable precision. I must digress to say more about these terms, because in astronomy they are all-important.

We may begin by supposing the sky to be solid – the *celestial sphere*, with its centre coincident with the centre of the Earth. Northward, the Earth's axis points to the North Celestial Pole, approximately marked by Polaris; there is no bright south polar star, the nearest naked-eye object being the fifth-magnitude Sigma Octantis.

On Earth, the latitude of a place is reckoned according to its angular distance north or south of the terrestrial equator. For example, my own

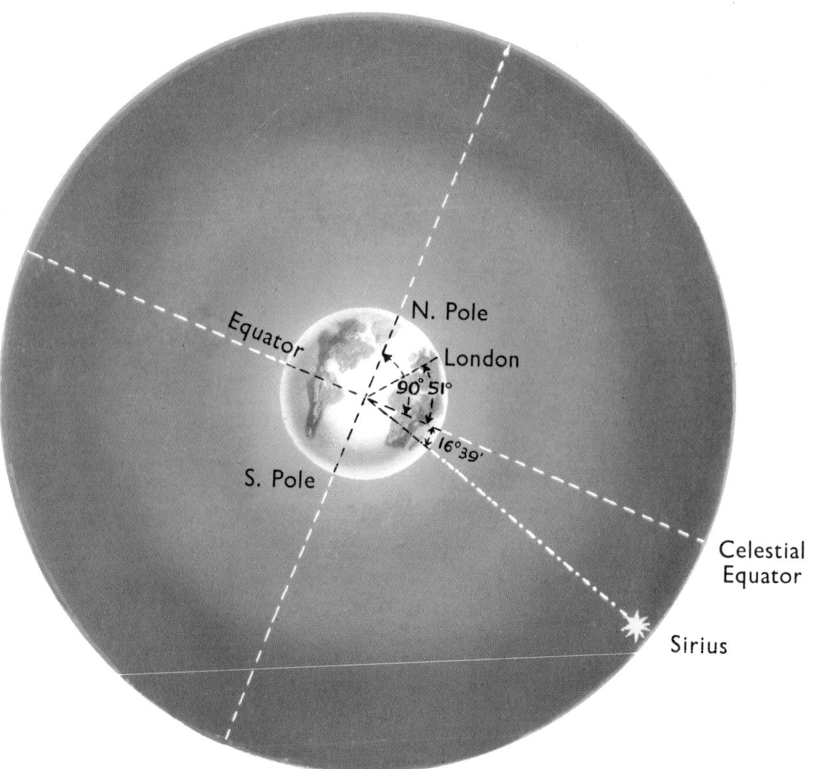

DECLINATIONS OF CELESTIAL OBJECTS. *The declination of a body in the sky is the angular distance of the body north or south of the celestial equator; the celestial equator is the projection of the Earth's equator on to the celestial sphere. Sirius, shown in the diagram, has a declination of south 16 degrees 39 minutes. The North Celestial Pole has, of course, a declination of N. 90 degrees; Polaris, the present Pole Star, lies within one degree of this. There is no conspicuous star within this distance of the South Celestial Pole. The declinations of the stars change very slowly owing to the effects of precession; the declinations of the Sun, Moon and planets naturally alter much more rapidly.*

ARIES, THE RAM. *The first constellation of the Zodiac.*

home at Selsey, in Sussex, has a latitude of 50·7 degrees North. If we project the plane of the Earth's equator on to the celestial sphere, we will have the celestial equator, dividing the sky into a northern and a southern hemisphere. The angular distance of a star north or south of this line is called the star's declination. Sirius has a declination of 16 degrees 41 minutes South (−16°41′), Betelgeux in Orion 7°24′ North, and so on.

If Polaris lay exactly at the polar point, its declination would be +90°. Actually it is +89°10′, so that it is 50 minutes of arc away from the pole – something which air and sea navigators have to be careful to remember. Note also that the altitude of the pole above the observer's horizon is equal to the observer's latitude. Thus from Selsey, the altitude of the pole is 50·7 degrees. From the equator, the pole lies on one horizon and the south celestial pole on the other. South of the equator, Polaris can never be seen.

Declination, then, corresponds to latitude on the Earth. Now let us turn to the celestial equivalent of longitude.

The Sun seems to travel through the Zodiac; its apparent path among the stars is called the *ecliptic*. Each March, about the 21st of the month (the date is not absolutely constant, owing to the vagaries of our calendar), the Sun crosses the celestial equator, moving from south to north. This moment marks the beginning of spring in the northern hemisphere, and the position where the Sun crosses the equator is known as the *vernal* or *spring equinox*. It is also known as the First Point of Aries, since in ancient times it lay in the constellation of Aries, the Ram. This is no longer so. As we have noted, the Earth's axis shifts slightly according to precession, so that the vernal equinox shifts too. By now it has moved out of Aries into the adjacent constellation of Pisces (the Fishes), though the old name is still used.

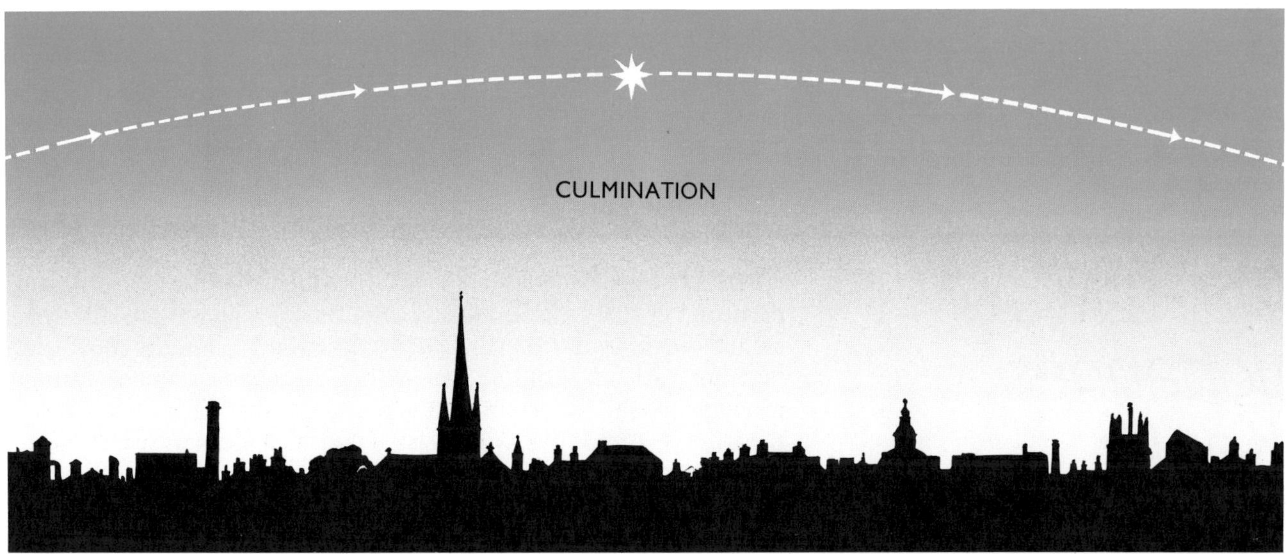

The vernal equinox is used as the zero for our measurements of a star's right ascension. First, we must refer to the *meridian* of the observer, which is a great circle in the sky passing through both poles as well as the *zenith* or overhead point. The vernal equinox crosses the meridian once every 24 hours – often, of course, it does so in daylight – and is then at its greatest height above the horizon, so that it *culminates*.

The difference in time between the culmination of the vernal equinox and the culmination of any particular star gives the star's right ascension. Sirius, for instance, crosses the meridian 6 hours 44 minutes after the vernal equinox has done so; its right ascension is therefore 6h 44m, while that of Betelgeux is 5h 54m, and so on.

It may sound confusing to measure the equivalent of a star's 'longitude' in units of time instead of in angular measure, and various other systems are in use, but on the whole astronomers have found that this is the most convenient method. Once we know a star's right ascension and declination, we can fix its position in the sky as accurately as we can fix a position on the Earth by quoting latitude and longitude.

The stars are so far away that they seem to keep in almost the same relative positions. The right ascensions and declinations change slowly, because of the effects of precession, but this is due only to the shift of the Earth's axis. We can also give the right ascensions of the Sun, Moon and planets, but obviously these will change rapidly.

Today, it is possible to give the positions of the stars to within a tiny fraction of a second of arc. Tycho Brahe had to be content with an accuracy of one or two minutes of arc. Yet this is no reflection upon his skill. Considering that he had no telescopes, it is amazing that he was as correct as he actually was.

Kepler, in particular, had great faith in Tycho – and it is ironical that he used the great Danish astronomer's work to prove the truth of the Copernican system which Tycho himself had so decisively rejected.

CULMINATION OF A STAR. *A celestial body is said to* culminate *when it reaches its greatest apparent height above the horizon.*

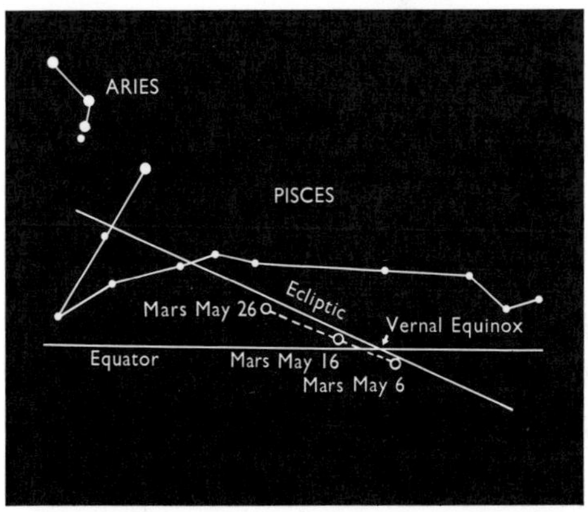

APPARENT PATH OF MARS IN MAY 1960. *Between May 6 and May 16, Mars passed by the First Point of Aries (Vernal Equinox) and also moved across the equator from the southern into the northern hemisphere of the sky.*

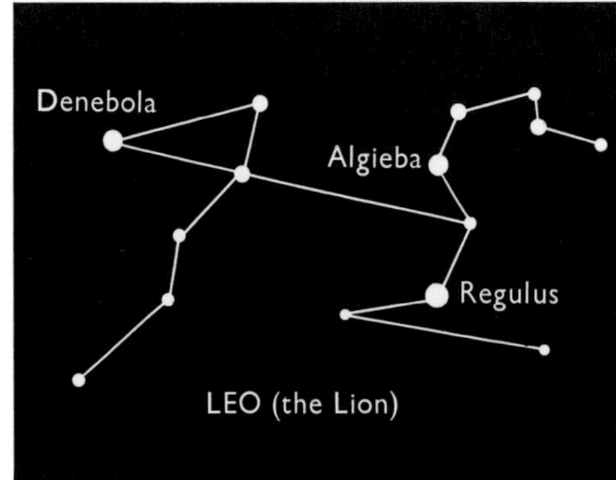

LEO. *Leo, the Lion, is one of the Zodiacal constellations. It contains one first-magnitude star (Regulus), and the 'Sickle', shaped rather like a question-mark twisted the wrong way round, is very easy to identify. Algieba and Denebola are of the second magnitude.*

8

The Laws of Johannes Kepler

THE first of the tragedies resulting from the Copernican theory took place on February 17, 1600. Giordano Bruno, who was not strictly an astronomer but who nevertheless possessed deep scientific knowledge together with an inquiring mind, had been going round Europe teaching that the Earth is in motion round the Sun. This did not please the Church authorities, who insisted that such ideas were dangerous and heretical. Finally Bruno came to Rome, and was captured by the dreaded Inquisition. After being kept in prison for seven years, he was burned at the stake.

It would be wrong to suppose that Bruno was burned because he had supported Copernicus. This was merely one of his many crimes in the eyes of the Church. Yet it showed that to put forward such theories was decidedly unsafe, and that any far-sighted scientist would be lucky to avoid serious trouble.

Tycho, of course, was in no danger at all, if only because he never doubted that the Sun moves round the Earth. He undoubtedly hoped that his assistant, Johannes Kepler, would use the Hven observations to verify the clumsy and artificial-looking Tychonic theory. Kepler began his work with an open mind. Fortunately, he was as brilliant a theorist as Tycho had been an observer.

It was a strange quirk of fate which drew Tycho and Kepler together, even though briefly, because no two men could have been less alike. Tycho was a nobleman, intensely practical, bombastic, unashamedly egoistic, and a superb observer. Kepler came of a somewhat disreputable family background, and had no practical skill at all, quite apart from the fact that a boyhood illness had left him with poor eyesight. Some of his views were far in advance of their time, while others were purely mediæval. It is fair to describe him as a curious mixture.

He was born on December 27, 1571, at Weil der Stadt in Württemberg. His grandfather had been the local burgomaster, and no doubt a worthy citizen, but the family reputation had declined. Johannes' father was shiftless and irresponsible; he had a chequered career, and finally vanished altogether. Katharine, his wife, was a busybody with a malicious tongue and a leaning toward the occult, which was to cause endless trouble in later life. For Johannes, the eldest of the six Kepler children, home life was the reverse of happy, but one episode is worth recalling; in 1577 his mother took him outdoors to see the bright comet of that year, and it may well be that this sparked off the boy's interest in astronomy.

Kepler meant to become a Lutheran pastor, and in 1589 he arrived at the University of Tübingen, where he studied mathematics as well as theology. It was while he was here that he became a convinced Copernican, inasmuch as he was sure that the Sun, not the Earth, lay in the centre of the Solar System. Events took an unexpected turn in 1594, when the authorities of the main school at Graz, in Styria, wrote asking whether Tübingen could supply a promising graduate to teach

JOHANNES KEPLER, *who drew up the famous Laws of Planetary Motion.*

mathematics and elementary astronomy. Kepler was nominated, and accepted, but he was not a success; he was an appallingly bad teacher, and his classes were badly attended, though he did his best.

During his spell at Graz he continued with his investigations into the design of the Solar System, and in 1596 he published a book, the English title of which may be given as the *Cosmographic Mystery*. Frankly, much of it was valueless, because Kepler was obsessed with a peculiar theory involving five regular solids to be fitted in to the spaces between the orbits of the planets, but Kepler sent a copy to Tycho Brahe, who was impressed. Meanwhile, religious problems were coming to the fore. In 1598 all Lutheran teachers were ordered to leave Graz, and Kepler was glad to accept Tycho's invitation to join him in Prague.

It would have been too much to hope that two men, both of genius and yet so utterly unlike, could work happily together. Actually the matter was never really put to the test, because, as we have seen, Tycho died suddenly in 1601; and Kepler succeeded him as Imperial Mathematician – at a greatly reduced salary.

Things were not easy. Kepler's meagre salary was often left unpaid; Tycho's relations claimed all the priceless manuscripts, and Kepler managed to secure them only after prolonged wrangles; and the situation in Prague was deteriorating steadily, due mainly to the Holy Roman Emperor's blundering incompetence. Yet Kepler did have one asset which was all-important: the Hven observations. In particular, Tycho had made a long series of measurements of the movements of Mars, and it was these which gave Kepler the vital clue.

His task was to work out a system which would fit all the facts. Unfortunately, nothing seemed to work. The theory of an Earth-centred universe was rejected out of hand, but neither could the movements of Mars be fitted into a system based upon the Sun – unless Kepler were prepared to accept that Tycho's observations had been inaccurate. This he refused to do. Finally, Kepler stumbled on the truth. The planetary orbits were not circular at all; they were ellipses.

The idea of elliptical orbits was not entirely new – it had been proposed, rather tentatively, by Arzachel of Toledo as early as 1080 – but it was revolutionary in every sense of the term, and even some of Kepler's admirers found it hard to accept. (Galileo, for instance, never did believe it.) But the breakthrough had been made, and at once Tycho's observations of Mars fell beautifully into place.

It is important to note that the planetary orbits are not very different from circles. For instance, the Earth's distance from the Sun ranges between 147 million and 152 million kilometres – only five parts in 150; the path of Venus is even more circular. Fortunately for Kepler, that of Mars has a higher eccentricity; the limiting distances are 249 million kilometres and only 207 million. But for this, he might never have found the answer.

Once the old idea of circular orbits had been abandoned, Kepler could draw up his three famous Laws of Planetary Motion. The first two were published in 1609, and the third in 1618. Kepler knew nothing about gravity, and believed that the planets were whirled around by some unknown force exerted by the Sun, but this made no difference to his Laws, which have formed the basis of all subsequent work.

Law 1 states that a planet moves round the Sun in an ellipse, the Sun

SUNDIAL, *constructed by Erasmus Habermeel, a mechanician at the Court of Kepler's benefactor the Emperor Rudolph II.*

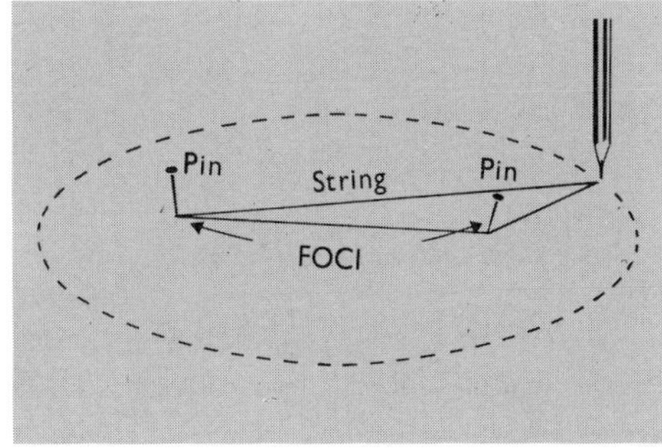

DRAWING AN ELLIPSE. *An ellipse may be drawn quite simply. Fix two pins into a piece of board, and join them with a thread, leaving a certain amount of slack. Then put a pencil through the thread, and trace out a curve. This will be an ellipse, and the pins will mark the* foci. *The wider apart you put the pins, the more* eccentric *will be the ellipse. If the two pins come together at a single point, the eccentricity is zero, and the ellipse becomes a circle. The orbits of the main planets known in ancient times are almost circular, but not quite – and it was this slight departure from circularity which enabled Kepler to prove the truth of the heliocentric theory as opposed to the Ptolemaic.*

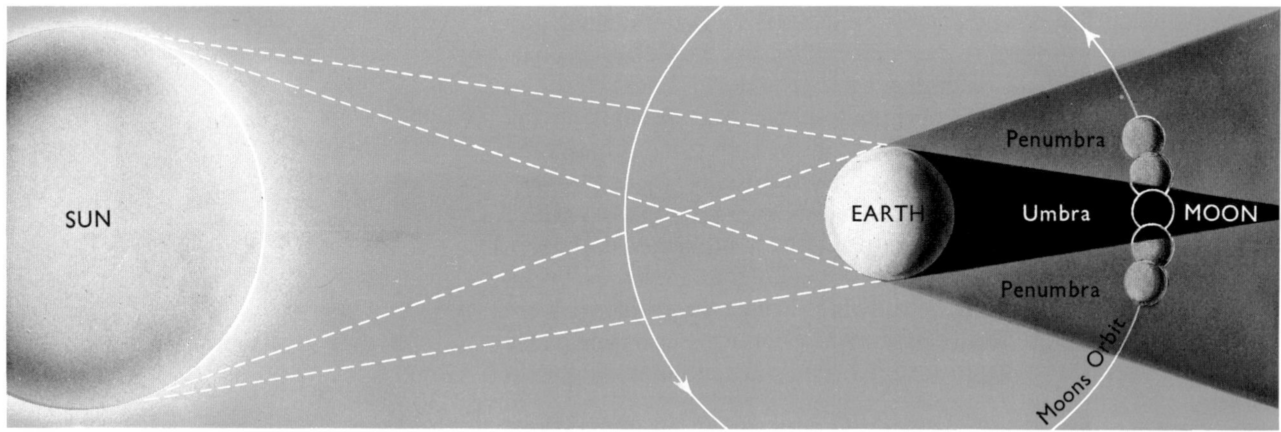

THEORY OF AN ECLIPSE OF THE MOON. *An eclipse occurs when the Moon passes into the shadow cast by the Earth. [This diagram is not to scale]*

occupying one focus of the ellipse while the other is empty. (If the two foci coincide, the ellipse becomes a circle.)

Law 2 states that the *radius vector* of the planet sweeps out equal areas in equal times. The radius vector is an imaginary line joining the centre of the planet to the centre of the Sun. In the diagram, the Sun is lettered **S**. Suppose that the planet moves from **A** to **B** in the same time that it takes to move from **C** to **D**; then the area **ASB** must be equal to the area **CSD**. In other words, a planet moves quickest when it is closest to the Sun. This naturally applies to the Earth; we move rather faster in December than in June. (To make the diagram clear, I have had to draw the orbit as highly eccentric, whereas, as we have noted, the real orbits of the planets are not far from circular; but the principle involved is exactly the same.)

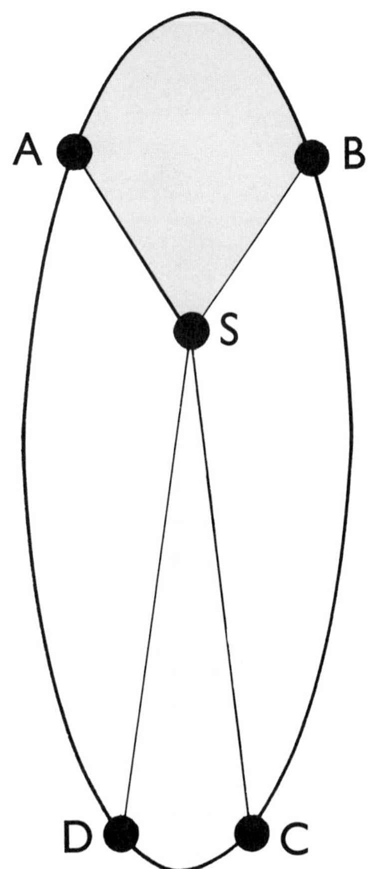

KEPLER'S SECOND LAW. *S represents the Sun; A, B, C and D a planet in four different positions in its orbit. According to Kepler's Law, a planet moves at its quickest when at its closest to the Sun.*

Law 3 states that for any planet, the square of the *sidereal period* (the time taken to complete one revolution; for the Earth, 365·25 or 365¼ days) is proportional to the cube of the planet's mean distance from the Sun. This may sound confusing, but all it means is that there is a definite relationship between the Earth-Sun distance and the "year". Therefore, as soon as we know a planet's revolution period, we can find out its distance from the Sun relative to that of the Earth, and this leads on to a complete scale model of the Solar System.* But you have to know one *absolute* distance, and Kepler did not. He believed that the Sun could be no more than 23 million kilometres from us. This was an improvement on the estimates of Tycho (8 million kilometres) and Copernicus (3·2 million), but it was still much too low, and the first reasonably good measurement was not made until 1672 – by the Italian astronomer Cassini, who gave 138 million kilometres.

The first two Laws came out in a book called *Astronomia nova* (New Astronomy) and the third, nine years later, in *Harmonices mundi* (Harmonies of the World), which, as usual, contained sound science mixed up with mysticism and astrology. But things in Prague had become increasingly difficult, and in 1612 Kepler moved to Linz, in

---

*Do you want a little mathematics? Very well.
Taking the Earth's mean distance from the Sun as 1 unit, that of Mars is 1·523. (The real mean distances are 149·6 and 227·8 million kilometres respectively; and if you multiply 149·6 by 1·523, you will find that the answer is indeed 227·8.) The periods are in the ratios of 1 to 1·88. (The real periods are 365·25 and 687 days respectively; in round figures, 365·25 multiplied by 1·88 comes to 687.) The cubes of 1 and 1·523 are 1 and 3·54, and the squares of 1 and 1·88 are also 1 and 3·54.

Austria, to lecture at the university there. His wife and one of his sons died, and altogether the situation was depressing in the extreme. To make matters even worse, Kepler's mother was arrested on a charge of witchcraft, and in view of the period in which she lived this was hardly surprising; she seems to have been a sinister kind of person, and Johannes had to fight energetically to secure her acquittal. He succeeded, though his mother did not live for long after her release from prison.

His last major book, *Epitome Astronomiæ Copernicæ* (Epitome of Copernican Astronomy) was really in the nature of a text, and summed up the views which Kepler held towards the end of his life. The Sun is unique, and occupies the centre of the whole universe; round it move the planets, and much further away lies the sphere of the stars. The stars are at different distances from us, as is indicated by their range in brightness, but they cannot be infinite in number, and so can occupy only a finite part of space. Nothing lies beyond, for one cannot have space where there are no bodies; the outer boundary of space may be a solid crystal sphere. The variation in a planet's velocity as it moves round the Sun, whirled along by the unknown 'pushing' force, can be explained if the core of the planet is divided into two hemispheres, one of which is attracted by this force and the other repelled; these hemispheres maintain a fixed direction in space, so that each hemisphere is turned alternately towards and away from the Sun, so that the planet is alternately pulled in and thrust outwards.

One final task remained: the completion of the Rudolphine Tables, dedicated to the old Emperor, which Kepler and Tycho had planned together. They finally appeared in 1628. By then he had moved, first to the city of Ulm and then to Silesia, but again his salary was often left unpaid, and in 1630 Kepler set out on a journey to collect some of the

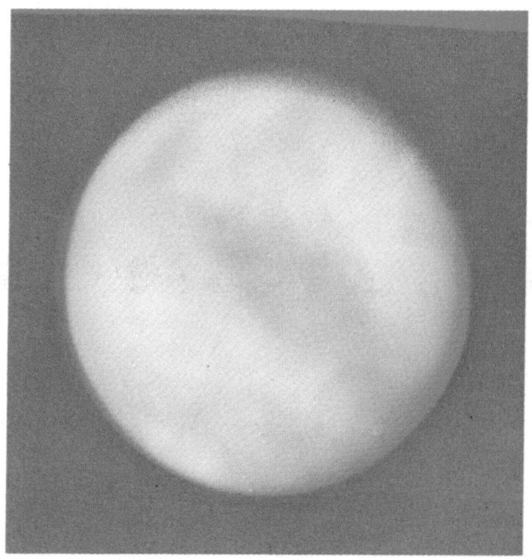

MARS, *photographed with the 152·4-cm reflector at Mount Wilson. The dark areas and the polar cap are well shown.*

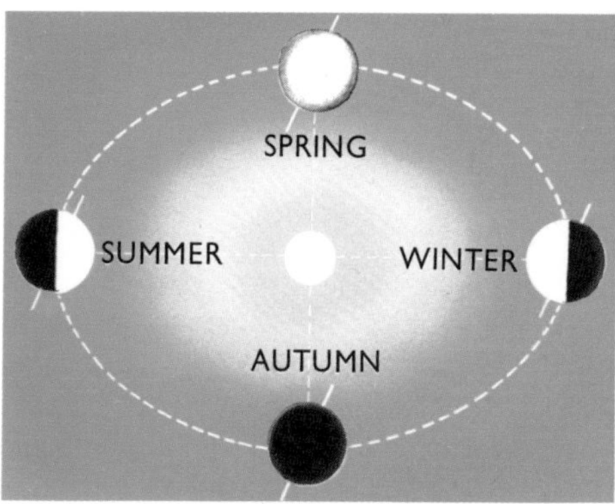

THE SEASONS. *This drawing relates to the northern hemisphere of the Earth; in the southern hemisphere conditions are of course reversed. During northern summer, the northern hemisphere of the Earth is tilted toward the Sun, and the rays strike this hemisphere more directly. At this time the Earth is actually at its greatest distance from the Sun (over 151,000,000 km).*

PHOTOGRAPHS OF AN ECLIPSE OF THE MOON. *Totality occurred between the fifth and sixth photographs. Generally the Moon is plainly visible even when totally eclipsed, though it is on record that at some eclipses the Moon has disappeared completely for a short period.*

money he was owed. He never arrived; in the town of Regensburg he collapsed with fever and exhaustion, and on 15 November he died. It was a sad end to a sad life.

Though Kepler was not an observer, he did make a few practical contributions, notably an improved kind of eyepiece to the newly-invented telescope. He observed the supernova which blazed out in 1604 in the constellation of Ophiuchus, the Serpent-bearer, and he also wrote an important book about comets. Moreover, he explained why the Moon turns a coppery or reddish colour during a lunar eclipse. Some of the Sun's light is bent or *refracted* on to the Moon by way of the blanket of atmosphere surrounding the Earth, which is why the Moon does not disappear completely when its supply of direct sunlight is cut off.

No description of Kepler would be complete without referring to his science-fiction novel, the *Somnium* (Dream), which was published posthumously in 1634. It was in effect a spirited defence of the Copernican theory, and the purely scientific footnotes are longer than the book itself, but it is still well worth reading. It describes how a young Icelander named Duracotus was taken to the Moon by friendly demons, and fiends a world populated with beings of varied kinds – some serpentlike, some with fins, and others covered with fur. In Kepler's day this did not sound so fantastic as it does now; there was no real reason to doubt that the Moon might be inhabited, and in the footnotes at least Kepler was keeping strictly to seventeenth-century science.

This, in its way, is typical of him. He, more than any other pioneer, had one foot in the ancient world and the other in the modern. Had he been born of a rich and noble family, his thoughts might have been directed along different lines – but then, he would not have been Johannes Kepler.

9

Telescopes and the Stars

THE PLEIADES, *photographed in red light with the 45·7-cm Schmidt telescope at Palomar.*

GALILEO. *A portrait of Galileo Galilei, the great Italian scientist who was the first great astronomer to use a telescope. Galileo was also a pioneer of experimental mechanics.*

NEXT in our list of the men who made the Copernican revolution possible we come to Galileo Galilei, who was not only one of the greatest of all scientific pioneers, but was also one of the most colourful – even if he did not have a Tycho-like artificial nose. His career overlapped Tycho's; he was born in Pisa in 1564, and during the period when Tycho was at Hven Galileo was busy lecturing first at the University of Pisa and then at Padua, but apparently the two never corresponded at all. I have often wondered what would have happened if they had met face to face. Galileo was not noted for his tact, and a confrontation might well have ended in a verbal explosion of supernova violence, if only because Galileo was as firmly in favour of the Sun-centred system as Tycho was against it. Galileo did have some later correspondence with Kepler, but it was very limited, and the fact that it lapsed was not Kepler's fault. The two were very different in character; Galileo was a born experimenter, and he is generally regarded as the real founder of the science of experimental mechanics.

When Galileo was ten his family moved back to their original home, Florence. The boy's early upbringing was conventional enough; his father, Vincenzio, was well-educated and talented, particularly as a musician. Then, when Galileo was seventeen, he entered the University of Pisa. He originally meant to study medicine, but he soon showed a strong desire to change over to science, and Vincenzio was wise enough not to stand in his way.

While still a student, Galileo made his first notable discovery. During a service in the Cathedral, he watched a lamp swinging on the end of a long cord, and saw that no matter whether the swing were short or long, the time needed for a full back-and-forward motion was always the same. Needless to say, this could have been noticed by anybody, but Galileo began to wonder just what it meant. Typically, he carried out some experiments in order to prove his point, and then put it to use, working out a device to measure short time-intervals such as a man's pulse-rate. He may even have tried to design a pendulum clock, though it seems that he never actually built one.

Galileo's first spell at Pisa ended abruptly in 1585. He had no degree, but his father could no longer afford to pay the fees, and Galileo began to earn his living by teaching. He also made his mark in various other ways; in 1586 he invented a device known as a hydrostatic balance, and produced some technical studies about the centres of gravity of solids, which led him into correspondence with several learned professors. In rather different vein, he gave two public lectures about the shape, size and location of hell, turning the whole topic into a sort of geometrical exercise!

To be financially secure he needed a regular job, and he looked around to see what he could find. Rather surprisingly, he managed to obtain the chair of mathematics at Pisa University, only four years after he had left the university without taking any degree at all.

THE LEANING TOWER OF PISA. *It has been said Galileo dropped stones off the top of the famous Leaning Tower of Pisa, though as a matter of fact he never carried out this experiment. Photograph by Dominic Fidler.*

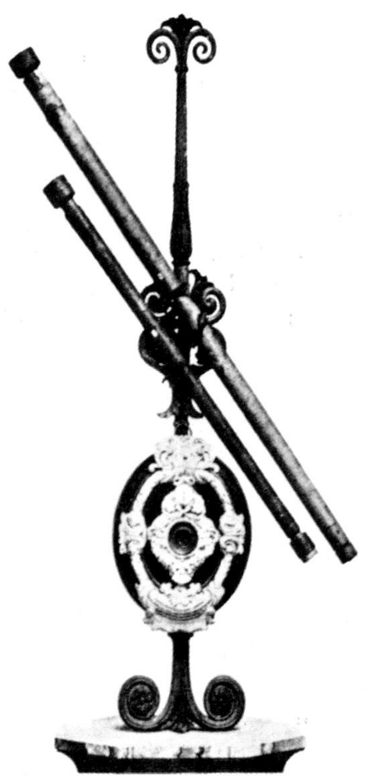

TWO OF GALILEO'S TELESCOPES, *preserved in the Tribuna di Galileo in Florence. A broken object-glass, with which the four satellites of Jupiter were discovered, is mounted in the centre of the ivory frame.*

Naturally he had to teach astronomy, and this meant accepting the Ptolemaic theory – outwardly at least; but it seems that at an early stage he had become a convinced Copernican, though it would have been folly for him to say so. During these years, also, he carried out his first pioneer researches into mechanics. There is a story that he went up the famous Leaning Tower to drop stones off the top – specifically in order to disprove Aristotle's statement that heavy objects would fall faster than light ones. Actually there seems no doubt that the story is untrue, but Galileo would have been quite capable of such an experiment.

In 1592 Galileo moved from Pisa to the University of Padua, and went on with his work. Astronomy remained his greatest interest; he observed the supernova of 1604, and even gave four public lectures about it, but it was in 1609 that he first heard of a discovery which altered the whole course of his life. This was the invention of the telescope.

It is rather surprising that despite the vast amount of research which has been carried out, nobody really knows who made the first telescope. The candidate usually favoured is Jan Lippershey, a lens-grinder who lived at Middleburg on the Dutch island of Walcheren. At any rate, Galileo lost no time in making a telescope for himself. Let me quote his own words, written later in his book *Sidereus Nuncius* or the Sidereal Messenger:

> 'About ten months ago a report reached my ears that a Dutchman had constructed a telescope, by the aid of which visible objects, although at a great distance from the eye of the observer, were seen distinctly as if near... At length, by sparing neither labour nor expense, I succeeded in constructing for myself an instrument so superior that objects seen through it appear magnified nearly a thousand times, and more than thirty times nearer than if viewed by the natural powers of sight alone.'

Telescopes of this kind are known as *refractors*. The light from the Moon, or whatever object is to be studied, falls upon a glass lens known as an *object-glass* or *objective*. This bends the light-rays, and brings them together at a point called the *focus*, where an image is formed. This image is then magnified by a second lens termed an *eyepiece*. The distance between the object-glass and the focus is the *focal length*.

Galileo's telescope was feeble judged by modern standards. Its object-glass was only 2·5 centimetres in diameter, as against the 101·6 centimetres of the largest refractor now in use, that at Yerkes Observatory in the United States. Yet it was good enough to enable Galileo to make a whole series of spectacular discoveries, and it is his work from 1610 to 1619 which marks the true beginning of telescopic astronomy.

He was not the first to turn a telescope towards the sky. In July 1609, some time before Galileo even heard of the Dutch discovery, an Englishman named Thomas Harriot even drew a map of the Moon as seen through a telescope. Harriot was at one time tutor to Sir Walter Raleigh, and he was certainly a skilled scientist; his lunar chart is remarkably good, but he never seems to have followed up his pioneer work. Another pre-Galilean observer of the Moon was a Welshman, Sir William Lower, who compared the surface with a tart that his cook had made; 'here some bright stuff, there some dark, and so confusedly all over'. But Harriot, Lower and others of the time were mere dabblers compared with Galileo, and the fact that the great Italian was not the first in the field is neither here nor there.

Naturally enough, one of the first objects he looked at was the Moon. At once he saw that the lunar surface is rough and mountainous. Along the *terminator*, or boundary between the daylight and the night hemispheres, he saw shining points, and realized that they must be mountain-tops catching the sunlight, so giving the terminator a jagged appearance. He saw the grey plains which later became known as seas (in Latin, *maria*), though whether he believed them to be waterfilled is not certain. He described the valleys, and also the walled circular formations which are always called craters. He made some drawings of the lunar surface, which were the first apart from Harriot's, and he tried to measure the heights of some of the lunar peaks. For this he used the shadow method. If you know the angle at which the sunlight is striking, and you also know the length of the shadow cast by the lunar mountain, it is a relatively simple geometrical problem to work out the height of the mountain itself. Galileo deduced that the peaks were as lofty as any on the Earth. Actually he concentrated upon the range known today as the lunar Apennines, and his estimates were too high, but at least they were of the right order. It was this kind of experiment which made Galileo stand head and shoulders above his contemporaries in telescopic astronomy.

He also found that the stars did not show obvious disks in his telescope. No matter what the magnification, they still appeared as points of light. Of course they seemed more brilliant, and more were visible; he could go down well below the sixth magnitude. As a typical example, he cited the famous star-cluster of the Pleiades, in Taurus (the Bull). To the naked eye seven stars are visible on a clear night, but Galileo's telescope showed him at least forty, and we now know that the cluster contains several hundreds.

It is not surprising that Galileo failed to see stellar disks. Even with modern telescopes a star still appears as a dot. This is because all the stars are so remote. If you turn a telescope towards a star and see a shimmering balloon of light, you may be quite sure that there is something wrong with the focusing.

Galileo next turned his attention to the Milky Way, the glorious band which stretches across the sky. He found that it was made up of 'a mass of stars', apparently crowded close together. Again he was correct, as any modern binoculars will show.

Yet it was with the planets that Galileo was most concerned. By now he was an open Copernican, but he was well aware that he would need positive proof before he could hope to make any inroad upon orthodox beliefs. Copernicus himself had been unable to provide any, quite apart from the fact that his original theory had so many faults. Galileo determined to do better, and it was with special interest that he turned his telescope towards the brilliant Venus.

According to Ptolemy, both Venus and the Sun move round the Earth, Venus being the nearer of the two. In the diagram – which, I admit, is very over-simplified – D is the deferent of Venus (the point round which it moves), and Venus is shown in four positions on its epicycle. The line Earth-D-Sun must always be straight. Now, since the Sun can illuminate only half a planet at a time, it is quite obvious that if Venus behaves in this way it can never be seen as a full disk, or even a half. Part of the time it will be invisible altogether, because its dark side will be turned towards us; at other times it will be a crescent.

This was not what Galileo found. Venus showed a full range, from

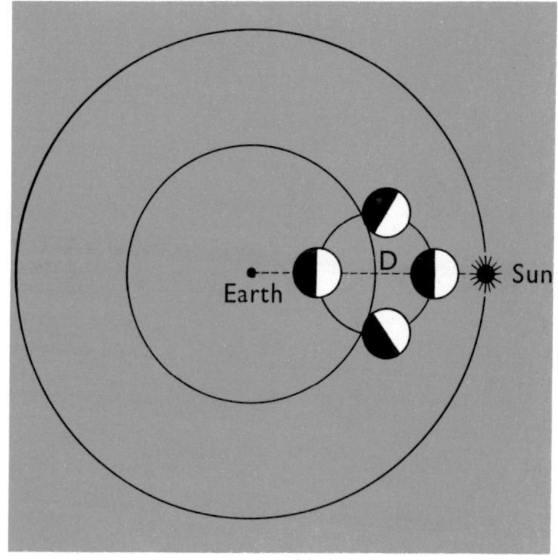

MOVEMENTS OF VENUS ACCORDING TO PTOLEMY'S THEORY. *According to Ptolemy, both Venus and the Sun move round the Earth, and the line joining the Earth, the Sun, and the deferent of Venus is always straight. This would mean that Venus could never show as a half or full disk; it would always be a crescent when visible at all. Though there have been many reported cases of the crescent shape of Venus being visible to the naked eye, it is certain that nobody could follow the changing phases without optical aid, and until the invention of the telescope there could therefore be no certain proof. When Galileo was able to use a telescope to study Venus, he saw at once that the old theories must be wrong, since they could not possibly account for Venus' behaviour.*

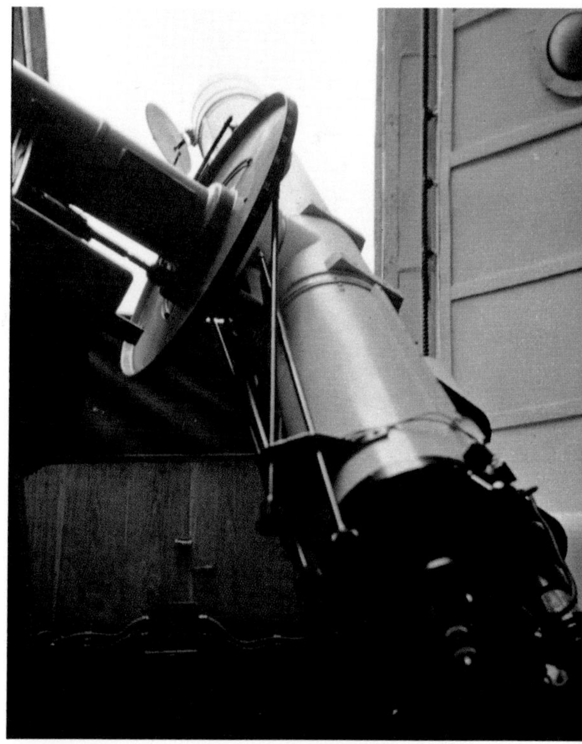

A MODERN REFRACTOR. *This is the refractor at the Sternberg Institute in Moscow; the object-glass is 20·3-cm across. Photograph by Patrick Moore, 1960.*

THE LUNAR MARE CRISIUM. *The Mare Crisium, or Sea of Crises, is one of the smaller but most conspicuous of the lunar maria. It lies fairly close to the limb, and its appearance is therefore markedly affected by libration. Lick Observatory photograph.*

new (invisible) to practically full; and on any Sun-centred theory – circular orbits or not – this is perfectly logical. The experiment was decisive. If Galileo had had any lingering doubts, Venus removed them.

In January 1610 Galileo turned his attention to Jupiter, and discovered that the planet was attended by four satellites. This was in absolute contradiction to the old belief that the Earth must be the centre of everything, and Galileo was quick to realize the significance of what he had found.*

The satellites were seen at about the same time by a German named Marius, who named them Io, Europa, Ganymede and Callisto. In a low-powered telescope they appear starlike; more powerful instruments show them as distinct disks, but it was only with the Voyager space-missions of 1979 and 1980 that we found out what remarkable worlds they really are. Incidentally, it seems that during his observations of the satellites Galileo also saw the planet Neptune, which happened to be in the same region of the sky. Of course he did not realize that it was anything but a star; Neptune was not identified until 1846.

In July 1610 Galileo managed to see the disk of the other giant planet, Saturn. The magnification was only thirty-two and the definition was poor, but Galileo realized that there was something very peculiar about Saturn's shape. At first he believed that he had discovered two satellites, but they showed no relative motion, and two years later he was astonished to find that the planet appeared single. Galileo was puzzled by Saturn; his telescope was not good enough to show that what he had actually seen was the famous system of rings, which disappeared in 1612 because the rings were edgewise-on to us.

Galileo, naturally enough, was elated by all these discoveries, and he felt secure enough to publish them in a manner which showed that he had finally abandoned the Ptolemaic system. Inevitably this led to adverse comment from the Church as well as from orthodox scientists. One professor at Pisa, Julius Libri, stated his profound disbelief in everything that Galileo had said, but stubbornly declined to look through a telescope at all. When Libri died, in late 1610, Galileo is reported to have said that he might at least have a good view of Jupiter's satellites as he passed by them on his way to heaven.

The first signs of real trouble came in 1615, when the Church authorities repeated that 'the doctrine that the Sun is the centre of the world and immovable is false and absurd, formally heretical and contrary to Scripture', and Galileo was officially warned to alter his views. Meanwhile he had resigned his post at Padua, and had settled in Florence as mathematician to the Grand Duke of Tuscany. For a while he was left in peace, and no doubt he felt secure in view of the fact that he had powerful friends – notably Cardinal Barberini, one of the most influential men in the Roman Catholic Church.

By the end of 1629 he had completed the book which we know as the *Dialogue Concerning the Two Chief World Systems*. It was written in Italian, not Latin, so that it could be read by anybody in the country. He

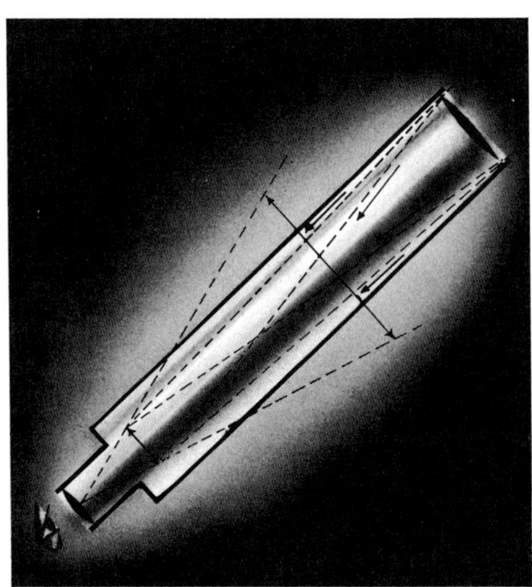

PRINCIPLE OF THE REFRACTOR. *The light is collected by the object-glass.*

*According to an old Chinese manuscript, an observer named Gan De saw a faint object close to Jupiter in 364 B.C. If this was a satellite – as seems likely – Gan De anticipated Galileo by over nineteen centuries!

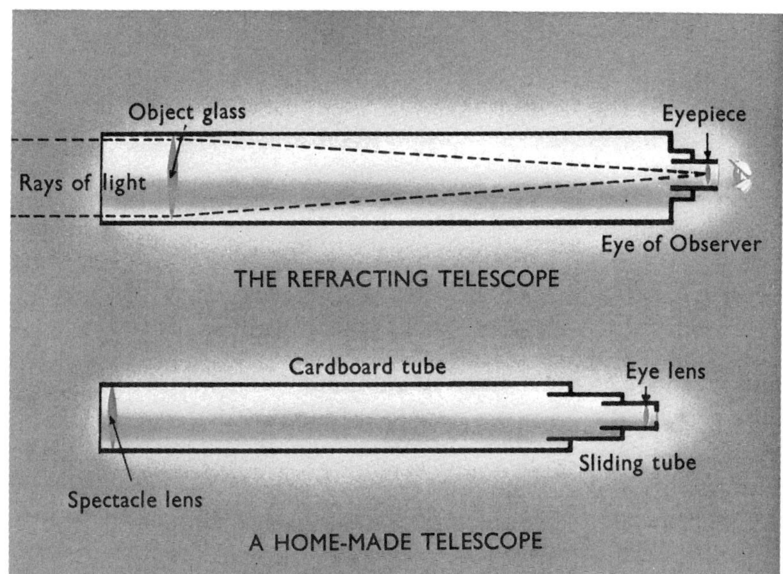

MAKING A SIMPLE TELESCOPE. *It is simple enough to make a small refractor out of spectacle-lenses and cardboard tubes. The cost is low, and the construction takes only an hour or two.*

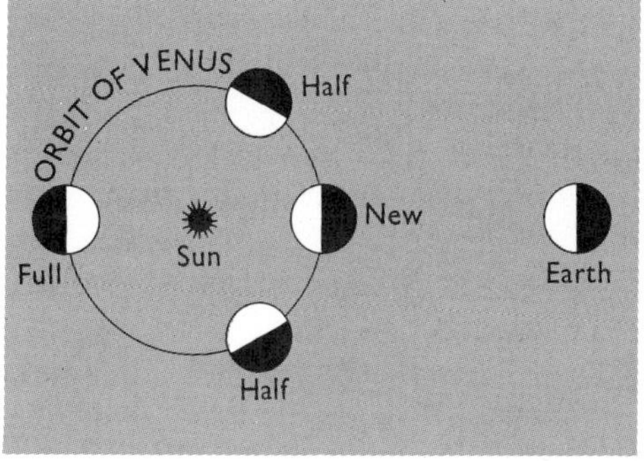

CORRECT THEORY OF THE MOVEMENTS OF VENUS. *The planet may be seen as a crescent, half or full disk. [This drawing is not to scale]*

submitted it to the censors, and with minor modifications it was passed for publication. It finally appeared in February 1632 – and at once the storm broke.

The book takes the form of a dialogue between three people. One of them was named Simplicio, and it was he who championed the Ptolemaic theory. Nobody who read the book could have the slightest doubt that it was purely Copernican, and that Simplicio was the stupid member of the trio of characters. Unfortunately the Pope believed – or was persuaded to believe – that he was being ridiculed. This was all the more startling because the Pope, Urban VIII, was none other than Galileo's old friend, Cardinal Barberini.

Galileo was summoned to Rome, brought to trial on a charge of heresy, and condemned. On June 22 he was forced to 'curse, abjure and detest' the false theory that the Earth moves round the Sun. He was too wise to protest; after all, it was less than forty years earlier that Giordano Bruno had been burned at the stake.

By now Galileo was an old man, and the rest of his life was passed at his villa at Arcetri. He was carefully watched by the Church authorities, and his visitors were carefully vetted. Moreover his eyes were failing, and during his last years he became totally blind. Yet he refused to give up; during this final period he even wrote what some regard as his greatest book, *Discourse on Two New Sciences*, which deals mainly with mechanics and so did not upset the Inquisition. In November 1641 he was taken ill, and on the night of January 8, 1642 he died. As a last piece of petty spite, Pope Urban VIII refused to allow a monument to be erected over his tomb.

THE MILKY WAY IN CYGNUS. *Photographed at Lowell Observatory with a 12·7-cm lens, exposure 3 hours, on April 13, 1930.*

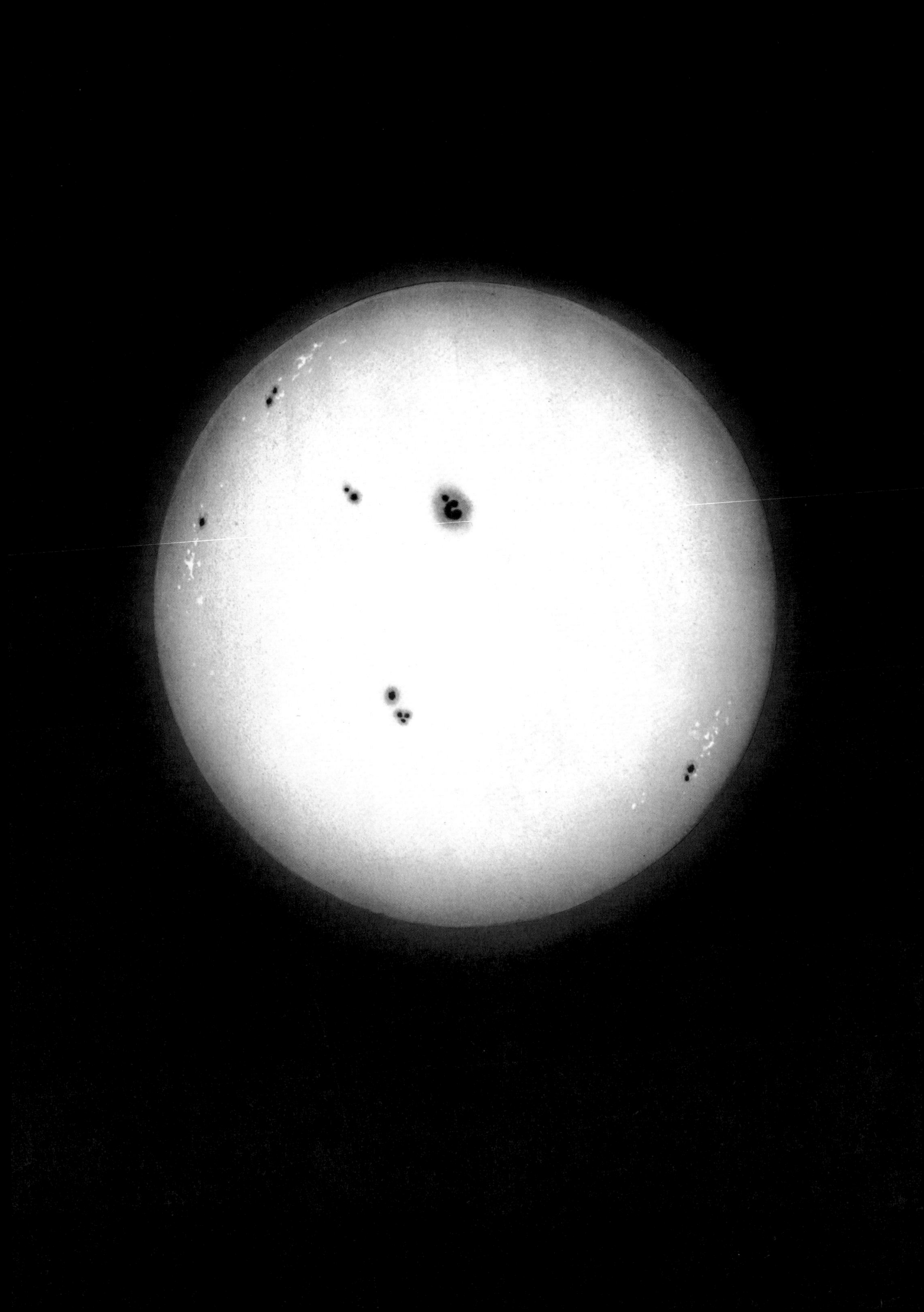

10

The Face of the Sun

ONE day during March 1611, Christoph Scheiner, professor of mathematics at Ingoldstadt in Germany, observed the Sun with his newly-acquired telescope. He saw that instead of being a blank yellow disk, the Sun had dark spots or patches on it. Scheiner was intrigued, and consulted his ecclesiastical superior, whose name was Busaeus. The discovery was not well received. Busaeus said, simply: 'I have read all the works of Aristotle several times from beginning to end, and I assure you that I have not found anything in them which could be what you are telling me. Go, my son, and calm yourself. I assure you that what you took to be spots on the Sun are only flaws in your glasses or your eyes.'

Scheiner was not satisfied, and when he had reports of similar observations by others he wrote some letters to his friend Mark Welser, who had them printed. A copy of the tract reached Galileo, who replied that he had been observing sunspots ever since November 1610. In fact both Scheiner and Galileo had probably been anticipated by a young Dutchman named Johannes Fabricius, whose book about sunspots came out in 1611, though – infuriatingly – he gives no dates for his drawings.

At any rate Galileo gave a correct explanation, whereas Scheiner did not. Remember, Scheiner did not (officially at least) believe in the Copernican theory, and he was disinclined to believe that a body as pure as the Sun could be spotted. So he preferred the idea that the spots were dark bodies close to the Sun, which would explain the observations without incurring the wrath of the Church. Galileo disagreed. In 1612 he wrote that 'Having made repeated observations I am at last convinced that the spots are objects close to the surface of the solar globe, where they are continually being produced and then dissolved, some quickly and some slowly; also that they are carried round the Sun by its rotation, which is completed in a period of about one lunar month. This is an occurrence of the first importance in itself, and still greater in its implications.'

PROJECTING THE SUN. *The only safe way to observe the Sun is to use a telescope to project the solar image on to a screen. This photograph was taken by W. M. Baxter in his observatory at Acton, London. The telescope is a 10·2-cm refractor.*

THE SUN, *(opposite) showing spot-groups and faculæ. The diminution in light near the edge of the disk 'limb darkening' is also apparent. From an observation made by Patrick Moore with a 10·2-cm refractor (projection method). Drawing by D. A. Hardy.*

As we have noted, modern measurements have shown that the distance of the Sun is 150,000,000 kilometres, with a slight range of 2½ million kilometres either way. The diameter of the globe is 1,392,000 kilometres – larger than the diameter of the Moon's orbit round the Earth. The volume of the Sun is well over a million times that of our own world. On the other hand the mass is not so great as might be thought, because of the lower density; only 1·4 times that of water. Remember, the Sun is made up of gas, though near the core, where the temperature is at least 14,000,000 degrees and the pressures are tremendous, the 'gas' behaves in a very unfamiliar manner.

The fact that the Sun is so hot makes it a potential source of danger to the unwary observer. To look straight at it with any telescope, or binoculars, is to invite disaster, and permanent blindness is almost inevitable – as has, unfortunately, happened on more than one

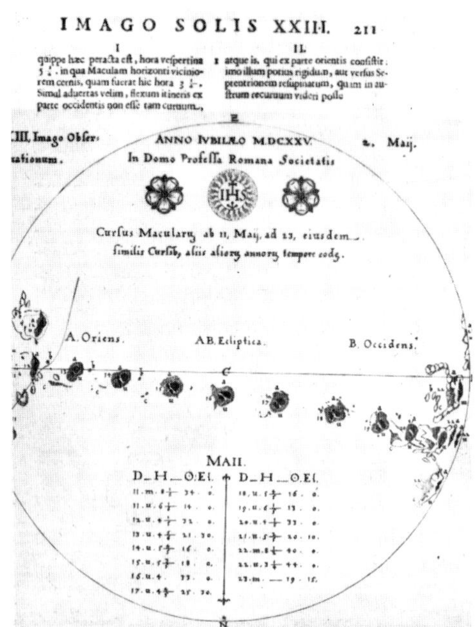

SUNSPOT DRAWING BY SCHEINER. *This series of observations was made in 1625 by Christoph Scheiner. Scheiner was one of the first telescopic observers to record sunspots, and there was an argument between him and Galileo with regard to priority – though in fact the question of priority is not in the least important.*

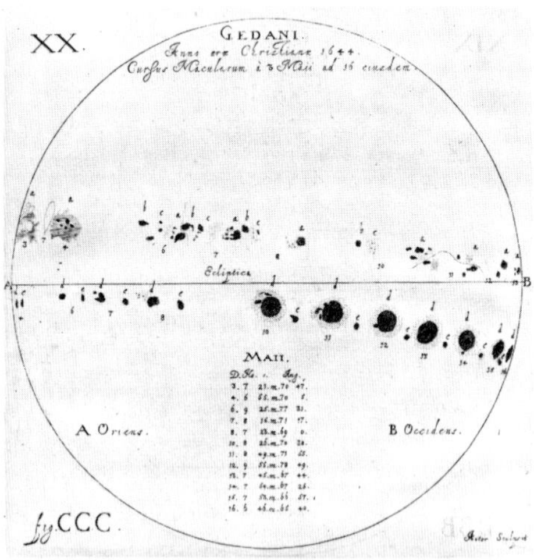

SUNSPOT DRAWINGS BY HEVELIUS. *Hevelius, one of the best telescopic observers of the seventeenth century, carried out most of his work from his private observatory at Danzig; his telescopes were among the best of their time, though very feeble judged by modern standards. His drawings of sunspots represented an advance on the work of Galileo, Scheiner and other pioneers.*

occasion. The only sensible method of observing sunspots was originally pointed out by Galileo's pupil Benedetto Castelli, and Galileo himself certainly used it. First, point the telescope towards the Sun by sighting along the tube, making sure that a solid metal or cardboard cap covers the object-glass. Then remove the cap, and project the Sun's image on to a white screen placed behind the eyepiece (a sheet of cardboard will do very well). Any spots present will be well seen, and there is no danger at all.

Most regrettably, some telescope suppliers provide dark filters which, they say, can be put over the eyepiece for direct observation of the Sun. My advice is – *never use them.* They can never give full protection, and are always apt to splinter without warning. I have emphasized this more times than I can count, in books, on the radio and on television, but I make no apology for repeating it yet again.*

It is also unsafe to stare straight at the Sun with the naked eye, even when the Sun is low down and looks deceptively mild and harmless. Naked-eye spots appear sometimes, and were recorded in ancient times; one Chinese record of 28 B.C. describes a patch as 'a black vapour, as large as a coin'. But one cannot be too careful.

There is only one observation which may be made without any kind of optical aid. Under favourable conditions, a glorious spectacle may be seen at the moment of sunset – a flash of vivid green as the last portion of the disk vanishes below the horizon. This 'Green Flash' is due purely to the Earth's atmosphere, but it is well worth looking for; it is best seen over a sea horizon – but again, never use binoculars or a telescope.

Since the Sun's bright surface or *photosphere* is gaseous, permanent features are not to be expected, and Galileo was right in claiming that spots are relatively short-lived. Small ones may last for only a few hours, while the record for longevity seems to be held by a spot-group which lasted for about two hundred days between June and December 1943. However, the group was not under continuous observation. As Galileo realized, the Sun spins on its axis, but it does not behave as a solid body would do. At the solar equator, the rotation period is about 25 days, but in higher latitudes the period is several days longer.

If you rotate a football which has specks of mud on it, the specks will seem to move round until they disappear over the edge of the football; if you go on spinning, the specks will reappear at the opposite edge. It is the same with sunspots. As the Sun rotates, the spots are carried from one side of the disk to the other, finally vanishing over the edge or *limb.* If they persist, they will reappear at the opposite limb about a fortnight later. Their appearance will be affected, too; near the limb a spot is foreshortened, and appears elliptical in form if the real outline is circular.

Suppose that we have a line of sunspots down the Sun's central meridian, as shown opposite. (In fact spots are never seen at the poles or right on the equator, but this makes no difference to the general principle.) Twenty-five days later the middle spot (A) will have come back to its original position, but in higher latitudes the spots lag behind,

*All sorts of strange things can happen. An old friend of mine was once projecting the Sun, in the conventional way, when he found that his beard was alight. He had forgotten to put a cap upon the small finder-telescope attached to the tube of the main instrument!

and the line will no longer be straight. This is termed differential rotation, and is of the utmost importance in modern theories of the Sun.

A spot is not a simple dark blob. There is an inner blackish *umbra*, and with larger spots a lighter surrounding *penumbra*. The shapes are often irregular, and one mass of penumbra may contain many umbræ. Yet even the darkest spots are not genuinely black. They are about 500 degrees cooler than the photosphere, and appear dark only by contrast. If a spot could be seen shining on its own, its surface brilliancy would be greater than that of an arc-lamp.

Spots usually appear in groups. Each group has its own peculiarities, but an 'average' two-spot group begins as two tiny pores at the limit of visibility. The pores grow into true spots, separating in longitude. Within a fortnight the group has reached its maximum length, with a fairly regular leading spot and a less regular follower – and, of course, various minor spots and clusters. Then a slower decline sets in. The leader is generally the last survivor, but eventually the whole group disappears.

Associated with larger groups are the so-called *faculæ*, which seem to have been first seen by Christoph Scheiner during 1611. Faculæ (Latin, 'torches') may be regarded as luminous clouds lying above the photosphere, and are now known to be made up chiefly of hydrogen. Faculæ often appear in regions where a spot-group is about to break out, and persist for some time after the group has vanished.

Even in 'quiet' areas, the Sun's surface is not calm. There is a granular structure, each granule being about 1000 kilometres in diameter and lasting for between six and ten minutes. They represent upcurrents, and it has been said that the whole of the Sun is continually 'boiling'.

Galileo never followed up his work on sunspots, but in a way he was lucky, because when he made his first observations, in 1610 or 1611, there were considerable numbers of spots; this was a period of solar maximum. To explain this, I must jump forward two centuries to the work of Heinrich Schwabe.

Schwabe, born at Dessau in Germany in 1789, was an amateur observer. In 1826 he became interested in astronomy, and began to make daily drawings of sunspots. He carried on this work for many years, and before 1850 he was able to show that there is a more or less

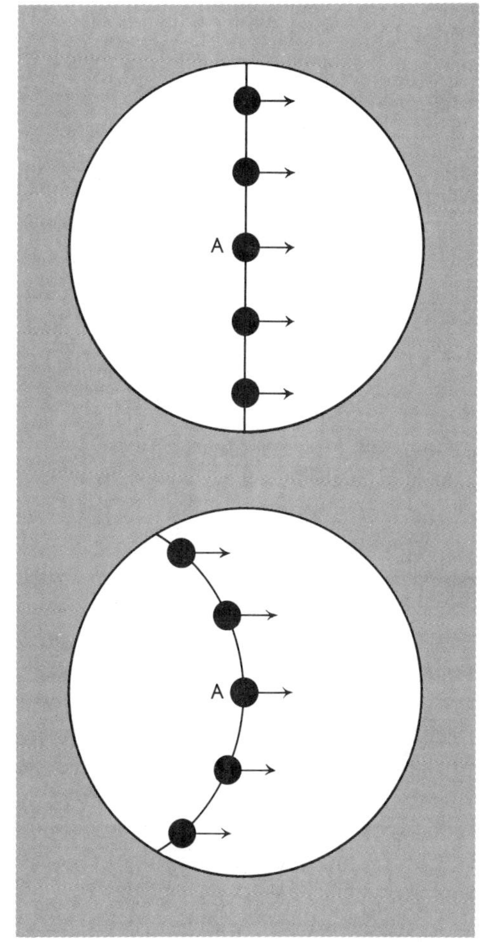

DIFFERENTIAL ROTATION OF THE SUN. *Region A, at the solar equator, completes its rotation quickest. In the polar regions, the length of the axial rotation period is appreciably longer.*

SUNSPOT PHOTOGRAPHS. *The same sunspot group photographed on three different days; August 19, 1959* (left), *21* (centre) *and 23* (right). *Photographs by W. M. Baxter.*

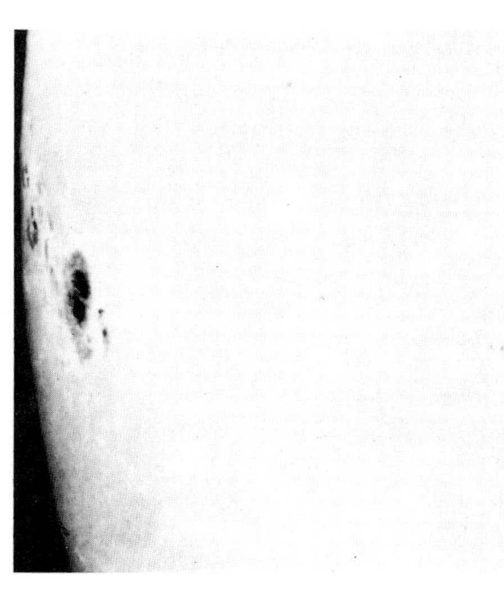

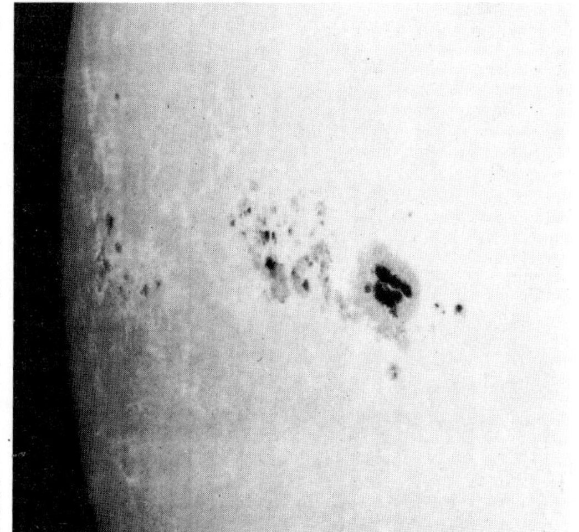

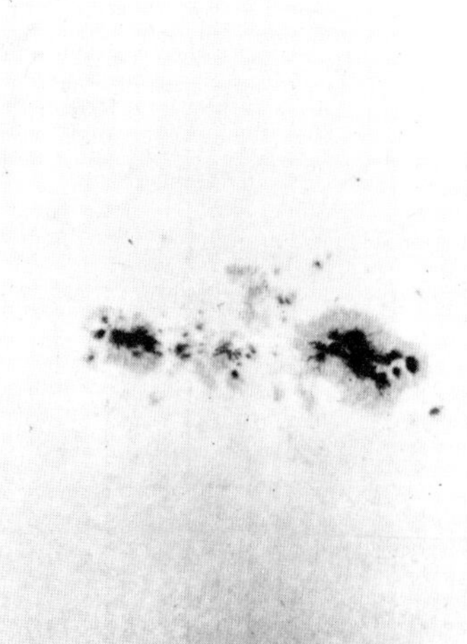

predictable cycle of activity. Every eleven years or so, the Sun is particularly active, and there are many groups of spots. Then activity dies down, and during minimum the disk may be completely clear for weeks at a time. Subsequently, activity builds up again until the next maximum is reached.

The cycle is not perfectly regular, and the eleven-year period is only an average, but it is at least a general guide. And not all maxima are equal in intensity. During 1947-8, for example, some giant spots were seen; the largest group ever recorded, that of April, covered an area of 18,130,000,000 square kilometres. The maximum of 1958 was also energetic; that of 1969 much less so. The last maximum occurred in 1980, so that at the time of writing (March 1983) we are on the way to the minimum which may be expected around 1985 or 1986.

Sunspots are never visible at the Sun's poles, and in 1861 another German observer, Friedrich Spörer, discovered a curious 'law' which still bears his name. (Spörer was originally a schoolmaster who made a hobby out of astronomy, though later in his life he worked at the Potsdam Observatory.) During the early part of a solar cycle the spots appear some distance from the Sun's equator; as the cycle progresses they invade lower and lower latitudes. As the cycle draws to its end, and

SOLAR DOMES AT PULKOVO. *The old Pulkovo Observatory at Leningrad, U.S.S.R., was destroyed during the war, and the present buildings are modern. The photograph, taken by Patrick Moore in 1960, shows domes housing equipment for studying the Sun.*

THE GREEN AND RED FLASHES. *These and other low-sun phenomena are due to the effects of the Earth's atmosphere. Photographs by Father D. J. K. O'Connell, S. J., Vatican Observatory.*

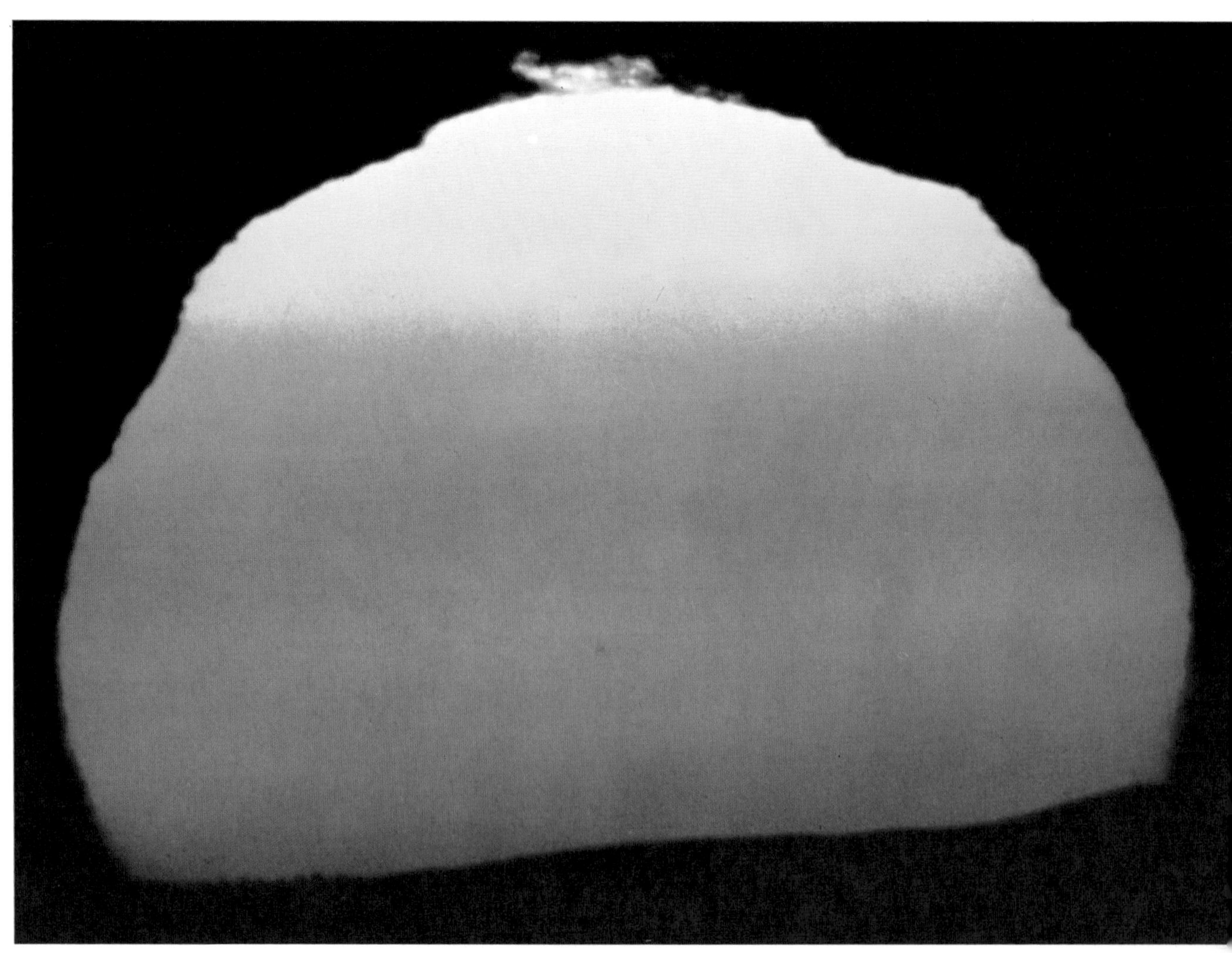

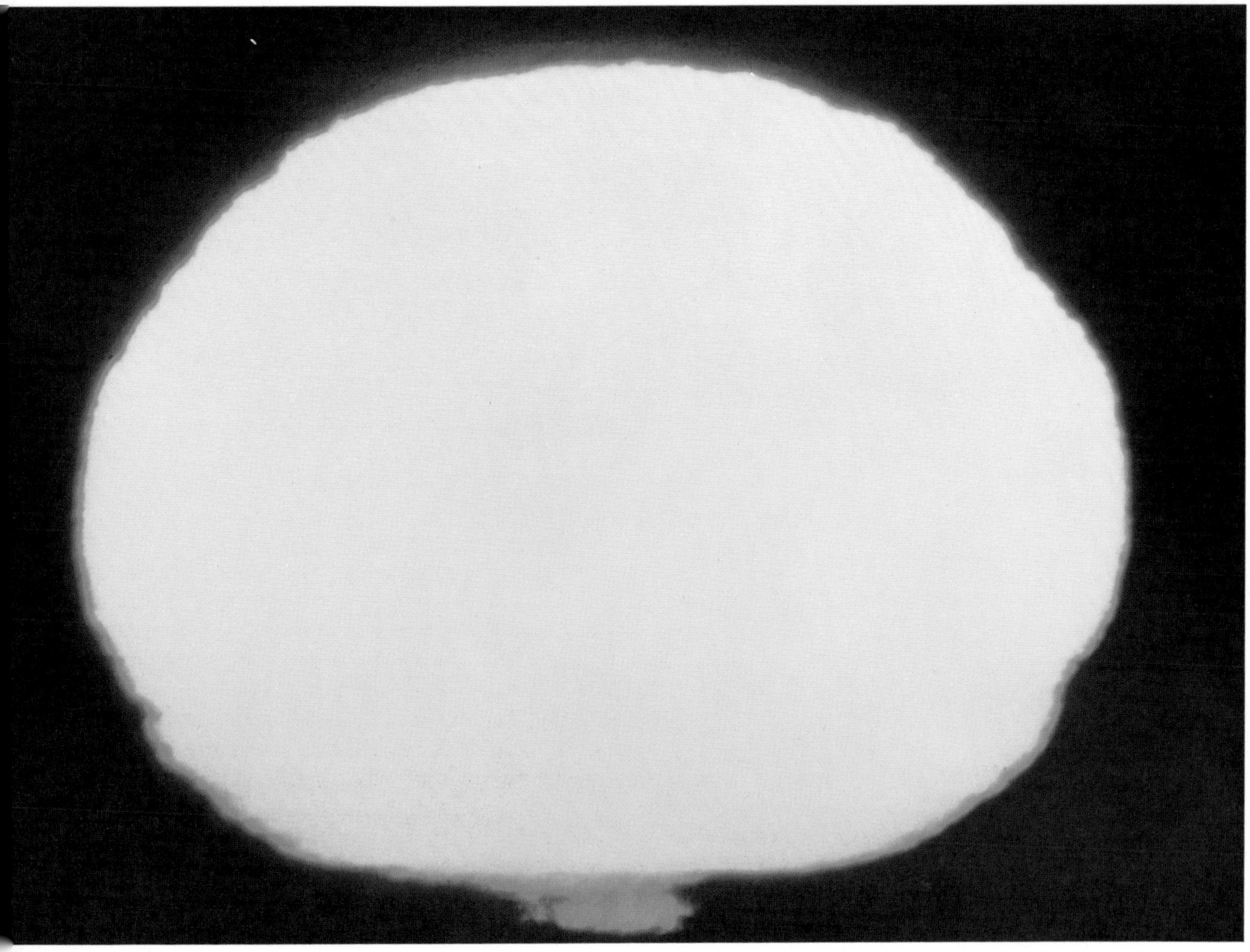

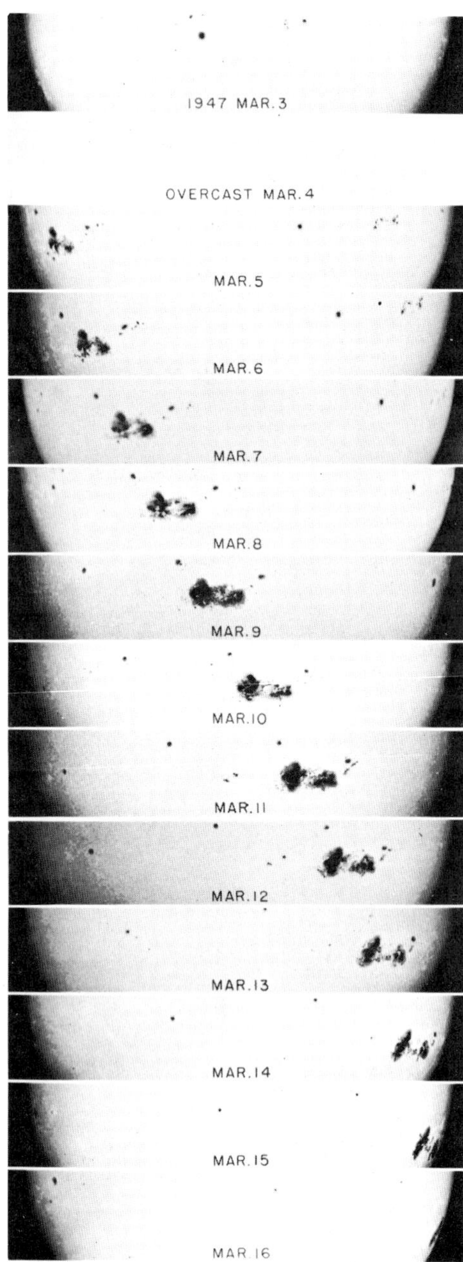

THE GREAT SUNSPOT GROUP OF 1947. *A day-by-day record from March 3 to March 16. This was the largest spot-group observed since accurate records were begun. Photographs by Mount Wilson and Palomar Observatories.*

its groups die away, small spots of the new cycle start to appear in high latitudes, and for a while there are two spot-prone regions; near the equator, due to the last spots of the dying cycle, and in higher latitudes, marking the early stages of the new cycle.

Whether the cycle is a permanent phenomenon is less certain than used to be thought. In 1890 E.W. Maunder, at Greenwich, examined old records, and found that to all indications there were virtually no spots between 1645 and 1715, so that the cycle was suspended. This 'Maunder Minimum' may not be unique, but there seems no doubt about it, and it may be significant that this was a period of cold weather over much of the world – during the 1680s the Thames froze regularly, and 'frost fairs' were held on it. There was also a lack of auroræ, which are brilliant glows in the upper atmosphere caused by electrified particles sent out from active regions on the Sun, particularly the violent, short-lived outbursts known as *flares*. Earlier records are infuriatingly fragmentary, but there is some evidence of another prolonged minimum between 1400 and 1510.

At the moment we have to admit that we do not know the reason for this behaviour, but there is recent evidence that some of our current theories about the Sun may be wrong. In particular, it had been expected that there would be a strong stream of elusive particles called neutrinos, but the numbers of neutrinos are unexpectedly small. We may have to do some drastic re-thinking.

What exactly are sunspots?

Many of them are low-lying. In 1774 the Scottish astronomer Alexander Wilson found that when a regular spot nears the limb, the penumbra to the limbward side seems to be broadened as compared with the opposite side, which can only happen if the umbra is depressed. But today it seems that the basic cause is magnetic – which brings us on to modern times. The magnetic fields associated with sunspots were discovered in 1908 by George Ellery Hale, about whom more anon, and the modern theory is based on the work of H. Babcock, carried out in the United States in 1961.

It is assumed that the solar magnetic lines of force run from one magnetic pole to the other, below the bright surface. The differential rotation means that the lines are distorted, and become looped. Eventually the lines are coiled right round the Sun, and a loop of magnetic energy breaks through to the surface; the upper gas is calmed and cooled, producing two spots. After about eleven years the knots in the magnetic lines break; the Sun 'snaps back' to its original state, but overshoots, so that the whole sequence of events begins anew. It has also been found that the magnetic polarities of the individual spots follow a definite law. If the leader is a 'north polarity' spot, the follower is a 'south polarity'. Conditions in each hemisphere are the same, but every alternate cycle there is a complete reversal – so that the real cycle may be twenty-two years instead of eleven.

I apologize for leaping backward and forward in time, but in tracing the story of solar observation there is really no alternative. Let us now go back to the seventeenth century, when attention was drawn to an entirely different type of phenomenon which was only indirectly connected with the Sun itself.

On the Copernican system there are two planets, Mercury and Venus, which move round the Sun at a distance less than that of the Earth. If, therefore, they pass directly between the Earth and the Sun, they will be visible; the planet will appear as a black spot passing across the solar disk, taking an hour or more to do so (of course, the actual period depends upon whether the planet crosses the centre or the edge of the Sun). This is known as a *transit*. The remaining planets are further out – though it is worth noting that an observer on, say, Mars could observe transits of the Earth.

If Mercury and Venus moved round the Sun in the same plane as the Earth, they would transit every time they reached inferior conjunction – that is to say, midway between the Earth and the Sun. In fact this is not so. Mercury's orbit is inclined to ours by 7 degrees, and that of Venus by 3·4 degrees. These angles are small, but the result is that at most inferior conjunctions Mercury and Venus pass either above or below the Sun in the sky, and escape transit.

It will be remembered that the last great work of Johannes Kepler was the preparation of the Rudolphine Tables of planetary movements. Kepler actually completed them in 1627, and from them he predicted that both Mercury and Venus would transit the Sun in 1631 –Mercury on November 7 and Venus on December 6. By that time Kepler was dead, but the transit of Mercury was successfully observed by the French astronomer Pierre Gassendi.

Mercury is a small world, and as it passed across the Sun it was invisible without a telescope. The transit of Venus promised to be more interesting. Afraid that Kepler's prediction might be in error, Gassendi began watching the Sun on December 4, and kept his observations continuously until sunset on the 7th. To his great disappointment he saw nothing. We now know why. The transit did indeed occur, but it took place during the northern night of December 6-7, when the Sun was below the horizon in France.

Kepler had forecast no more transits of Venus until 1761, but fresh calculations were made by a young English clergyman, the Rev. Jeremiah Horrocks, curate of Hoole in Cheshire, showing that a transit would take place on November 24, 1639 (old style; the new style date is December 4 – this was before the final change-over to our modern calendar). Horrocks finished his calculations only a short while before the transit was due, and he had time only to alert his brother Jonas, who lived near Liverpool, and his friend William Crabtree, in Manchester. Both Jeremiah Horrocks and Crabtree were able to see Venus in front of the Sun, though Crabtree's view was limited to a few minutes when the clouds luckily cleared away just before sunset.

Horrocks was an astronomer of great promise. He was born in 1619, studied at Cambridge, and became a firm supporter of the Copernican system. He seemed destined for a brilliant career, but he died in 1641 at the early age of twenty-two.

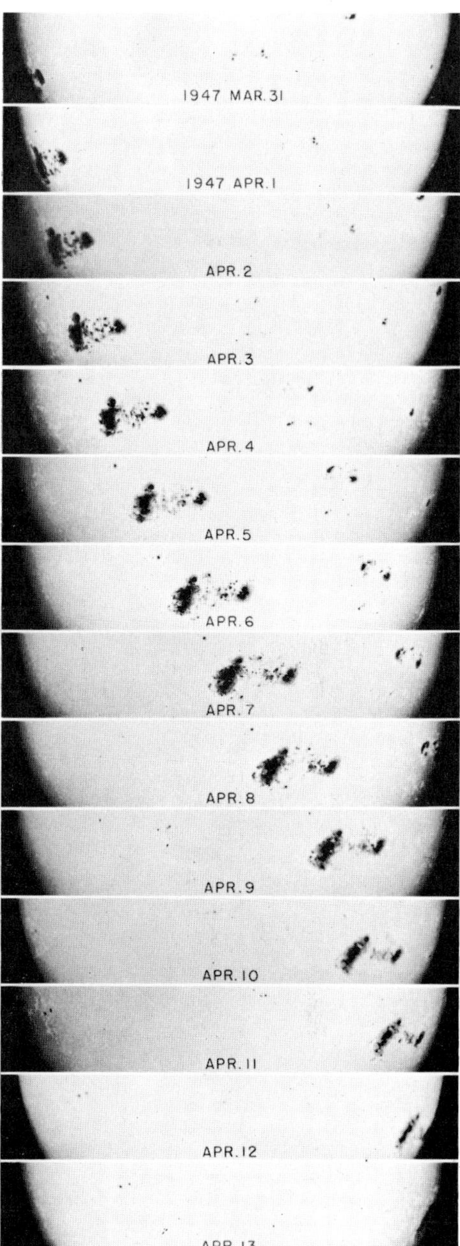

THE GREAT SUNSPOT GROUP OF 1947. *(Above) A continuation of the record, from March 31 to April 3 – the rotation following that shown on page 60. The two series of photographs may be compared, showing the changes in the spot-group. At its greatest extent, in April, the group was easily visible to the naked eye. It persisted for several more rotations of the Sun before it finally disappeared.*

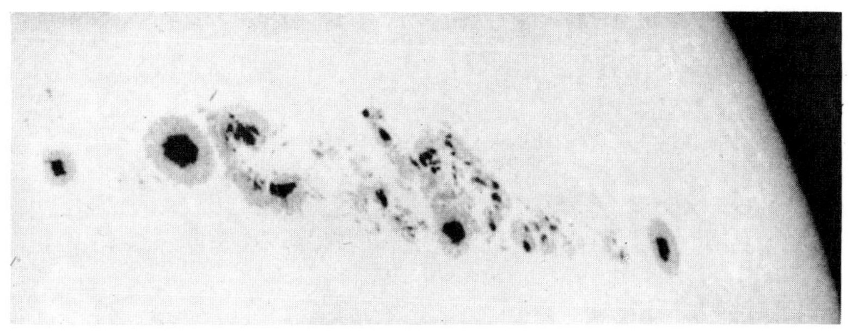

GIANT SUNSPOT STREAM, *February 20, 1956, (left) photographed at Mount Wilson and Palomar Observatories.*

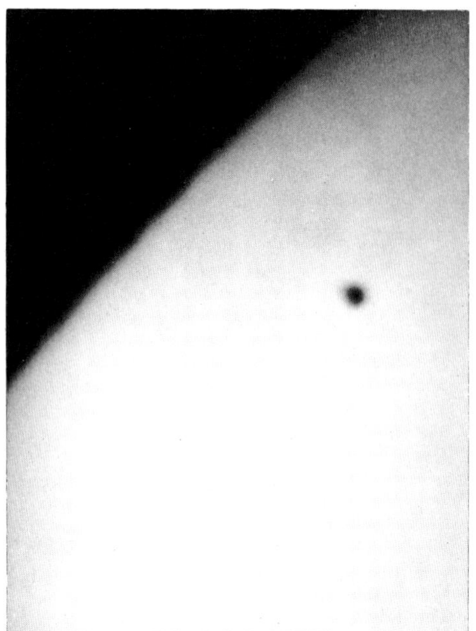

TRANSIT OF MERCURY, NOVEMBER 7, 1960, *photographed at 15 hours with the 25·4-cm reflector of the Norwich Astronomical Society.*

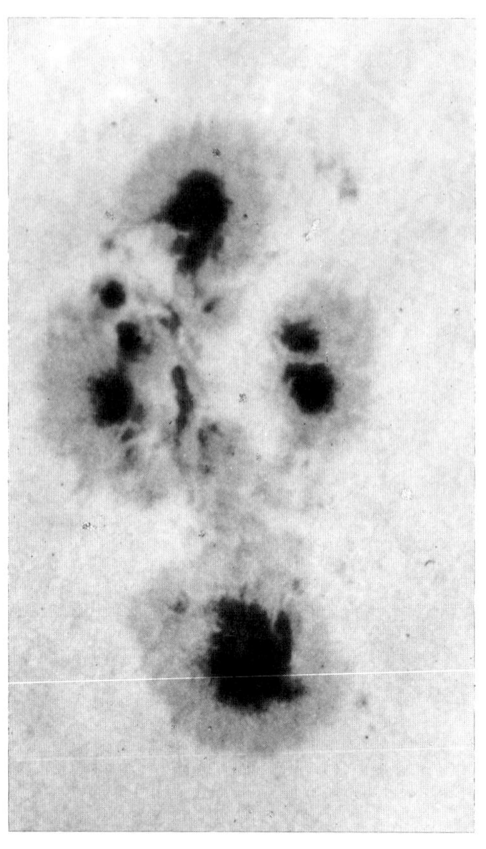

LARGE SUNSPOT GROUP, MARCH 31, 1960, *(right) photographed with a 10·2-cm refractor by W. M. Baxter.*

Transits of Mercury are not too infrequent; the last took place in 1973, and there will also be transits in 1986 and 1999. The more interesting transits of Venus are much rarer. They occur in pairs separated by an interval of eight years, after which no more occur for over a century. Since Horrocks' time the only transits have been those of 1761, 1769, 1874 and 1882; the next will be in 2004 and 2012. During transit, Venus is easily visible with the naked eye. It cannot be mistaken for a sunspot, because it moves right across the solar disk in a few hours.*

It was tragic that Jeremiah Horrocks died so young. Carr House, from which he observed the transit, still stands, and if you go to Hoole Church, you will see a beautiful stained-glass window set up in his honour. His career was cut short, but he will always be remembered as being the first man to see Venus as a black spot against the dazzling face of the Sun.

*According to an old manuscript, an Eastern scholar, Al-Faraki, 'saw Venus like a mark on the face of the Sun' from Kazakhstan in A.D. 910, but whether or not this is authentic is rather uncertain.

11

Exploring the Solar System

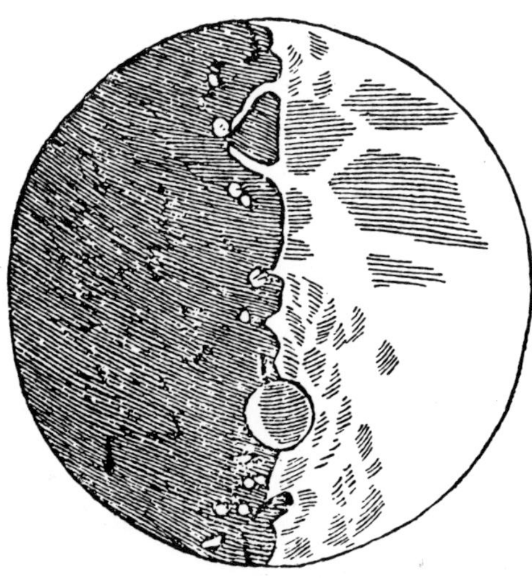

GALILEO'S MAP OF THE MOON. *This was the second lunar chart drawn with the aid of the telescope; the first was Harriot's. It is naturally rough, but various features are identifiable.*

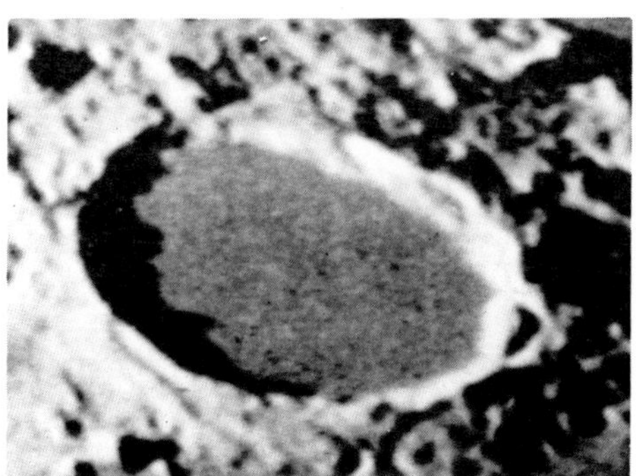

THE LUNAR CRATER PLATO. *This was the formation which Hevelius called 'the Greater Black Lake'; its floor is very dark in hue, and is relatively smooth. The full diameter is 96 km. The crater is circular, but appears elliptical due to foreshortening effects. Any small telescope will show it, and the darkness of its interior means that it is recognizable under all conditions of illumination.*

GALILEO died in 1642. Isaac Newton, whose work may be said to have laid the foundations of present-day theoretical astronomy, published his greatest work – the *Principia* – in 1687. Before going on to discuss Newton and gravitation, I feel that we must pause to say something about the other astronomers of the period, who were busy using telescopes to explore our nearest neighbours, the members of the Solar System.

As we have noted, two of the earliest observers of the Moon were Thomas Harriot and Sir William Lower; but Harriot never followed up his pioneer lunar chart, and Lower seems to have made no serious observations. (He is a somewhat shadowy figure; apparently he was expelled from Oxford for drunken behaviour, and later sat as M.P. for Lostwithiel in Cornwall before retiring to Treventy, in the calm of his native Wales. He and Harriot corresponded for some time.)

Maps of the Moon were soon produced. Christoph Scheiner drew one in 1614, Gassendi another in 1640, and so on; they were inevitably better than the only surviving pre-telescopic chart, drawn by William Gilbert in 1601. But it was not until 1647 that a major advance was made. It was due to a brewer who lived in Danzig, the modern Polish town of Gdańsk. His family name has been given variously as Hewelcke, Höwelcke, Höfelius and so on, but he signed himself by the Latinized version of Johannes Hevelius, which is how we remember him.

Hevelius was a brewer by trade, but he was also a city councillor, and an important man in Danzig. He travelled widely in his younger days (he visited London in 1631), and corresponded voluminously with most of the leading astronomers of the time. He taught himself how to make instruments, including telescope lenses, and he set up an elaborate observatory on the roof of his house; at that time it was probably the best in Europe. Sadly, it was burned down in 1679, with the loss of almost all his instruments, books and manuscripts. He made a noble effort to rebuild, but the new instruments were not so good as the old ones had been. Hevelius died in 1687, about six months before the publication of Newton's *Principia*.

Hevelius took great care in making and maintaining his telescopes, but they were quite unlike modern instruments. They had small object-glasses, and were immensely long, so that they were remarkably cumbersome.

There was a good reason for this. All the early observers had been painfully aware that refracting telescopes of the type built by Galileo had the unfortunate effect of producing false colour round any bright object. A star would seem to be surrounded by gaudy rings, which may have looked impressive, but were the last thing that the observer wanted. The cause was known, but at the time nobody could suggest a remedy.

What we normally call 'white' light is not white at all; it is a blend of all the colours of the rainbow, from red to violet. When a beam of light

45.7-M TELESCOPE USED BY HEVELIUS. *One of the clumsy 'aerial telescopes' of the seventeenth century.*

enters a lens it is bent or refracted, but the various colours are affected unequally. Red light, for example, is refracted more than blue, and so is brought to focus at a different point. The diagram here is out of scale, since the real difference between the red focus and the blue focus is very small – but it was more than enough to make the early refractors very inconvenient.

Even Newton could not see a cure, and it was not until later that a partial remedy was discovered by an English amateur, Chester More Hall. Meanwhile, Hevelius and others found that the false colour nuisance could be reduced by making telescopes of very long focal length.

Hevelius' best telescope had an objective only a few centimetres in diameter, but the focal length was nearly 47 metres! It was clearly hopeless to fit it into a solid tube, and so the object-glass had to be fastened to a mast, in this case 27 metres high. When we look at old drawings of the instrument, we wonder how Hevelius managed to use it at all, but somehow or other he did, and even produced a good star catalogue. Neither was his the longest of the so-called aerial telescopes. Huygens used a 64-metre refractor, and the French astronomer Adrien Auzout is said to have planned a 183-metre instrument, though it was never built.

Hevelius is best known for his star catalogue, his excellent list of comets, and of course for his lunar mapping. He knew that the dark areas are plains while the brighter regions are uplands, and he made measurements of mountain heights which were better than Galileo's. He also named the various features, and drew a complete map a third of a metre in diameter.

Hevelius' scheme was to give terrestrial names to the lunar craters and plains. One large crater was named by him 'Etna'; another was 'the Greater Black Lake', and so on. The system was not convenient, and it was soon abandoned. Less than half a dozen of Hevelius' names are still used. Neither does his map survive. After his death, it is said that the copper plate was melted down and made into a teapot. Fortunately, we have copies.

Meanwhile, a Jesuit professor of mathematics, Francisco Grimaldi, had been observing the Moon with the aim of drawing up a map. The map was actually published by another Jesuit, Giovanni Riccioli, who was considerably older than Grimaldi (he was born in 1598) and taught first at Padua and then at Bologna. Riccioli was no theorist; he had no faith in the Copernican system, and preferred Tycho's hybrid, but he

QUADRANT MADE AND USED BY HEVELIUS, *dated 1659. From an old woodcut.*

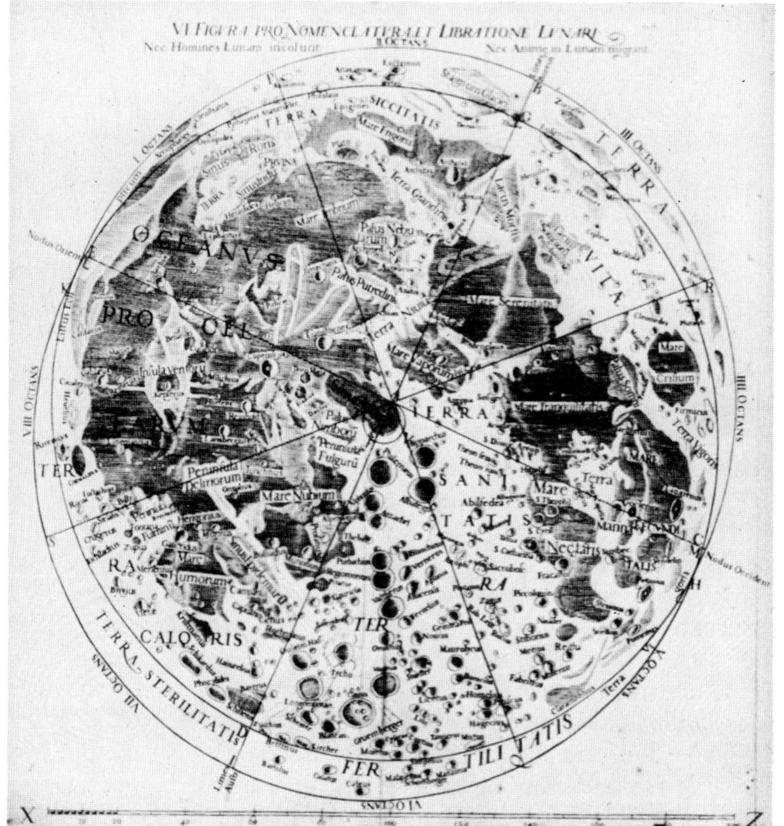

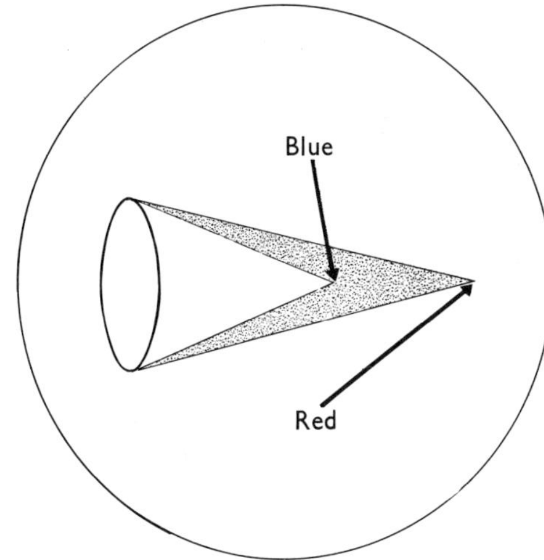

UNEQUAL REFRACTION OF LIGHT. *(Above) Blue light is bent or* refracted *more than red, and is brought to a different focus. In the diagram, the differences have been considerably exaggerated, for the sake of clarity.*

LUNAR MAP BY RICCIOLI. *(Right) Riccioli's lunar map appeared in 1651; it was based mainly upon observations made by his pupil Grimaldi. His system of nomenclature soon superseded the rather poor geographical names of Hevelius.*

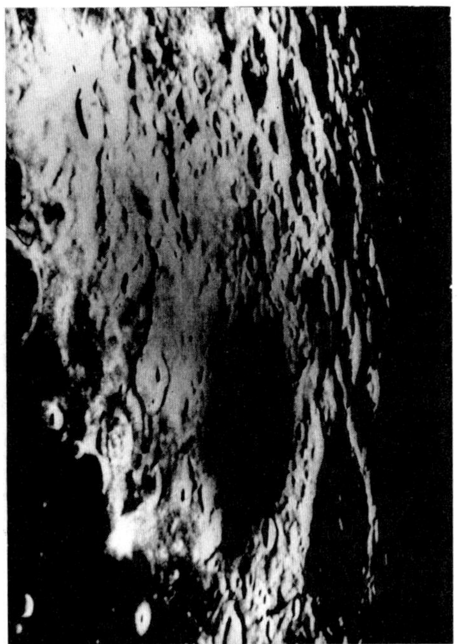

LUNAR CRATERS: REGION OF GRIMALDI AND RICCIOLI. *Grimaldi is the very large formation with an almost black floor; indeed, this is the darkest point on the Moon's surface. The walls of the crater are relatively low, but Grimaldi is always easy to recognize; had it been placed nearer the centre of the Moon's visible disk it might well have been classed as a minor 'sea'. Riccioli, below and to the right of Grimaldi, is rather smaller, but has a diameter of about 161 km; its floor contains one very dark area. Note the interesting 'valley' entering Riccioli from the west (left), to which attention has been drawn by the Japanese observer Miyamori.*

SATURN. *(Opposite) From an observation by Patrick Moore. Drawing by D. A. Hardy.*

was a reasonably accurate observer. The final map was based on Grimaldi's studies, though Riccioli may have made some contributions to it.

After careful thought he decided to jettison Hevelius' names, and introduce a completely different system. The dark plains, which he believed to be seas, were given picturesque names such as the *Oceanus Procellarum* (Ocean of Storms), *Mare Serenitatis* (Sea of Serenity), *Sinus Iridum* (Bay of Rainbows) and so on. For the craters, he used the names of famous personalities, usually those who had been connected with astronomy in some way.

Since Riccioli was a warm admirer of Tycho Brahe, he gave the great Danish observer the most prominent crater on the whole Moon – an 87-kilometre formation in the southern highlands, which is the centre of an extraordinary system of bright streaks or rays, and dominates the whole scene near the time of full moon. Copernicus – together with Aristarchus, who had proposed the 'moving Earth' theory so long before – was, as Riccioli commented, 'flung into the Ocean of Storms', but at least both craters are very conspicuous. Copernicus, like Tycho, is the centre of a ray system, while Aristarchus is the brightest object on the whole surface. It is often easy to see even when illuminated only by earthshine, and unwary observers have often mistaken it for an erupting volcano.

All the large craters were renamed. Hevelius' 'Greater Black Lake' became Plato; Julius Cæsar and Sosigenes were allotted craters near the centre of the disk, because of their efforts in connection with the calendar; Riccioli took care to give himself a very large walled plain, and Grimaldi another, though Galileo was rather grudgingly given a small, low-walled crater near the edge of the Ocean of Storms.

Riccioli's system is still in use. Later astronomers have added to the list, and some of the names are outwardly surprising. For instance, there is a Birmingham on the Moon, named after the nineteenth-century Irish amateur John Birmingham. We also meet with Billy (after a French professor of mathematics who lived from 1602 to 1679), a couple of Olympians (Atlas and Hercules) and even a Hell. However, the lunar Hell is not noted for any great depth. It honours Father Maximilian Hell, of Hungary, who made useful observations of the 1769 transit of Venus. And since 1959, when the Moon's far side was first photographed from a space-craft, names have had to be allotted there too.

So far as the planets were concerned, the most skilful observer of the period was certainly a Dutchman, Christiaan Huygens, who was born in 1629 and studied at Leyden. For some years he lived in Paris, and visited England, where he met Newton. He became an expert telescope-maker, and invented a new type of eyepiece which we still call the Huygenian. More important still were the improvements which he made in the construction of watches, and the invention of the pendulum clock. Galileo had considered the use of the pendulum as a timekeeper, but it was certainly Huygens who made the first clock of this sort.

His first major astronomical discovery was made in connection with Saturn. Galileo had been frankly puzzled; he had originally believed the planet to be triple, but when the two attendant bodies disappeared he was at a loss. Various other theories were proposed, some of them really weird. The French mathematician Gilles de Roberval believed

Saturn to be surrounded by a hot zone which gave off vapours; Giovanni Hodierna, of Sicily, thought that Saturn might be a globe with two dark patches on it, and so on. Sir Christopher Wren (who, remember, was an astronomer before turning to architecture) worked out a theory according to which Saturn had an elliptical corona, meeting the globe in two places and rotating with Saturn once in each sidereal period (in the case of Saturn, over twenty-nine years). But Wren never published his theory, because before he was ready to do so he heard of Huygens' work – and realized that Huygens was right, while he was wrong.

Huygens solved the problem in 1655, though he did not announce his result until some time later. In his *Systema Saturnium*, published in 1659, he stated that Saturn is surrounded by 'a thin, flat ring, which nowhere touches the body of the planet'. The ring-system is circular, though from Earth it appears elliptical; the main part of it has an overall diameter of over 270,000 kilometres, but it is only a few kilometres thick at most, and so when the system is edgewise-on to us, as happens regularly, it can be seen only as a thin line of light. In small telescopes it disappears altogether. The diagram here shows the changing appearance. Since then the rings have been at their widest in 1975, but edge-on in 1980. Of course this does not indicate any real change; everything depends upon the angle from which we are observing. In 1675 G.D. Cassini, in Paris, discovered that there is a gap in the ring system, known still as Cassini's Division. It separates the two main bright rings, and was thought to be empty, though the Voyager space-craft results of 1980 and 1981 have given us a very different picture of it.

In 1705 J.D. Cassini (son of G.D.) made the suggestion that the rings were not solid. He was right. The rings are made up of vast numbers of small, icy particles, each of which is whirling round Saturn in the manner of a dwarf moon. No solid or liquid ring could exist so close to Saturn; it would be quickly broken up by the powerful gravitational pull. Theoretical confirmation of Cassini's theory was given much later: by James Clerk Maxwell, in 1875.

In 1655 Huygens made another interesting discovery: that of Saturn's largest satellite, Titan, which is almost as large as the planet Mercury. A small modern telescope will show it, but Huygens did well to detect it with his clumsy aerial refractor.

Huygens next turned his attention to Mars, which had at least a solid surface. In 1659 he recorded a marking on the disk – the V-shaped

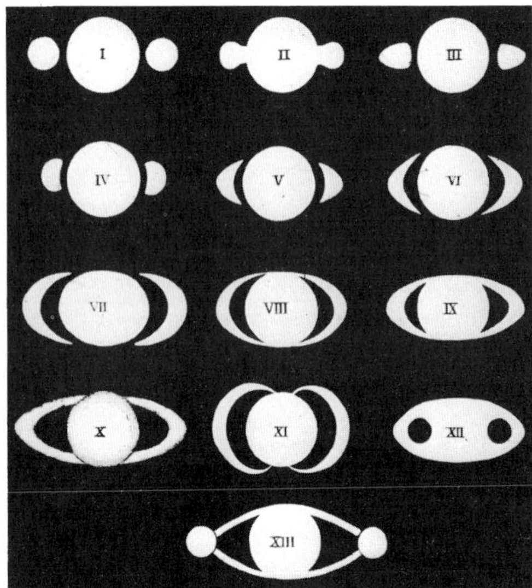

DRAWINGS OF SATURN MADE BY HUYGENS

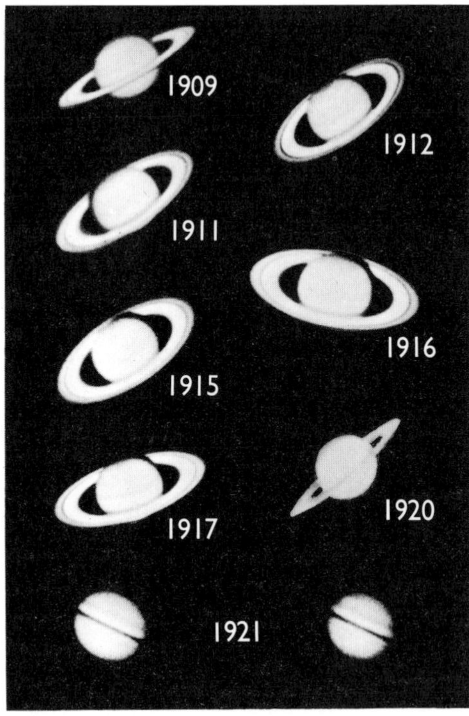

PHOTOGRAPHS OF SATURN, *showing the ring-system under different angles. Lowell Observatory photographs.*

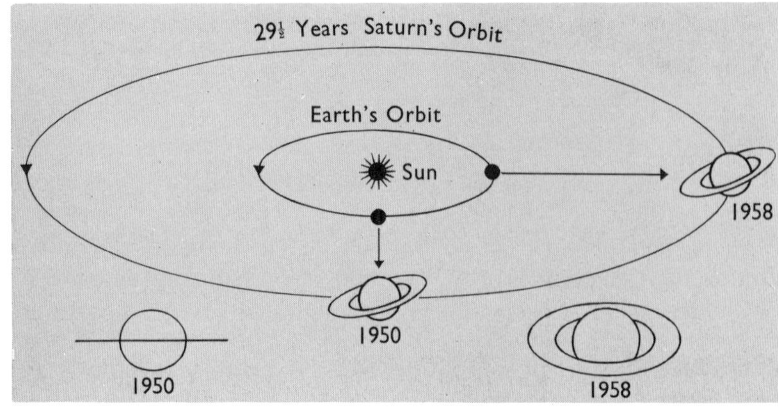

THE CHANGING ASPECTS OF SATURN'S RINGS *(right)*

feature which we now call the *Syrtis Major* (it used to be known as the Hourglass Sea). By watching the drift of this feature as it was carried across the disk by virtue of Mars' axial rotation, Huygens was able to show that the rotation period or 'day' there is about half an hour longer than our own.

Huygens made these discoveries when he was living in his native Holland, but in 1665 he went to France, at the invitation of Louis XIV, and stayed there for sixteen years. It was during this period that he invented the pendulum clock. Unfortunately there were religious troubles to be faced; Huygens was a Protestant, and in 1681 all Protestants in France became unpopular. Huygens accordingly returned home, and died at The Hague in 1695.

Another astronomer busy in France at this time was Giovanni Domenico Cassini. Cassini was born in Italy in 1625, and in 1650 became a professor at Bologna. Like Huygens, he was a careful observer of the planets; he measured the rotation period of Mars, and also that of Jupiter, whose globe is obviously flattened at the poles, and whose rotation period is less than ten hours long – though, as with the Sun, the equator has a faster rotation than the poles.

In 1666 Cassini discovered that the Martian poles are covered with whitish deposits which look very like snow or ice. These polar caps grow and shrink with the seasons on the planet, and have vitally important effects upon Martian meteorology.

It was also in 1666 that Adrien Auzout pointed out that it was high time to establish a national observatory, equipped with the best instruments available. Up to then there had been only one such establishment – at Copenhagen, completed in 1637 (it was re-established in 1656). All the great observers, including Galileo, Hevelius and Huygens, had built and set up their own equipment. Auzout wanted to alter all this, and fortunately the French King, Louis XIV, was a patron of science. The result was the founding of the famous Academy of Sciences in Paris, and no time was lost in starting the construction of an observatory. Cassini seemed to be the obvious man to direct it, and he accepted an invitation to come to France.

Difficulties arose almost at once, mainly because the King wanted the observatory to look magnificent while Cassini was more interested in its scientific value. As soon as Cassini arrived and saw the half-finished building, he told Louis that unless it was drastically altered it would be of no use whatsoever. The King was far from pleased, and a serious quarrel was only narrowly avoided. Eventually Cassini carried

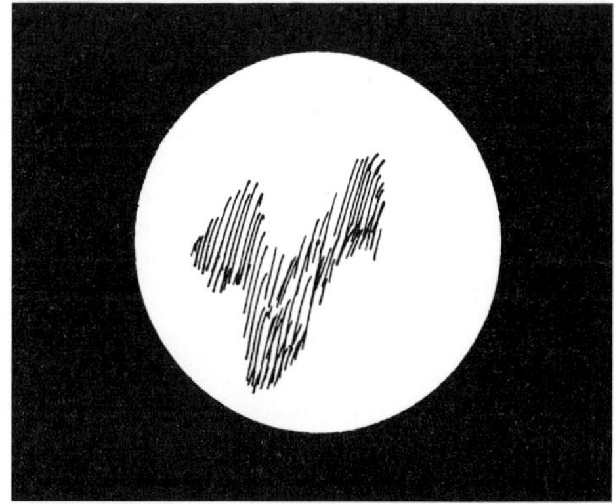

HUYGENS' ORIGINAL DRAWING OF MARS. *This, the first telescopic drawing of Mars to show surface detail, was made by C. Huygens in 1659. The Syrtis Major, the most prominent of the dark features, is clearly recognizable.*

JUPITER. *A photograph taken in red light with the 508-cm Hale reflector at Palomar. The Red Spot is shown. The black disk above is the shadow of Jupiter's third satellite, Ganymede, and Ganymede itself is seen to the right of the planet.*

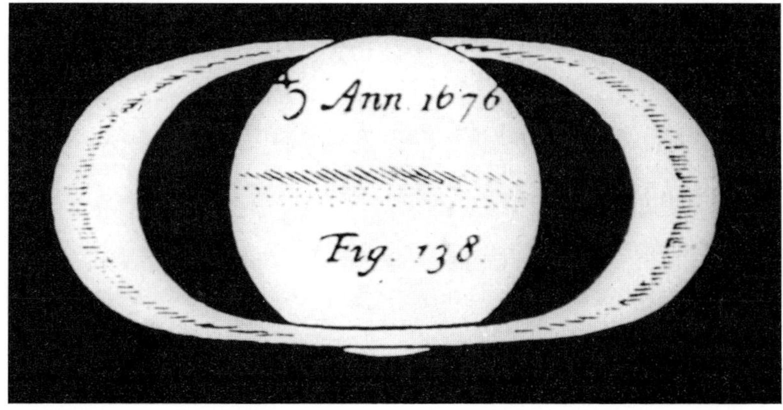

CASSINI'S DRAWING OF SATURN. *(Left) This is one of the drawings made in 1676 by G. D. Cassini, at Paris. It shows the famous division in the rings which Cassini discovered, and which is still known by his name. The drawing is naturally rough by modern standards, and the shape of the ring-form is not accurate, but it was of course made with one of the clumsy 'aerial telescopes', and it is a tribute to Cassini's skill as an observer that the rings are shown in recognizable form. One of the belts crossing the surface of the planet is also shown. Belts on Saturn resemble those of Jupiter, but are considerably less prominent.*

APPARENT MOVEMENT OF MARS AMONG THE STARS, 1960. *At this time Mars was near opposition, and the constellation of Gemini, with the 'twins' Castor and Pollux, is shown; the shift of Mars is very obvious. The dates of the photographs are: (upper) October 26, (right) November 19, (lower) December 26. Colour photographs taken by K. S. G. Stocker.*

his instruments into the open air outside the observatory. This was by far the best plan, since he was still forced to use aerial telescopes of small aperture and enormous focal length. However, it was at least a beginning.

The French skies are less clear than those of Italy, but Cassini was able to carry on with his work. He discovered four more satellites of Saturn, now known as Iapetus, Rhea, Dione and Tethys; he refined the value of Mars' rotation period as being 24 hours 40 minutes (less than three minutes too long), and he re-measured the distance of the Sun. He gave a value of 138,000,000 kilometres, which was not so very wide of the mark. He never went back to Italy, and when he died in 1712 his son, J.D. Cassini, succeeded him as Director of the Observatory.

One of Cassini's colleagues at Paris was a Dane, Ole Rømer, who arrived in 1671. It was Rømer who made another important discovery: that of the velocity of light.

Cassini had made careful measurements of the satellites of Jupiter, and had produced tables of their movements which were much the best in existence. In particular, it was possible to work out when the four moons should be eclipsed by Jupiter's shadow, just as our Moon is sometimes eclipsed by the shadow of the Earth. However, he found that his predictions were often wrong; sometimes the eclipses were too early, sometimes too late. There seemed to be nothing wrong with the tables, and the eclipses of the Galilean satellites are not hard to observe. Anyone using a small telescope can watch them, and time them accurately.

Cassini wondered whether the speed of light might have anything to do with these discrepancies, but he did not follow up the idea. Rømer did. The errors were due to the fact that Jupiter's distance from us is always changing, so that the light from the planet does not always take the same time to reach us. Rømer calculated that instead of moving instantaneously, light travels at a rate of 300,000 kilometres per second. This was an excellent result; the modern value is 299,792.456 kilometres per second.

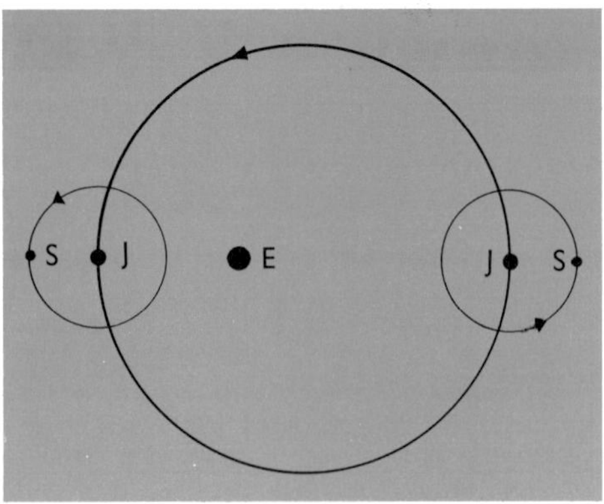

MEASURING THE VELOCITY OF LIGHT. *This shows the method adopted by Ole Rømer. When Jupiter* (J) *is at its closest to the Earth* (E), *the light from the satellite* (S) *has a lesser distance to travel, and so its eclipses are earlier than had been predicted.*

Rømer's work did not end there. In 1681 he went back to Denmark as Professor of Astronomy at Copenhagen, where he was responsible for various improvements in equipment. In particular, he invented the transit instrument and the meridian circle, both of which contributed largely to the development of precise astronomy. Rømer lived on until 1710.

Clearly, then, observational astronomy had made great strides since Galileo had first turned his tiny telescope to the sky in the winter of 1609-10. But theoretical work had been making even more striking progress, and this brings us on to the greatest figure in the whole history of science: Isaac Newton.

12

The Genius of Newton

ISAAC Newton was born at Woolsthorpe, near Grantham in Lincolnshire, in 1642. His father died before he was born, and his mother was left in charge of the family farm. There was little money to spare, and to make matters even more difficult England was in a disturbed state; this was the time of the Civil War between King Charles I and Cromwell's Roundheads. Isaac went to the village school, but showed no signs of unusual intelligence. All we really know about his early school career is that he was particularly fond of making models.

When he was twelve he was sent to King's School, Grantham, which had been founded over a century earlier by Henry VIII. For the first few terms he remained near the bottom of his class. Then – according to a story which may or may not be true – a bigger boy, who was above Newton in form, jeered at him and kicked him. Newton had no intention of bearing such treatment, and a fight followed, ending only when Newton had beaten his rival and rubbed his nose against the wall. From that time on he worked hard, and reached the top position in the school.

Meanwhile his mother had married again, but in 1656 her second husband died, and Mrs Newton brought Isaac back from Grantham to help her run the farm. The experiment was not a success. The boy was not in the least interested in farming; it is said that when he was supposed to be keeping an eye on the labourers he used to spend his time sitting behind a hedge, working out problems in mathematics. Wisely his mother sent him back to Grantham, and in 1661 he went to Cambridge University, enrolling as an undergraduate at Trinity College. Here he met Professor Isaac Barrow, who became his tutor in mathematics.

Barrow was the son of a linen draper. Like Newton he had done little work during his first years at school, and had distinguished himself only by his fondness for fighting. As he was enormously strong, he won far more fights than he lost, and was held in respect by the other boys. Later he went to Cambridge, and became a Fellow of Trinity College. During the Civil War he left England, and travelled around Europe and Asia Minor. Although he was by now a well-known scholar, his character had not changed, and on one occasion he showed his courage very plainly. During a sea voyage from Leghorn to Smyrna his ship was attacked by pirates. Barrow was not to be frightened; he stayed on deck, and fought so fiercely that the pirate vessel sheered off.

STATUE OF NEWTON. *This photograph, taken by Patrick Moore in 1961, shows a statue of Newton which has been set up in the Lincolnshire town of Grantham. It was in this town that Newton received his education, and his birthplace, Woolsthorpe, is not far off. It was in Lincolnshire, too, that Newton carried out much of his important work – notably during the Plague period, when Cambridge University was closed, and all the students had been sent back to their homes.*

PRODUCTION OF A SPECTRUM. *When a beam of apparently 'white' light is passed through a prism, it is split up into its component colours, and in this colour photograph the production of a rainbow or continuous spectrum is shown. This effect was first noted by Newton.*

Barrow was a Royalist, and came back to England after the restoration of Charles II in 1660. In 1663 he was once more at Cambridge, and was appointed to the Lucasian Chair of mathematics – so called because the money to found the appointment had been provided by the will of a Mr Lucas.

Barrow was strong-minded as well as clever, and he was an ideal tutor for the rather shy and retiring Newton. The two worked very happily together. Newton took his degree in 1665, and four years later Barrow resigned the Lucasian Chair so that Newton could succeed him. Clearly, then, he had faith in his pupil's ability, and this faith was more than justified later on.

Meanwhile, fresh disasters had come to England. In 1665 the Great Plague struck London; 17,000 people died in August alone, and 30,000 in September. The King and Court left London, but cases of plague were reported from other parts of the country, and there was no real safety anywhere. When the disease reached Cambridge, the authorities wisely closed the University. All the students left Cambridge to go back to their homes, and Newton accordingly returned to Woolsthorpe.

THE 304·8-CM LICK REFLECTOR. *This is one of the most modern of large reflectors, and is one of the largest telescopes in the world. The tube is a skeleton, and the large size of the instrument means that various optical systems can be used.*

To most men such an interruption would probably have been a handicap, but to Newton it was an advantage. He was able to work quietly, on his own, and with no money troubles. For the next year or so he was left in peace, and he managed to carry out a tremendous amount of original research.

One subject which interested him was the nature of light. We have seen that the object-glasses of early refractors produced false colour, and that this was due to the unequal bending of the different colours which merge together to make 'white' light. Newton demonstrated this by an ingenious experiment. He passed a beam of sunlight through a prism, and split it into the usual rainbow. He then passed one particular colour – violet, for instance – through a second prism. No second rainbow was produced, showing that the violet light was 'pure'. Had Newton but known it, he had taken the first steps in the science of spectroscopy.

His aim was to produce an *achromatic* object-glass – that is to say, a lens which would produce no false colour. Unfortunately he could see no way of doing it, and after considering the matter carefully he decided that there was only one solution. He must make a telescope which did not use an object-glass at all.

A 30·5-CM NEWTONIAN REFLECTOR. *This photograph shows a typical Newtonian reflector on an equatorial mounting, The focal length is 279 cm. The telescope was used by an English nineteenth-century observer, W. F. Denning, for his studies of planetary surfaces. Photograph by Patrick Moore.*

Some years earlier, in 1663, the Scottish mathematician, James Gregory, had suggested using a mirror instead of a lens to collect light. Gregory never built such an instrument – as he admitted, he had no practical skill – but Newton developed the idea, and produced the first *reflector*. His optical system was not exactly the same as Gregory's, but it was more convenient, and is more widely used today.

On the Newtonian system, the light from the object to be observed passes down an open tube until it hits a mirror at the lower end. This mirror is curved, and reflects the light back up the tube on to a smaller mirror or *flat*, placed at an angle of 45 degrees. The flat sends the rays into the side of the tube, where they are brought to focus and the image is magnified by an eyepiece in the usual way. In a Newtonian reflector, then, the observer looks into the tube instead of up it.

Since a mirror reflects all parts of the spectrum equally, the false colour trouble does not arise, and Newton's first telescope was a definite success. It had a metal mirror 2.5 centimetres in diameter, and was presented to the Royal Society of London in 1671. It caused a major sensation, and led to some controversy, because some of the members of the Society had different ideas about the nature of light.

Newton's reflector looks very small compared with modern instruments. Today, many amateurs have reflectors of incomparably greater power; my own, at Selsey, has a 39-centimetre mirror, and the largest telescope in the world today (the Russian reflector in the Caucasus) has a mirror no less than 600 centimetres across. But Newton had shown the way, and his tiny 2·5-centimetre telescope was the direct ancestor of those we now use.

Gregory's original idea was to reflect the light back down the tube, passing it through a hole in the centre of the main mirror. There are unpleasant disadvantages here, though the Cassegrain arrangement, which also makes use of a central hole, is better. Meanwhile, let us turn to the laws of gravitation – beginning with the tale of Newton and the apple, which, unlike those of Canute and the waves, Alfred and the cakes, and for that matter Galileo and the Leaning Tower, is probably true.

According to the story, Newton was sitting in his Woolsthorpe garden one afternoon when he saw an apple fall from a tree branch to the ground. A very common sight, but he began to wonder just why the apple had fallen. There must be some definite force which pulled it to the ground; but what was this force, and how far did it extend? Newton realized that the force which pulled on the apple was the same as the force which keeps the Moon in its path round the Earth – or the Earth in its orbit round the Sun. This led him on to the concept of *universal gravitation*, according to which every particle of matter attracts every other particle with a strength which becomes weaker with increasing distance.

Put in this way, the story is over-simplified, and to reduce the theory to mathematical terms needed a remarkable amount of hard work. But the apple was the starting-point, so let us follow the reasoning further.

Suppose that instead of being around 6 metres high, the tree had risen to 60 metres, or even 60 kilometres? The apple would still have dropped to the ground, gathering velocity as it fell. This would even apply to a tree 376,000 kilometres in height. The Moon is 376,000 kilometres away from us – so why does it not fall, just as the apple did?

Newton found the answer. The reason why the Moon does not drop is because it is moving. It is not easy to give an everyday analogy, but some idea of what is meant can be gathered from taking a cotton-reel and whirling it round on the end of a string. The reel will not fall so long as it continues moving quickly enough to keep the string taut, and the Earth's pull on the Moon may be said to act in much the same way as the string on our cotton-reel. And the Moon goes on moving simply because there is nothing to stop it. (I realize that in this analogy I am neglecting the mass of the Earth. I give it here merely because I have never been able to think of anything better.)

Now suppose that the holder of the string lets go suddenly. The reel will fly off in a straight line. If we forget about the pull of the distant Sun, it is easy to see that the Moon also would move off in a straight line if the Earth were not pulling upon it, and Newton concluded that any moving body will continue moving in a straight line unless some outside force is acting upon it. This is the famous *law of inertia*.

NEWTON'S FIRST REFLECTOR. *The photograph shows a replica of the first reflecting telescope, demonstrated by Newton to the Royal Society. The mirror, 2·54-cm in diameter, was made of speculum metal. Though it was so small, the telescope functioned perfectly, and demonstrated the soundness of Newton's arrangement.*

SIR ISAAC NEWTON

THE MOUNT WILSON 152·4-CM REFLECTOR, *(right) seen from the north-west. This was the first of the large Mount Wilson reflectors, made by Ritchey at the instigation of George Ellery Hale; it is interesting to make a comparison between this vast instrument and the original small telescope made by Newton. Large reflectors such as the 152·4-cm are seldom used for visual observations, and nearly all their work is carried out by means of photography.*

A 21·6-CM NEWTONIAN REFLECTOR. *This has an equatorial mounting made in the late nineteenth century by Browning. The focal length is 132-cm. Photograph by Patrick Moore.*

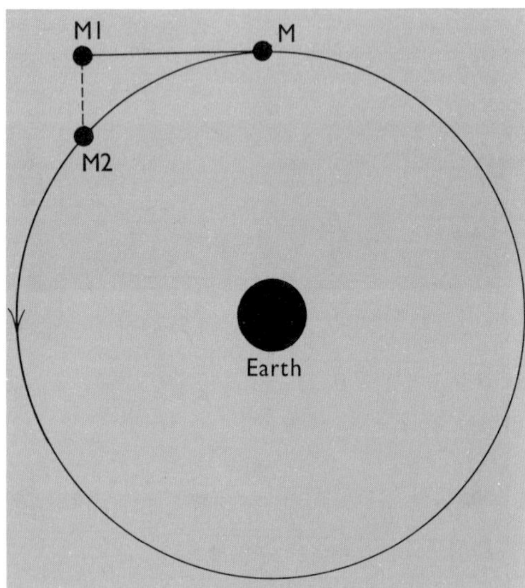

GRAVITATIONAL EFFECTS OF THE EARTH UPON THE MOON. *This diagram, which is not to scale, shows the Moon's path round the Earth. But for the presence of the Earth, we may suppose that the Moon would move from M to M1 in one minute. However, the Earth's gravitational pull means that instead of moving uniformly from M to M1, the Moon is 'pulled down' to M2. It may be said that the Moon has 'fallen' from M1 to M2 in one minute, and it goes on 'falling' all the time, though it does not drop any closer to the ground.*

Newton knew the force of the Earth's pull at ground level, since this was the force affecting the apple, and he worked out the law according to which force weakens with increasing distance from the Earth (or, more correctly, from the Earth's centre). According to his calculations, the Moon should 'fall' at five metres per minute, which would be the distance between **M1** and **M2** in the diagram.

This did not agree with observation. The distance 'fallen' in one minute is not five metres, but only four. As Newton said, the figures 'agreed pretty nearly', but not well enough to satisfy him.

In this sort of calculation, a body such as the Earth behaves as though all its mass were concentrated at a point at the exact centre of the globe. As Newton had to make his observations from the Earth's surface, he had to know the distance from the surface to the centre of the globe. There is a story that the discrepancy in the fall of the Moon was due to Newton's having used a wrong value for the Earth's radius. Apparently this is not so; one link in the mathematical argument was still missing, and it was not until some time later that Newton found out what was wrong. Meanwhile he realized that in order to make his calculations he had to develop an entirely new branch of mathematics. He called it the 'method of fluxions', but it corresponds to what we now term the calculus.

Newton returned to Cambridge after the end of the Plague danger, and in 1672 he was elected a Fellow of the Royal Society, a scientific body which had been founded early in the reign of Charles II. (One of the prime movers was Prince Rupert – 'Rupert of the Rhine', Charles I's dashing cavalry leader.) Newton was still comparatively unknown; he had said nothing at all about his researches into gravitation, and his first official contribution to astronomy was the reflecting telescope. Unfortunately he had an immediate disagreement with another Fellow of the Society, Robert Hooke, who was to play an important rôle in Newton's life.

Hooke was seven years older than Newton, and a man of a different type. He was physically weak, and it was said of him that 'his figure was crooked, his limbs shrunken; his hair hung in dishevelled locks over his haggard countenance. His temper was irritable, his habits solitary.' He was violently jealous and suspicious, and was not above claiming credit for work which had actually been done by others. Yet he was a brilliant scientist, and was of an inventive turn of mind. For instance, he developed the 'universal joint' known to every mechanic, and he also constructed new meteorological instruments, such as the hygrometer for measuring the moisture in the atmosphere. A list of his inventions would take up many pages. On the other hand his range was so wide that he seldom followed any particular investigation through to the end, as Newton did, and though he was an excellent mathematician he was by no means Newton's equal.

Hooke admitted that the new reflecting telescope was effective, and hinted that he had himself made one as long ago as 1664, though he failed to produce it. He was also highly critical of Newton's theories about light, and Newton was never disposed to be questioned. Newton was touchy and sensitive, and was always reluctant to become involved in public arguments. He did have some angry exchanges with Hooke and others, but on the whole he preferred to say as little as possible, and he even refused to publish some of his results. Indeed, his most important work on the nature of light, contained in his book *Opticks*, remained unpublished until 1704, after Hooke's death.

Other leading Fellows of the Royal Society at that period were Edmond Halley, of comet fame, and Sir Christopher Wren. Wren is best remembered as the great architect who was responsible for the design of St. Paul's Cathedral and other buildings following the Fire of London, but he was also an astronomer, and had at one time been Professor of Astronomy at Oxford. In 1684 Hooke, Halley and Wren discussed the problem of gravitation, and came to the conclusion that what we now term the *inverse square law* must be true.

Hooke had published some work on gravitation, including some of the same conclusions as Newton had reached during his spell at Woolsthorpe. Hooke knew, then, that the force between any two bodies will become weaker if the bodies are moved further apart, and he believed that he had discovered the relationship between pull and distance. It can be explained by simple arithmetic, so let us take the convenient case of two planets which move round the Sun at distances of 2 million kilometres and 5 million kilometres respectively. (No planet is anything like as close-in as this, but let us make matters as easy as possible!). Two squared, or 2×2, is 4. Five squared, or 5×5, is 25. Then according to the inverse square law, the Sun's force on the two planets will be in the ratio of $\frac{1}{4}$ to $\frac{1}{25}$, so that the force on the more distant planet will be only $\frac{4}{25}$ of that on the nearer planet. It can then be shown that each planet will move in an elliptical path.

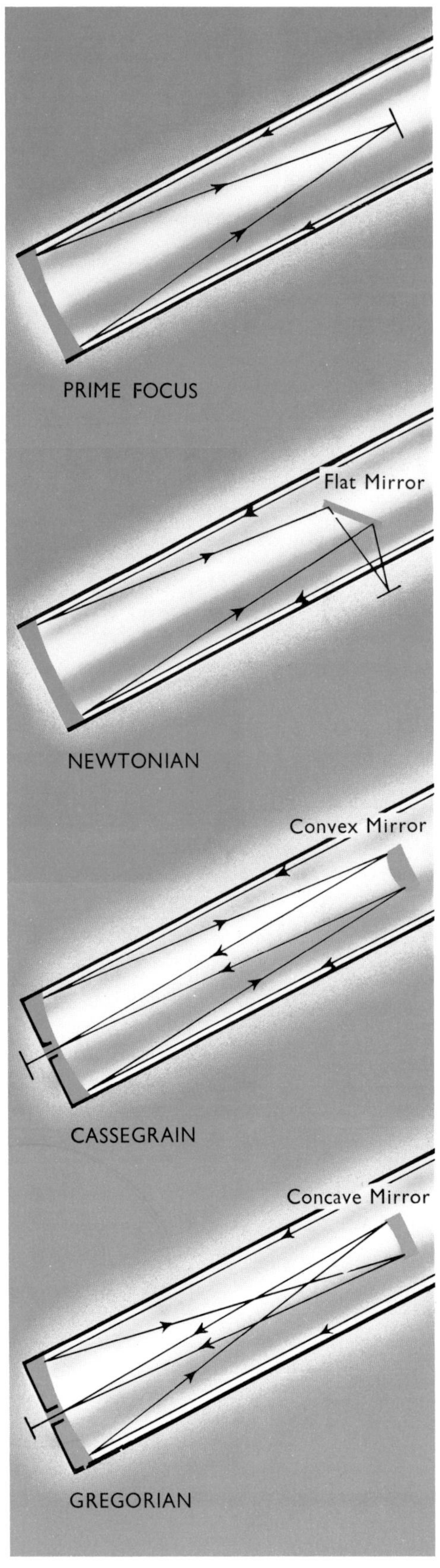

FOUR TYPES OF REFLECTORS

HYGROMETER MADE BY HOOKE. *(Right) This was one of the many scientific instruments developed by Hooke; indeed, his researches extended into almost all branches of science.*

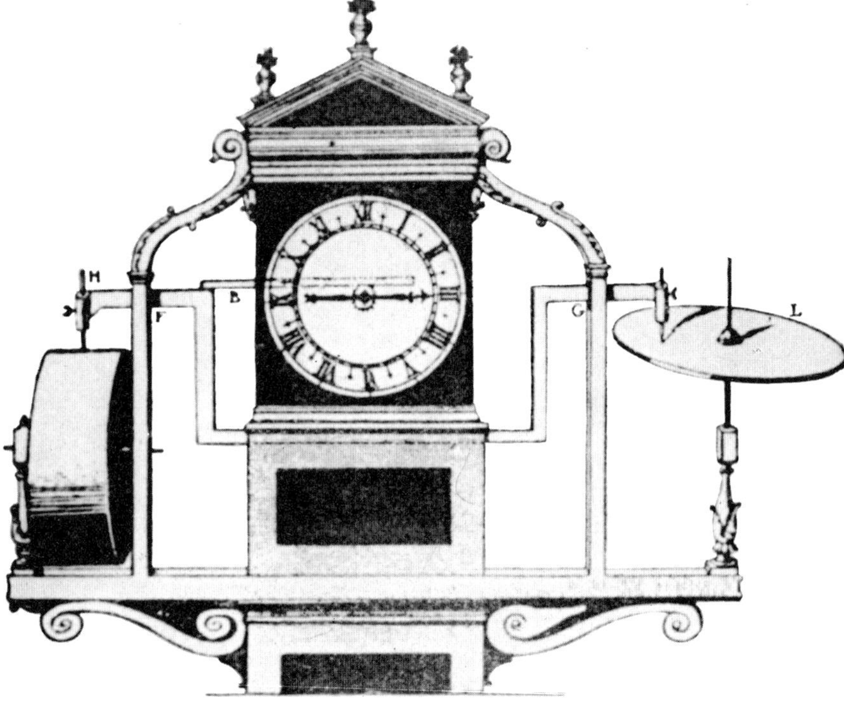

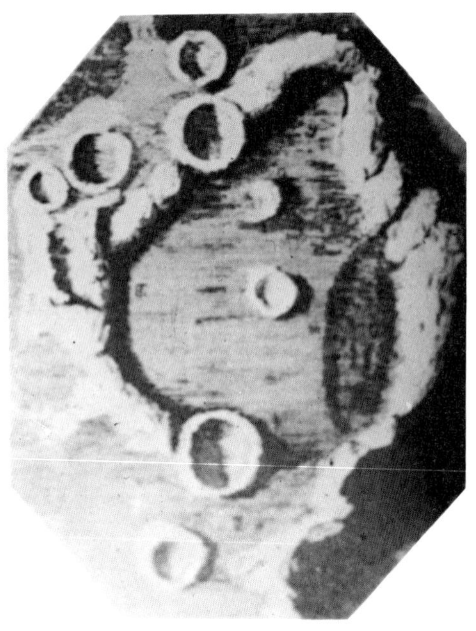

LUNAR DRAWING BY HOOKE. *Robert Hooke was one of the first observers to study the Moon in detail. This drawing shows some craters in recognizable form, and is remarkably accurate considering the low-powered telescope with which it was made.*

Hooke guessed this instinctively, but he was not a good enough mathematician to prove it. Neither were Halley and Wren, and evenutally Halley went down to Cambridge to consult Newton. He was astounded to learn that Newton had solved the problem years earlier, but had not announced it, and had even lost his notes!

Halley persuaded Newton to rework the calculations, and to allow them to be published. Newton agreed, rather reluctantly, and during 1685 and 1686 he worked on the book which he called the *Philosophiæ Naturalis Principia Mathematica* (Mathematical Principles of Natural Philosophy), but which is known to everyone simply as the *Principia*.

Halley's idea was that the book should be published by the Royal Society, but there were money difficulties; the Society had just issued a tremendous book by Francis Willughby called *The History of Fishes*, and the book had been a financial failure. (Later, when Halley was a salaried official of the Royal Society, he was presented with fifty copies of *The History of Fishes* instead of being given the £50 which the Society owed him. Halley is known to have had a strong sense of humour, but whether he saw the joke or not remains uncertain!)

There was the added trouble that Newton was being constantly irritated by his disputes with Hooke and others, and he was quite capable of changing his mind and withdrawing the *Principia* altogether. Halley realized this, and generously offered to pay for the publication out of his own pocket. This was done, and the book finally came out in 1687.

The *Principia* had taken Newton fifteen months to write, and has been described as the greatest mental effort ever made by one man; it has ensured that Newton's name will live for all time. Many of the age-old problems of astronomy were solved at a stroke. As well as dealing with gravitation and the movements of the planets, the *Principia* contained sections dealing with matters such as the tides, whose cause had not previously been understood. It is impossible to describe

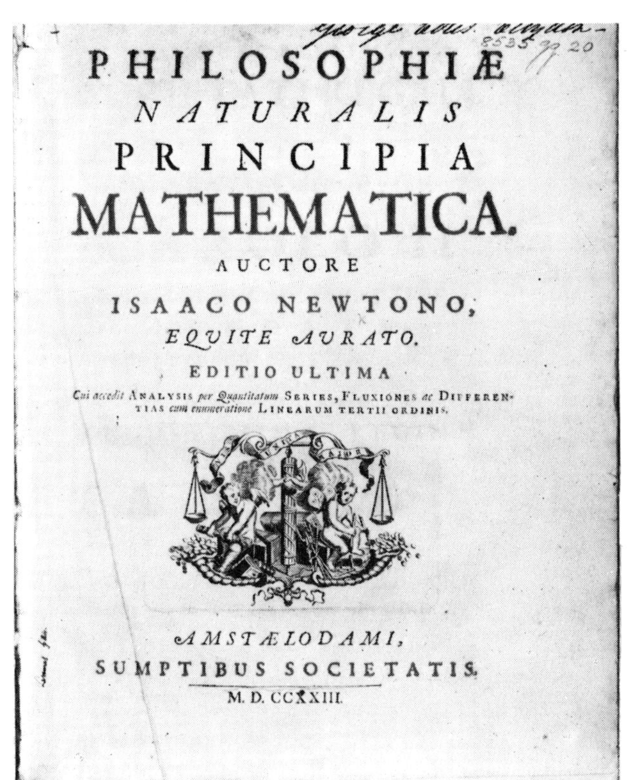

PHILOSOPHIÆ
NATURALIS
PRINCIPIA
MATHEMATICA.
AUCTORE
ISAACO NEWTONO,
EQVITE AVRATO.
EDITIO ULTIMA
Cui accedit ANALYSIS per Quantitatum SERIES, FLUXIONES ac DIFFERENTIAS cum enumeratione LINEARUM TERTII ORDINIS.

AMSTÆLODAMI,
SUMPTIBUS SOCIETATIS.
M. D. CCXXIII.

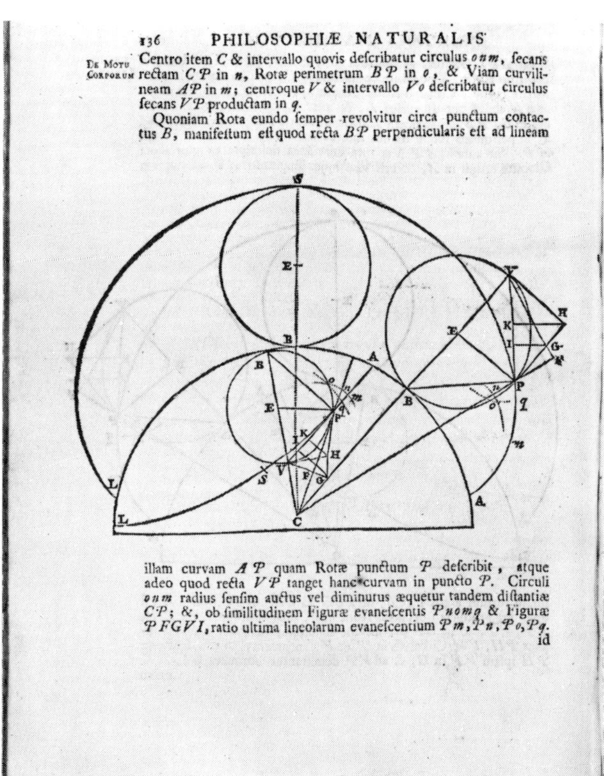

136 PHILOSOPHIÆ NATURALIS

DE MOTU CORPORUM

Centro item C & intervallo quovis deſcribatur circulus onm, ſecans rectam CP in n, Rotæ perimetrum BP in o, & Viam curvilineam AP in m; centroque V & intervallo Vo deſcribatur circulus ſecans VP productam in q.

Quoniam Rota eundo ſemper revolvitur circa punctum contactus B, manifeſtum eſt quod recta BP perpendicularis eſt ad lineam illam curvam AP quam Rotæ punctum P deſcribit, atque adeo quod recta VP tanget hanc curvam in puncto P. Circuli onm radius ſenſim auctus vel diminutus æquetur tandem diſtantiæ CP; &, ob ſimilitudinem Figuræ evaneſcentis Pnomq & Figuræ PFGVI, ratio ultima lineolarum evaneſcentium Pm, Pn, Po, Pq. id

TITLE-PAGE AND SPECIMEN PAGE FROM THE *Principia – or, to give the book its full title* Philosophiæ Naturalis Principia Mathematica. *This has often been regarded as the greatest scientific work ever produced.*

the book in a few lines, but those who have studied it can only wonder how a man still less than fifty years old could have accomplished so much.

Newton lived for forty years after the publication of the *Principia*. He was active in many fields as well as science; for instance he entered Parliament for a time, and then became Master of the Royal Mint, helping to revise England's coinage – which badly needed attention, since many of the coins in circulation had been reduced in value by having pieces chopped off them. In 1705 he was knighted by Queen Anne, and he also became President of the Royal Society. Two further editions of the *Principia* were issued before Newton died in 1727, and was – fittingly – buried in Westminster Abbey.

Newton was certainly not infallible. He made mistakes, and some of his ideas sound strange today. For instance, he spent a great deal of time experimenting in alchemy, and he made careful studies of old writings in the hope of finding hidden truths in them; he left a mass of notes and manuscripts which, frankly, have no value except to the scientific historian. There was also an unhappy quarrel with the great German mathematician, Gottfried Leibniz, as to who had invented the calculus. The truth of the matter is that Newton's 'method of fluxions' and Leibniz' calculus had been developed at about the same time; very probably Newton was the first to make use of it, but Leibniz' system was better, and became generally adopted. The dispute was long and bitter, and was still going on when Leibniz died in 1716.

Yet Newton's scientific errors were relatively few, and his *Principia* did more for astronomy than any other book before or since. Those who follow its reasoning, and who bear in mind that it represented only part of Newton's work, will hardly question that he was the greatest scientific genius not only of his own time, but perhaps of all time.

13

The Royal Observatory

THERE is still a widespread impression that astronomy is a purely academic science, interesting enough, but with no practical value. In fact this is not true, and never has been. For example, astronomy is the basis of all timekeeping. It is also the basis of navigation, and this leads me on to the story of the Royal Greenwich Observatory, which was founded in 1675, at the express order of King Charles II, to help British sailors.

Britain has always been a seafaring nation. By the seventeenth century prolonged journeys were being undertaken, but maps were by no means accurate, and to make matters worse navigation was still in the hit-or-miss stage. After a voyage lasting for a week or two, sailors far out of sight of land seldom had much idea of where they were.

To fix one's position on the surface of the Earth, it is necessary to know both latitude and longitude. Finding latitude is easy enough, because all that has to be done is to measure the altitude of the Pole Star and then make a slight correction to allow for the fact that the Pole Star is nearly a degree away from the actual pole. True, there is no bright south polar star, but a sailor could always work out his latitude with reasonable accuracy. The real problem was to find his longitude.

A ship's longitude is the difference between the meridian of the ship itself, and a standard meridian such as that of Greenwich. Local noon is easily found; it is the moment when the Sun reaches its greatest height above the horizon. A reliable clock will give Greenwich time at the same moment, and the longitude of the ship can be calculated.

Unfortunately, sailors of Charles II's day had no good clocks. Huygens had tried to make a timekeeper which would be accurate at sea, but without success (the pendulum clock was useless, for obvious reasons). For the moment the problem seemed to be almost insoluble. Accordingly, a Board of Longitude was set up and energetic efforts were made to find some way out.

As long ago as 1474 Regiomontanus had suggested that one way to measure longitude would be to find the position of the Moon against the stars. Since the Moon is so close to us, it moves quickly – about thirteen degrees per day – and can be used as a clock-hand to show time; and once the time is known, longitude can be found. The need, then, was to have a very accurate value for the Moon's position. It was also essential to have a really reliable star catalogue.

The best available catalogue was Tycho Brahe's. However, Tycho had had to work with the naked eye alone; he had no telescopic sights, and the errors in his catalogue were still unacceptably large. (It is ironical that Tycho died less than ten years before the invention of the telescope. He, of all people, would have made superb use of it.)

A Frenchman visiting England, Saint-Pierre, put forward an alternative scheme which also involved using the Moon and the stars. His theory was not practicable, and Saint-Pierre himself seems to have been a somewhat disreputable individual, but at least his proposal

HERSTMONCEUX CASTLE, *site of the present-day Royal Greenwich Observatory. The Castle, which is near Hailsham in Sussex, was photographed by Patrick Moore while alterations to it were still being made.*

OLD GREENWICH. *The original Royal Observatory as it must have appeared in Flamsteed's time.*

caused official interest. A committee of Fellows of the Royal Society was appointed, and presented a report, giving their opinion that longitudes could indeed be found by using the Moon as a 'clock' – provided that there were good enough star positions.

King Charles II may have had his little weaknesses, but one of his undoubted qualities was his interest in science. He decided that the only course was to build a new observatory so that a proper star catalogue could be compiled.

The site selected was the Royal Park at Greenwich. In those days Greenwich was a small village outside London, and the choice seemed excellent. Sir Christopher Wren was called in to design the buildings, and by 1675 the observatory was ready. Typically, the King paid for it by selling 'old and decayed gunpowder' to the French, though his generosity did not extend to providing instruments; the astronomer in charge was expected to obtain these for himself.

The chosen astronomer was the Rev. John Flamsteed, who had been born at Denby, near Derby, in 1646 and had taken his degree at Cambridge, where he first met Isaac Newton. Flamsteed had been in charge of the original Royal Society investigation, and was known to be an excellent observer as well as a mathematician, so that the choice was a fairly obvious one, particularly as he was on good terms with the influential Sir Jonas Moore, Surveyor-General of Ordnance at the Tower of London. It was due to Moore that he was able to buy instruments good enough to allow him to start work on the new catalogue. Subsequently he was appointed Astronomer Royal (the first holder of this post), and remained in office until his death in 1719.

Unfortunately he had many difficulties to face. His salary was a mere £90 per year, and instead of skilled assistants he had initially to make do with a 'silly, surly labourer' named Cuthbert, who was much more interested in the local ale-houses than in the Observatory. Moreover,

REV. JOHN FLAMSTEED

PROSPECTUS INTRA CAMERAM STELLATAM.

THE ROYAL OBSERVATORY IN FLAMSTEED'S TIME. *An interior view, showing observers with quadrant and telescope.*

FLAMSTEED HOUSE – *the old Royal Observatory as it is today, showing the Octagon Room surmounted by the time-ball. Photograph by Patrick Moore, 1961.*

Flamsteed was not personally popular. He was irritable and touchy, due in part to his chronically frail health. He was also a perfectionist, and refused to publish any of his results without being sure that they were as accurate as humanly possible.

Trouble began when Newton asked for his observational data, and Flamsteed said bluntly that he was not ready to provide them. The quarrel was patched up for a while, and in April 1704 the two men met at Greenwich, apparently on friendly terms. At this meeting Newton asked for a report on the star catalogue for which astronomers all over the country were waiting. Flamsteed replied that he was almost ready for printing arrangements to be put in hand, and Prince George of Denmark, husband of the new sovereign Queen Anne, generously promised to pay for the whole publication.

Still Flamsteed was not ready, but he handed to the Royal Society committee a copy of the observations as well as an incomplete manuscript of the catalogue itself. He made it clear that the catalogue was not to be printed as it stood, but was to wait until it had been completed and checked; the observations, however, could go forward. Printing was duly begun.

Four more years went by, and still Flamsteed did not submit his finished catalogue. Other disputes arose, and came to a head in 1711 with the publication of Flamsteed's observations. They took the form of a large book containing not only the observations which Flamsteed had passed for publication, but also the star catalogue, which he had not. The Royal Society committee had become tired of waiting, and had asked Edmond Halley to make the best of things. Halley had therefore supplied whole pages of material on his own account, and had added a preface which could not be anything but harmful to Flamsteed's reputation.

Flamsteed was furious, particularly with Halley and Newton. He wanted to revise the catalogue and re-issue it, but Newton held some of the observations and refused to give them back. In 1715 a large number of copies of the book fell into Flamsteed's hands, and he publicly burned them 'that none might remain to show the ingratitude' of two of his countrymen. He never did see the eventual catalogue; it was completed by two of his later assistants, Crosthwait and Sharp, and published in 1725.

At least the final version, the *Historia Cœlestis*, proved to be well worth waiting for. It included nearly three thousand stars, and was far more accurate than Tycho's catalogue. It represented the first major contribution to science of the Royal Greenwich Observatory, and it proved that despite the long delays, the choice of Flamsteed as Astronomer Royal had been a good one.

EDMOND HALLEY, *the second Astronomer Royal.*

However, it did not by itself solve the problem of longitude-finding. As well as knowing the positions of the stars, one has also to have a sound knowledge of the movements of the Moon. This was the problem tackled by Halley, who succeeded Flamsteed as Astronomer Royal.

Halley had been born in 1656. His parents were well off, and he did not have to face money troubles. He went to Oxford, but left before taking his degree in order to sail to the island of St. Helena, where Napoleon Bonaparte was to be exiled a century and a half later. Tycho had catalogued the northern stars, and Flamsteed was just starting work at Greenwich, but the southern stars which never rise over Europe had been completely neglected. Halley's main object was to catalogue them as well.

St. Helena was not a particularly good choice, and Halley had to contend with bad weather, but he was able to draw up a catalogue of 381 southern stars, and he was even nicknamed 'the southern Tycho'. When he returned to England, Oxford University granted him an honorary degree.

Halley first came into close contact with Newton during the 'inverse square law' discussions, and the two men always remained on friendly terms; from all accounts Halley was of a jovial disposition. He paid for the publication of the *Principia*, and this may indeed have been his greatest contribution to science.

Just as Newton's name is always linked with gravitation, so Halley's is associated with the famous comet. In fact it was Halley's book about comets, published in 1705, which provided the most dramatic proof that the laws laid down in the *Principia* were correct.

A brilliant comet with a tail stretching half way across the sky is a glorious spectacle. Not unnaturally, the ancients found it frightening, and an indication that the gods were angry, so that disasters of all kinds were imminent. For instance the Roman writer Pliny, who was killed during the A.D. 79 eruption of Vesuvius which overwhelmed the towns of Pompeii and Herculaneum, once wrote that 'we have in the war between Cæsar and Pompey an example of the terrible effects which follow the apparition of a comet... that fearful star which overthrows

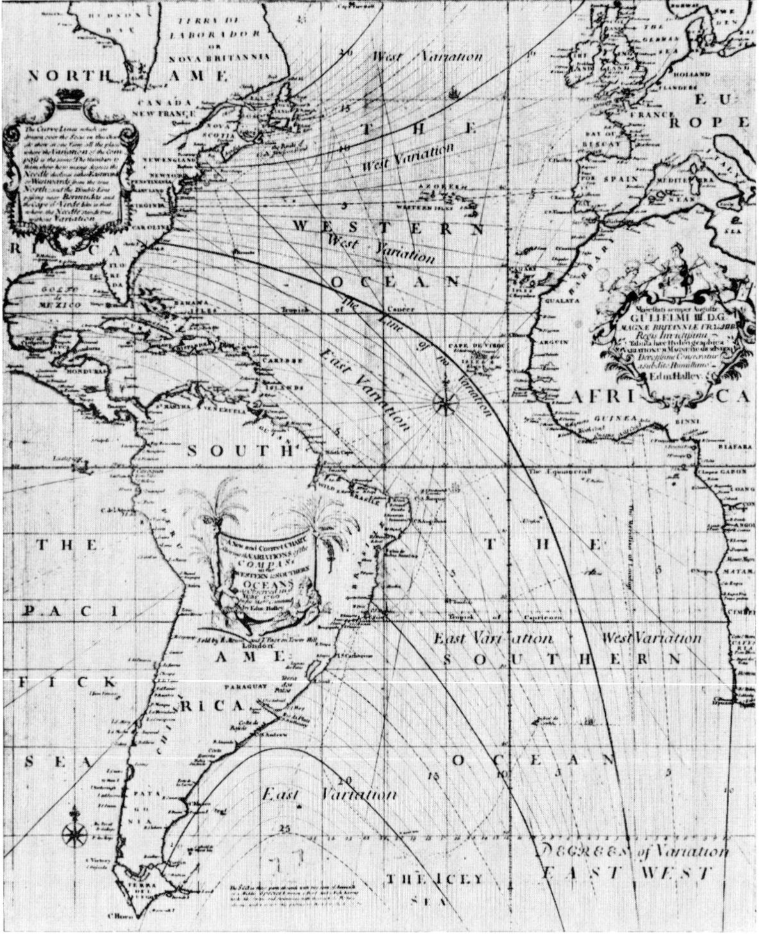

HALLEY'S MAGNETIC CHART. *Though astronomy was Halley's main study he also paid attention to other branches of science. During his voyage to St. Helena he noticed that the compass needle did not point due north, and from this he deduced – correctly – that the north magnetic pole is not situated exactly at the geographical pole. Years later, in 1698, he was given a commission in the Royal Navy, and sailed as captain of a ship, the Paramour, on a journey which enabled him to study the 'variation' of the compass – that is to say the difference between true north and magnetic north. In a second cruise, in 1699-1701, he went to the South Atlantic and encountered Antarctic icebergs. A year later he was able to publish a chart showing the magnetic variations over the whole world. Sea navigators found it remarkably useful.*

the powers of the Earth, showing its terrible locks.' And a bright comet led the Roman Emperor Vespasian to declare that 'this hairy star does not concern me; it menaces rather the King of the Parthians, for he is hairy, while I am bald.' Actually Vespasian died in the same year, though one must take leave to doubt whether the comet had anything to do with it!

The trouble was that nobody had much idea about the nature of a comet. Aristotle had taught that comets were hot, dry exhalations, rising from the ground and carried along by the motion of the sky. When they had become sufficiently hot, they caught fire, and either burned up rapidly as meteors or slowly as comets. Galileo disagreed – one can hardly imagine his taking anything that Aristotle had said at face value – but his own theory was no better; to him, a comet was simply sunlight refracted in the Earth's atmosphere. Tycho had shown that comets were further away than the Moon, but even Newton had originally believed that they might travel in straight lines rather than in elliptical orbits round the Sun.

Comets could not be predicted. They would appear without warning, and remain striking for a few nights, weeks or months before fading away. They did not seem to fit into any system, Copernican or otherwise.

A bright comet appeared in 1682. Halley observed it, and became interested in the whole problem. He therefore collected all the recorded observations of comets seen between 1337 and 1698, and

HALLEY'S CHART OF THE SOUTHERN SKY

began to analyse them, working out the orbits as well as he could, using Newton's principles. He realized that the comets of 1531 and 1607 had orbits strikingly similar to that of the comet of 1682. Could it be that the three were one and the same, so that the comet moved round the Sun in a highly eccentric ellipse in a period of seventy-six years? Halley believed so, and predicted that the comet would return in 1758. By that time he knew that he would be dead, but he added, modestly: 'If the comet should return according to our prediction, about the year 1758, impartial posterity will not refuse to acknowledge that this was first discovered by an Englishman.'

Halley died in 1742. Later, new calculations were made by three French astronomers, Lalande, Clairaut and Madame Lepaute, modifying the date of perihelion to 1759. Yet Halley was vindicated in every respect. On Christmas Night 1758, a German amateur, Johann Palitzsch, duly picked up the comet, and it passed through perihelion on March 12 of the following year.

Halley's Comet came back on schedule in 1835, and once more in 1910. Records of it date back to 240 B.C. and perhaps as far as 467 B.C., but it has not always been striking; it made a brave showing in 684 (when it was widely seen; a drawing of it as it appeared in that year was published in the Nürnberg Chronicle long afterwards), 1301, and 1456, when it was so spectacular that Pope Calixtus III publicly condemned it as an agent of the Devil. On the other hand it was much less prominent in 1145 and 1378. Unfortunately the coming return, that of 1986, will not be a good one, because the comet will be badly placed in the sky. To see Halley's Comet at its best, I am afraid that you will have to wait until the return of 2062. However, it will certainly be a naked-eye object for some time between December 1985 and the end of March 1986. It was recovered in October 1982 by E. Danielson and D. Jewitt, at Palomar Observatory in California. It was then of magnitude 24·2, and therefore one of the faintest objects ever recorded.

(Incidentally, the fear of comets is not dead even yet. At the 1910 return we actually passed through the end of the tail, and women in Chicago boarded up their windows because of a fear of carbon monoxide gas, while an enterprising gentleman in Atlanta made a large sum of money by selling what he called comet pills – though I have no idea what they were meant to do.)

There have been many brilliant comets in the past; those of 1811, 1843 and 1858 were magnificent by any standards. But these spectacular visitors take thousands or even millions of years to orbit the Sun, and so they cannot be predicted. In fact, Halley's Comet is the only bright comet which has a period of less than four or five centuries.

Another of Halley's discoveries was that three bright stars, Sirius, Procyon and Arcturus, had shifted slightly since the time when Hipparchus had drawn up his star catalogue, and Sirius had even shifted since Tycho Brahe's work at Hven, indicating that it must be close by stellar standards. This was the first indication of *proper motion*, and showed that the old term of 'fixed stars' was misleading. It also finally disposed of Aristotle's contention that the skies must be changeless.

Halley became Astronomer Royal in 1719, on the death of Flamsteed, but he had to face immediate difficulties. Mrs Flamsteed descended upon the Observatory like an east wind, and removed all the

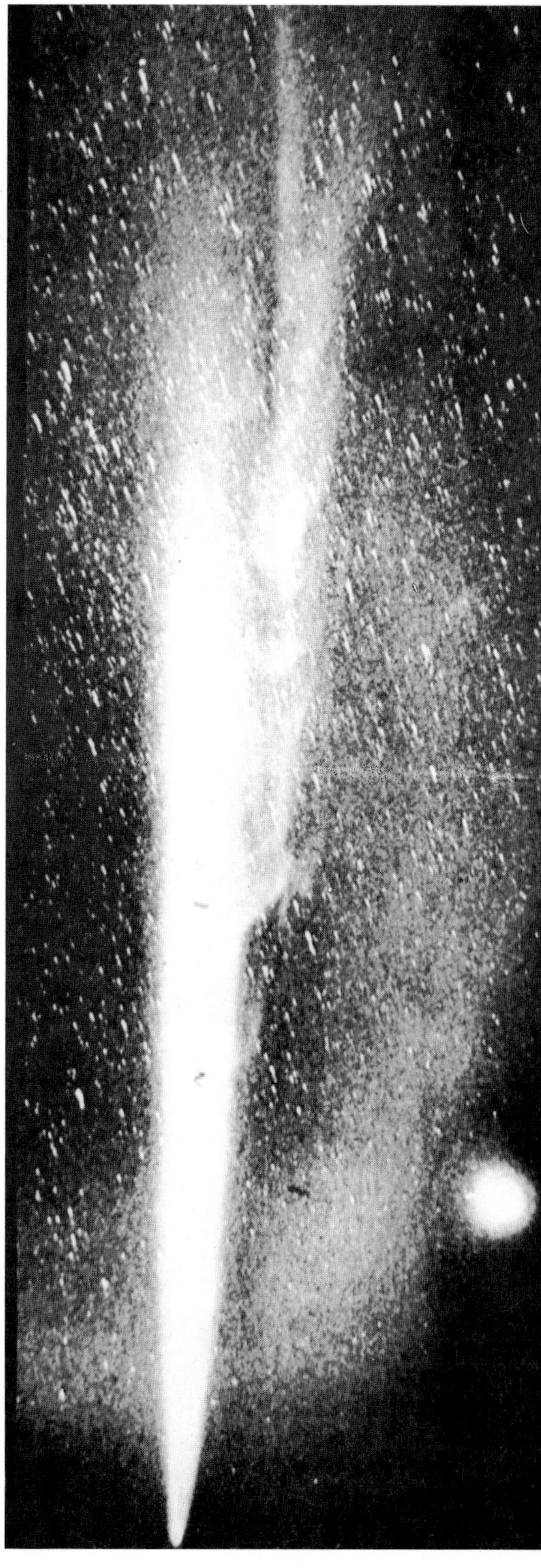

HALLEY'S COMET AND VENUS, *photographed by Slipher in 1910.*

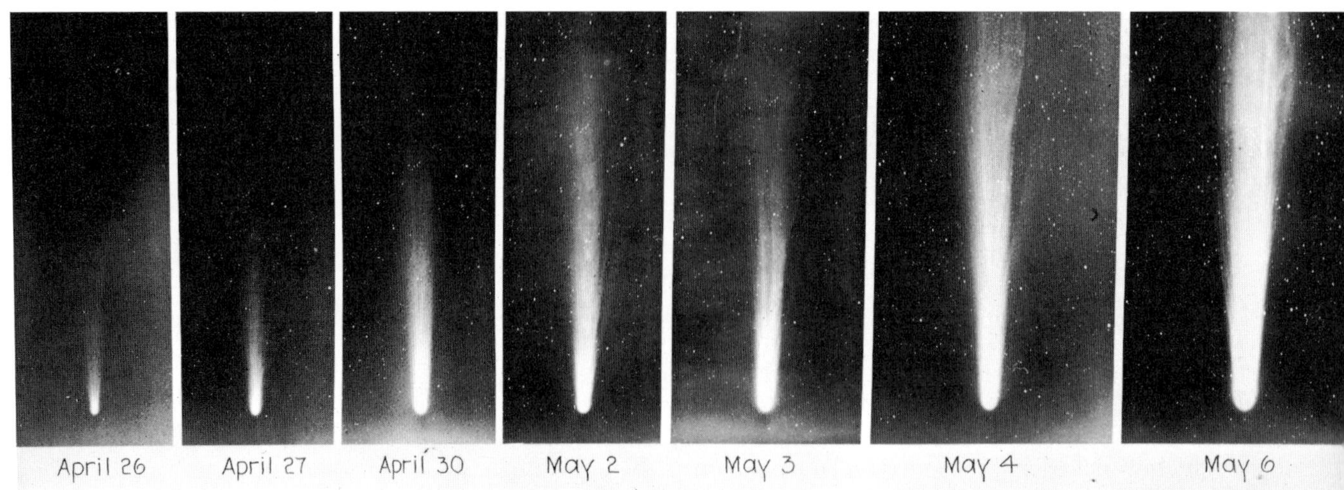

HALLEY'S COMET IN 1910, *(above) photographed at Mount Wilson on various dates.*

PROPER MOTIONS OF STARS IN ORION. *(Right) Orion at the present time. (Far right) the constellation as it will be in* A.D. *75,000.*

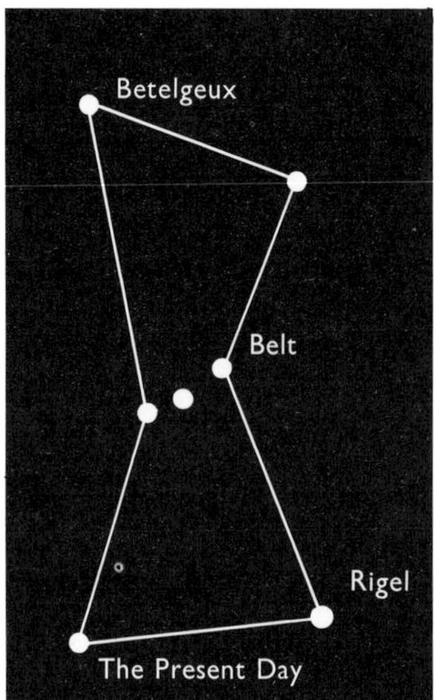

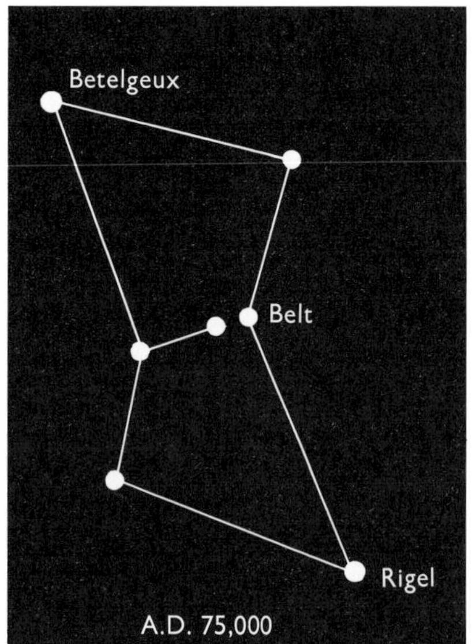

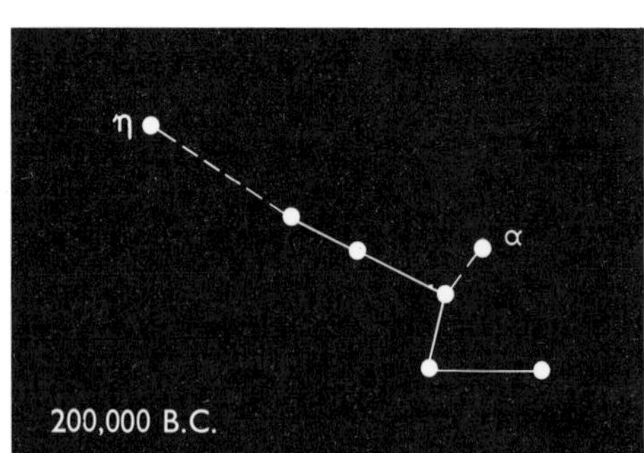

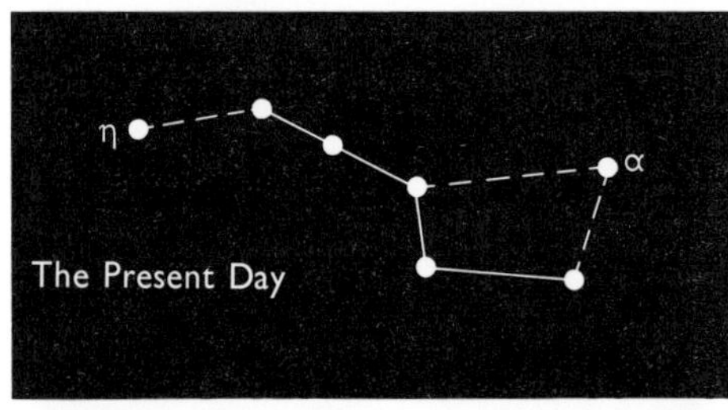

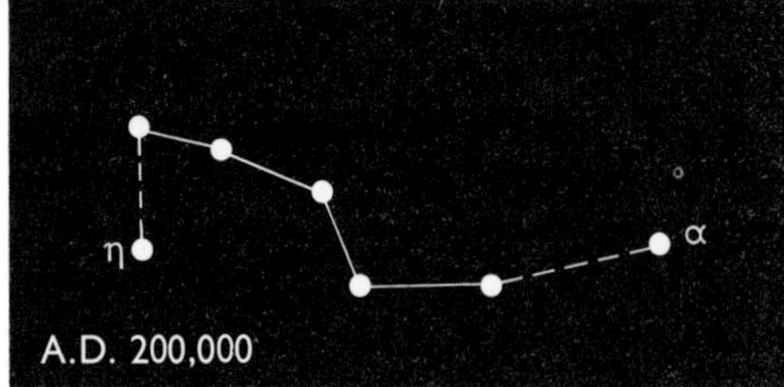

PLOUGH OR DIPPER. *The seven stars in the past, present and future. Two of the stars, α (Dubhe) and η (Alkaid) are moving in a different direction from the other five.*

instruments, which were legally hers. Unfortunately they have never been traced. Halley had to begin again, which he did. Some of his equipment is still to be seen in the Octagon Room, which had been designed by Sir Christopher Wren.

By this time Halley was well over sixty, but with his usual energy he returned to the problem of longitude-finding. Flamsteed had provided the star catalogue; Halley set out to study the movements of the Moon. It took him nearly nineteen years, but he completed the work, and it proved to be of immense value – though, ironically, it was never used for its original purpose; the longitude problem was solved by John Harrison, an English clockmaker, who developed the marine chronometer, a timekeeper which was good enough to meet all the requirements. Harrison did eventually win the substantial cash prize offered by the Board of Longitude, though not until shortly before his death in 1776.

Rather unfairly, Halley has always been overshadowed in history by Isaac Newton. Of course, Newton was the greater of the two, but Halley's own contributions to astronomy will never be forgotten – quite apart from his famous comet.

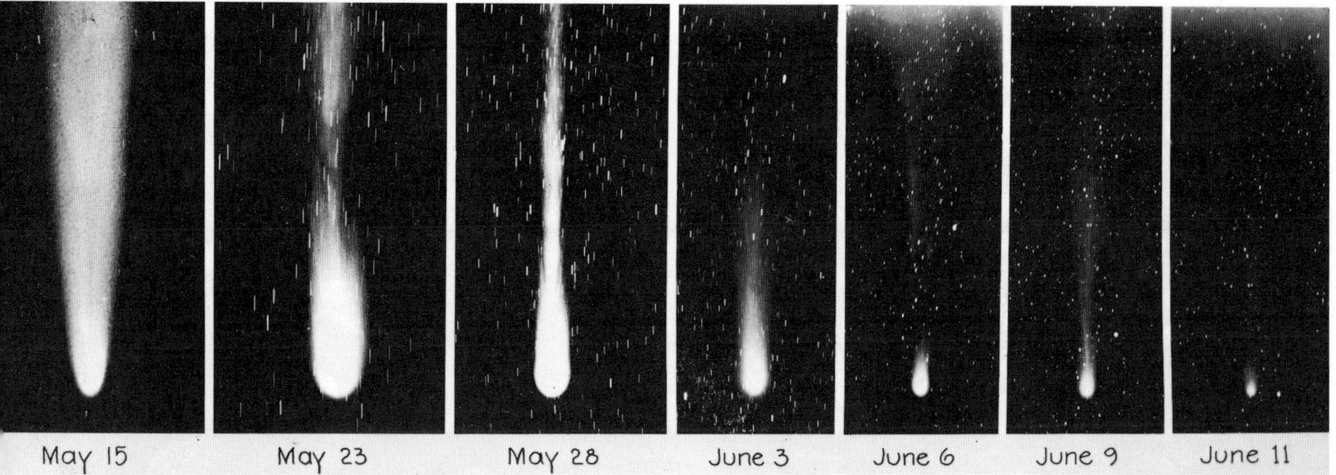

THE MERIDIAN LINE AT GREENWICH OBSERVATORY. *This line at the old Royal Observatory (now known as Flamsteed House) marks the boundary between the eastern and western hemispheres. Reckoning has not been altered by the shift of the Royal Observatory from Greenwich Park to Herstmonceux. Photograph by Patrick Moore.*

HARRISON'S NO.1 MARINE CHRONOMETER. *This was the first 'clock' which proved capable of keeping time accurately over long sea voyages.*

14

When Venus Crosses the Sun

UP to now I have been describing the lives and works of astronomers of Western Europe: Greeks, Britons, Italians, Germans and the rest. American science had hardly begun, and in the Far East there had been no real advances since the ancient Chinese had recorded comets and eclipses. What, then, of Russia?

Scientists of the modern U.S.S.R. are very much in the forefront of affairs, but during the eighteenth century Imperial Russia was frankly backward. Few people could read or write, and in the few schools and universities the Copernican theory was rejected both on scientific and on religious grounds, and as late as 1740 most Russian professors clung to the old idea that the Sun must move round the Earth. The change-over to a more enlightened view was due in part, at least, to the first of Russia's great astronomers: Mikhail Vasilyevich Lomonosov.

Lomonosov was born on an island off the coast at Archangel in 1711. His father, a free peasant, was a fisherman, and during his boyhood Mikhail went on several fishing expeditions into the White Sea, well inside the Arctic Circle. Like most Russian children of the time he had very little schooling, but by the age of fifteen he had learned how to read and write, and – according to a story which may well be true – he ran away from home and joined a train of sleds carrying frozen fish, bound for Moscow. He enrolled at a school attached to a large monastery, and his real education began. He did not confine himself to science; he also started to write poetry.

He made rapid progress, and came to the notice of the St. Petersburg Academy of Sciences, which had been founded by the Czar Peter I. In 1735 he was sent to the University of St. Petersburg, and a year later he and other students went on to the Marburg University in Germany principally to study chemistry and mineralogy.

It cannot be said that his period there was an unqualified success. To put it mildly, he was anything but a teetotaler, and some of the staid German professors were distinctly shocked. When he returned to Russia in 1741 he abandoned the wife whom he had married after arriving in Marburg, and back at St. Petersburg he insulted some of his colleagues, after which he was imprisoned for several months. While in gaol, he wrote two of his most famous poems. Evidently he was forgiven, and by 1745 we find him a full member of the Academy, and director of the main chemical laboratory, which he himself designed.

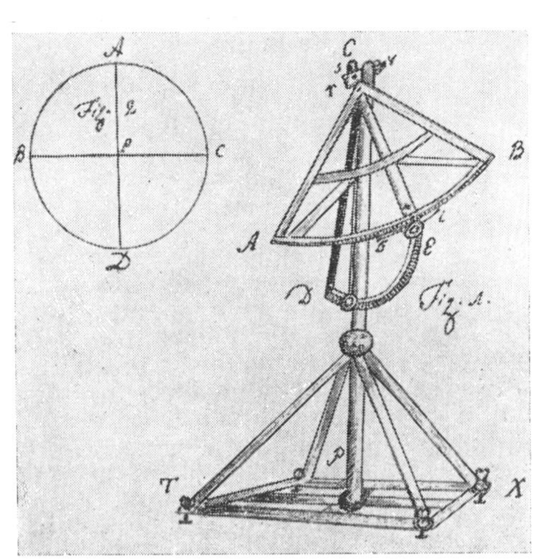

QUADRANT, *designed and used by M. V. Lomonosov.*

Lomonosov's activities covered a wide range. He was interested in problems of navigation, and with the aid of his own measuring instruments he determined the latitudes and longitudes of many of the main cities of the country, so drawing up the first accurate map of the Russian Empire. He was a pioneer of what we now term the kinetic theory of gases; he was a geologist and a meteorologist; he developed a solar furnace, using lenses and mirrors; he investigated electrical phenomena, including the spectacular auroræ or polar lights; and he also wrote a work on grammar which had a profound effect upon the

Russian language. In astronomy, he was a champion of the Copernican theory, to the disgust of his more conservatively-minded colleagues, and he also devised a modified optical system for a reflecting telescope. But he is perhaps best remembered astronomically for his 1761 observations of the transit of Venus.

When Edmond Halley had been in St. Helena, he had watched a transit – not of Venus, but of Mercury. Following up an earlier suggestion by James Gregory, he had realized that such transits might provide a way of measuring the distance of the Sun. The method was fairly straightforward, but it depended upon an exact determination of the moments when the transit began and ended. Mercury is small and elusive; but Venus, much larger and closer, held out more promise, and Halley certainly regretted that he could not live long enough to see the next transit, that of 1761 – the first since that of 1639 which had been observed by Horrocks and Crabtree.

When the time came, the transit was carefully studied by astronomers all over the world, and various special expeditions were dispatched. Unfortunately the observations were badly affected by a phenomenon which became known as the Black Drop. As Venus passed on to the Sun it appeared to draw a strip of darkness after it, and when this disappeared the transit was already in progress. Similar troubles were experienced at the next transit, that of 1769. Many years later the German astronomer Johann Encke published a final result based on the observations, and gave the Sun's distance as 153,304,000 kilometres, which is now known to be considerably too great.

No description of the 1761 transit would be complete without referring to the amazing misfortunes of the French astronomer Guillaume Legentil, who decided to observe it from Pondicherry in

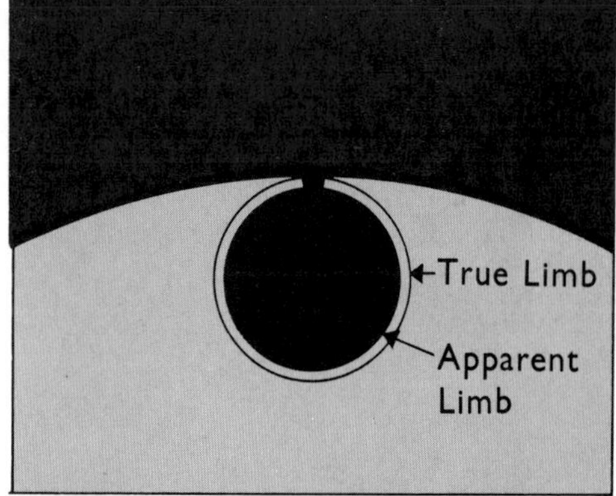

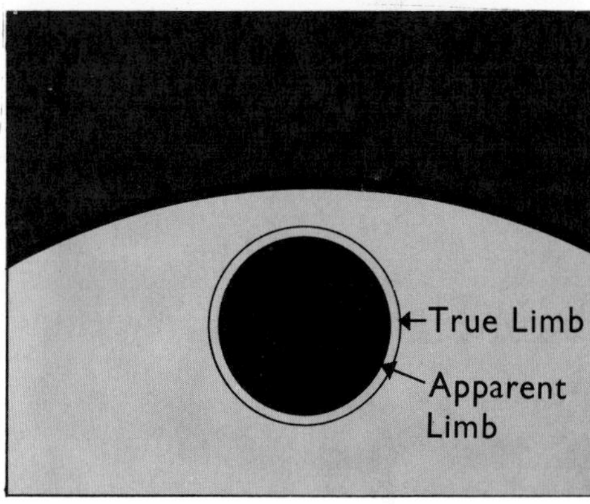

THE BLACK DROP. *This was the effect which ruined the accuracy of the transit of Venus method of determining the Sun's distance.*

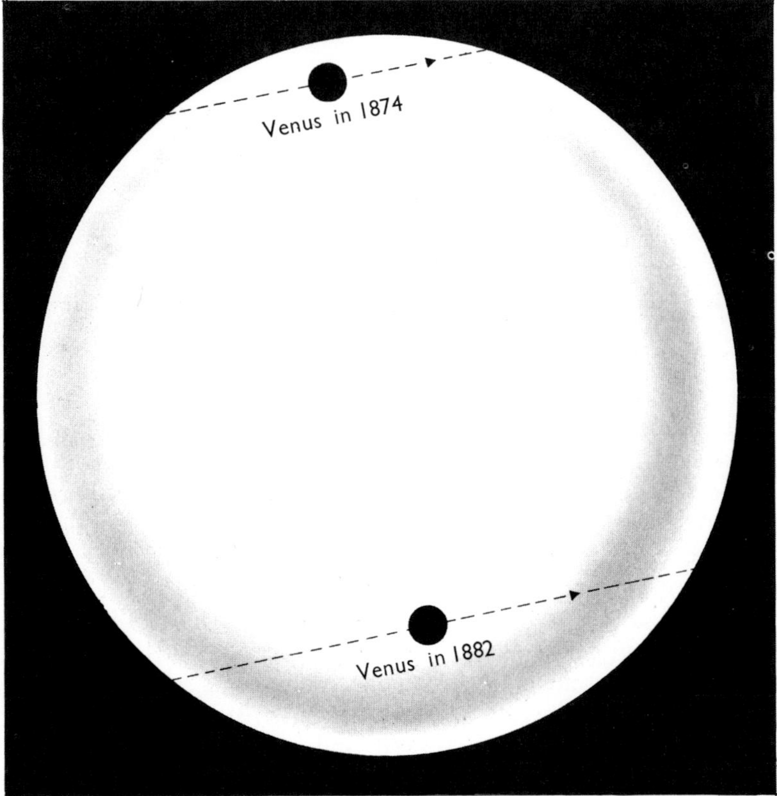

TRANSITS OF VENUS IN 1874 AND 1882. *The diagram shows the apparent track of Venus across the face of the Sun during these two transits. In neither case was the transit central, but it had been hoped to obtain measures which would be accurate enough to give a reliable estimate of the Sun's distance from the Earth. Great attention was paid to observing the transits, but the results proved to be very disappointing, and the next two transits (those of 2004 and 2012) will not be regarded as of much astronomical importance.*

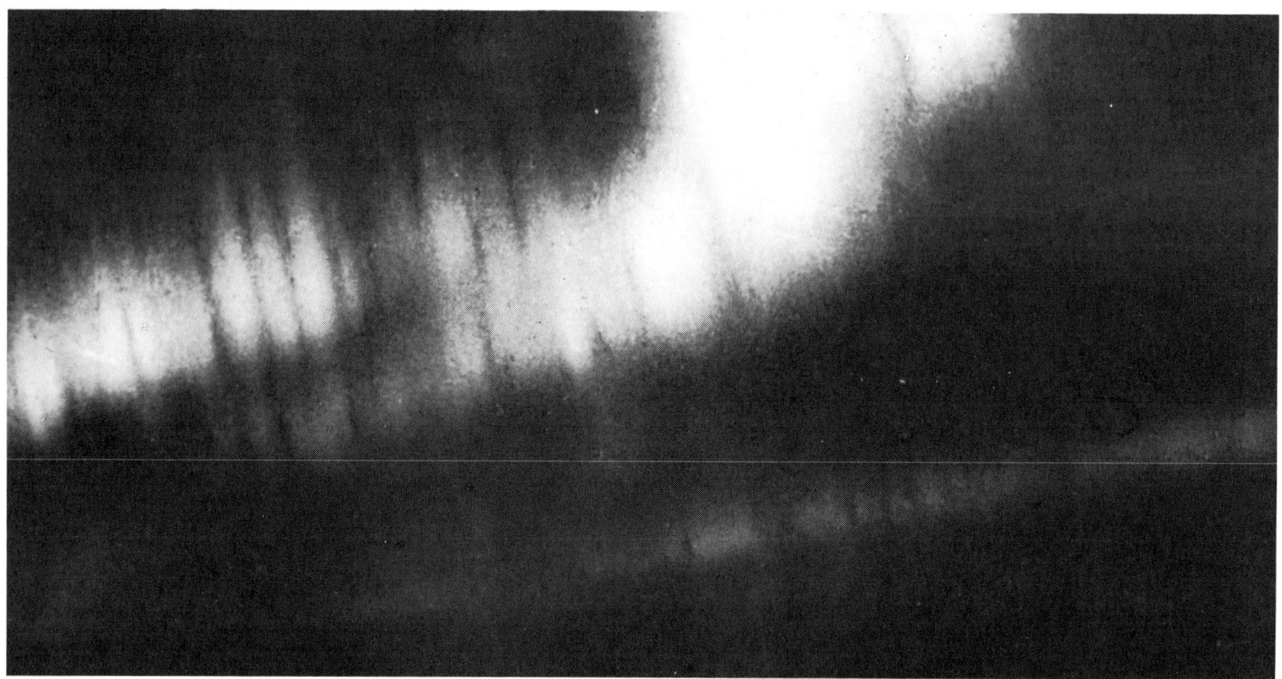

THE AURORA BOREALIS. *(Centre) Lomonosov was among those who studied the spectacular aurorae or Polar Lights.*

VIEW OF RUSSIA'S OLDEST OBSERVATORY. *(Foot) The Chamber of Curiosities in Leningrad, where the first observatory in Russia was set up and used by M. V. Lomonosov. The photograph was taken from the opposite bank of the river. There is no longer an observatory in the building, and the building itself is now used as a museum. Lomonosov's observations of the 1761 transit of Venus were not actually carried out from here, but from another site in St. Petersburg (now Leningrad). Photograph by Patrick Moore, 1960.*

India, where conditions were expected to be ideal. He sailed in a French frigate, but unluckily for him a war between England and France was going on, and about this time Pondicherry was captured by the English, so that Legentil had to turn back. Before he could reach land the transit was over, and all he could do was to make rough notes from the deck of his ship. Rather than risk a second delay he elected to stay in India for the next eight years, and observe the 1769 transit instead. Again he was unlucky; clouds covered the sky, and nothing could be seen at the critical moment. Since it was asking rather too much to wait for the next transit (that of 1874) Legentil set off for home. Twice he was shipwrecked, and reached Paris after a total absence of eleven years to find that he had been presumed dead and

OCCULTATION OF REGULUS BY VENUS, JULY 7, 1959. *(Above) It is very seldom that Venus passes in front of, and occults, a bright star, but such an event took place on the early afternoon of July 7, 1959, when Regulus in Leo was occulted. These drawings, made by Patrick Moore with the 31·8-cm reflector at Henry Brinton's observatory in Selsey, show the changing position of Venus and Regulus. For a brief period before immersion Regulus was shining through the atmosphere surrounding Venus, and appreciable fading was recorded, so yielding information as to the height of the atmosphere of Venus. It will be many centuries before Venus again passes in front of a 1st-magnitude star.*

that his heirs were just about to distribute his property. One can imagine that his enthusiasm for transits was somewhat dampened.

Mikhail Lomonosov watched the 1761 transit from his home in St. Petersburg, where he had erected a refracting telescope of focal length 1·4 metres. One thing interested him particularly. Just before the transit began, the Sun's limb seemed to become blurred, and a similar appearance was seen just after the transit was over. Moreover, there was a curious sort of 'blister' as Venus passed right on to the Sun.

Lomonosov could find only one explanation. In his own words: 'The planet Venus is surrounded by a considerable atmosphere, equal to, if not greater than, that which envelops our earthly sphere.' This strengthened his view that some of the planets, at least, might not be so very unlike the Earth, and he even suggested that life might exist upon them.

He was, of course, right in believing that Venus has an atmosphere. We now know that the ground pressure there is at least 90 times that of the Earth's air at sea-level, though the surface temperature is much too high for any life of our type to exist there.

Lomonosov died in 1763. Elaborate preparations were made for the 1769 transit, and one of the expeditions was under the command of Captain Cook, whose task was to take the astronomers to a favourable observing site in the South Seas. He went to Tahiti, where he and his colleagues – notably the chief astronomer, Charles Green – made good observations before continuing with the famous voyage of discovery. Sadly, Green died on the return voyage.

Cook's expedition had been organized at the instigation of Nevil Maskelyne, who had become Astronomer Royal (and who is also remembered for founding the *Nautical Almanac*). It was only one of many; but again the Black Drop ruined the observations. Neither were the transits of 1874 and 1882 much more fruitful. Of course, the whole method is now completely obsolete, and the next transits, those of 2004 and 2012, will be regarded as of no more than academic interest; but at least it is worth remembering that it was at the transit of 1761 that Mikhail Lomonosov discovered that Venus, like the Earth, is surrounded by a mantle of atmosphere.

15

The King's Astronomer

SIR WILLIAM HERSCHEL, *often regarded as the greatest observer in astronomical history.*

OBSERVATORY HOUSE, SLOUGH. *The house where three Herschels – William, Caroline and John – lived and worked. The great 40-foot reflector was erected in the garden. Eventually the house fell into disrepair, and was pulled down in 1960. This photograph taken by Patrick Moore was probably the last picture ever taken of it – three days before demolition was begun.*

THE GREAT NEBULA IN ORION. *(Opposite) Photograph taken by the 508-cm Hale reflector at Palomar.*

IT is fair to say that the 'Copernican revolution', begun with *De Revolutionibus* in 1543, ended with the work of Newton. Once the *Principia* had been published, no serious scientist could continue to believe in an Earth-centred universe. The new era had arrived. Then, almost a century later, came perhaps the greatest of all astronomical observers: William Herschel.

Of course, there were many famous astronomers and mathematicians during the intervening years. There was Alexis Clairaut of France, born in 1713, who studied higher mathematics when still a small boy, and sent a valuable paper to the French Academy of Sciences when he was only eleven. Clairaut later revised the predictions of Halley's Comet, and found that it would return to perihelion not in 1758, but in 1759; he was in error by only a month. There was the brilliant mathematician Leonhard Euler, who continued his work even after losing his sight; he made all his calculations in his head. There were also tragic figures such as Jean Sylvan Bailly and Bochart de Saron, who were guillotined during the French Revolution; it was said that 'the Republic has no need of wise men'. And there were many more; but in purely observational astronomy, Herschel was in a class of his own.

He was born in Hanover in 1738, and christened Friedrich Wilhelm. His father, Isaac, was a bandmaster in the Hanoverian Guards, and most of his children were musically talented. William (as he is always known) was no exception. In his own words: '1753 May 1; I was engaged as a Musician in the Hanoverian Guards, being then 14 years and some months of age, and I remember playing for this purpose on the Hautboy and on the violin before General Sommerfeld, who approved my performance.'

Yet he was not cut out for an Army career. In 1756 he was sent to England with his regiment to guard against the danger of a French invasion; the Seven Years' War was raging, and at that time, of course, England and Hanover were ruled by the same king (George II). The Guards remained for some months, which gave William the chance to learn English.

He was then sent back to Hanover, and was present at the Battle of Hastenbeck in July 1757, when the Duke of Cumberland's force was utterly routed by the French. William relates how he spent one night in a water-filled ditch, which he did not enjoy in the least. Subsequently he left the Army – quite legally, because he had never been formally enlisted; he had been too young – and in the autumn of 1757 he arrived back in England, together with his elder brother Jacob.

(Let me here dispose of the oft-quoted story that Herschel was a deserter. He was not. A document signed by General von Spörken, now on display in the Herschel Museum at Bath, makes it quite clear that the story is untrue.)

A 15·2-CM NEWTONIAN REFLECTOR. *The tube is a skeleton, which is perfectly satisfactory for instruments of this type.*

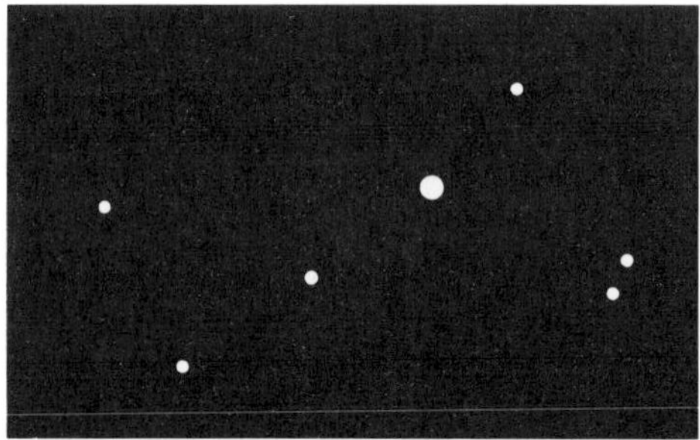

MOVEMENT OF URANUS AMONG THE STARS. *Observations made by Patrick Moore on March 4 and 6, 1960.*

By now Herschel was an accomplished musician. At first he earned his living by copying music, but then went to Richmond in Yorkshire to take charge of a small band in the Earl of Darlington's militia. In 1765 we find him in Halifax, having obtained the post of organist there in the face of stiff competition – notably from Dr. Wainwright of Manchester, known to be an expert player. It is said that the organ-builder, an elderly German named Snetzler, complained that Wainwright 'run about the keys like one cat', while Herschel's performance made him comment 'Very goot! Very goot! I vill luff dis man, for he gives my pipes room to schpeak!'

At Halifax William was an undoubted success, but he did not stay for long, and in 1766 set out for Bath, where he had been invited to become oboist in Linley's celebrated Pump Room Orchestra. At this time Bath was one of the most fashionable centres in England, and Herschel, with his good looks, pleasant personality and musical talent, fitted well into the scene. On October 4, 1767 the new Octagon Chapel was opened, with Herschel as organist, and he was joined by his brother Alexander, also an excellent musician. Then, in 1772, William paid a visit to Hanover. When he came back to England he brought with him his sister Caroline, who had been in danger of becoming a household drudge. This was a wise move; for the rest of his life Caroline remained his devoted assistant.

Herschel could have had a successful full-time career in music, but as early as February 1766 he had become interested in astronomy, and after borrowing a small telescope he became anxious to use something larger. He could not afford a powerful telescope, and so he decided to make one.

He experimented with refractors, but without much luck, and he soon turned to reflectors. In 1773 he was able to buy some equipment and half-finished mirrors from a neighbour, and proceeded to turn his house into a workshop, much to Caroline's consternation. Helped by Alexander, who was of a mechanical turn of mind, he started mirror-grinding. Failure followed failure, but at last he made a good mirror with an aperture of 13 centimetres and a focal length of 1·7 metres. On March 4, 1774 he turned his new telescope toward the Great Nebula in Orion, and his real life's work had begun.

He moved house several times, but finally arrived at No. 19 New

EXTERNAL GALAXY N.G.C.205 *(in Andromeda) – a system far beyond our Milky Way. Herschel was one of the first to suggest that such objects might be external. Photograph taken with the 508-cm Hale reflector.*

King Street, Bath. On 13 March, 1781 he set up his latest reflector, of 15·7 centimetres aperture, and was busy making routine observations when he came across something unusual. It was not a star; it showed a disk, which no star can do, and on succeeding nights it was seen to move slowly, which again no star can do. What was it? Herschel took it for a comet; he notified the Astronomer Royal at Greenwich, Nevil Maskelyne, and also Hornsby, the Radcliffe Observer at Oxford. On April 26 his friend Watson read Herschel's paper at a meeting of the Royal Society. It was headed *Account of a Comet*.

Herschel's description was accurate enough. His paper began:

> 'While I was observing the small stars in the neighbourhood of H Geminorum, I perceived one that appeared visibly larger than all the rest; being struck with its uncommon magnitude, I compared it to H Geminorum and the small star in the quartile between Auriga and Gemini, and finding it so much larger than either of them, suspected it to be a comet... The comet, being magnified much beyond what its light would admit of, appeared hazy and ill-defined with these great powers [magnifications 227 to 2010], while the stars preserved that lustre and distinctness which from many thousand observations I knew they would retain. The sequel has show that my Surmises were well founded, this proving to be the comet we have lately observed.'

Yet almost at once, doubts arose. The stars mentioned by Herschel were identifiable; his H Geminorum is our 1 Geminorum (No. 1 in Flamsteed's catalogue of the stars in that constellation), while the 'small star in the quartile' is 132 Tauri. But it soon became clear that the object was not moving in the way that a comet would do. Even before Herschel's paper was read, Maskelyne had written to him, saying that 'it is as likely to be a regular planet moving in an orbit nearly circular to the Sun'. Calculations made by several mathematicians – Andres Lexell of Finland, Boscovich of Italy, and Darquier and de Saron of France – showed quite conclusively that Herschel had found a new planet moving far beyond the orbit of Saturn, the most remote of the planets known in ancient times. We now call it Uranus.

It is quite true to say that Herschel was not looking for a new planet, and did not recognize its real nature even when he had found it. His aim was to 'review the heavens', and to find out the way in which the stars were distributed in space. On the other hand, it is unfair to say that his discovery was sheer luck. As he wrote to his colleague Dr. Hutton:

> 'It has generally been supposed that it was a lucky accident which brought this star to my view; this is an evident mistake. In the regular manner I examined every star of the heavens, not only of this magnitude but many far inferior, it was that night *its turn* to be discovered... Had business prevented me that evening, I must have found it the next.'

Actually Herschel was not the first to see Uranus. It had been recorded by Flamsteed in 1690, and even given a catalogue number: 34 Tauri. Pierre Le Monnier, a French astronomer, saw it half a dozen times between 1750 and 1771. It has often been said that if he had compared his observations he could not have failed to make the discovery, and there is also a famous story that because of his lack of order and method, one of his observations of Uranus was later found scrawled on the back of a bag which had once contained hair perfume. In fact this may be unfair to Le Monnier, because apparently his observations were made when Uranus was close to its stationary point, and the paper-bag story may well be untrue. It has also been said that

16·5-CM INCH REFLECTOR MADE BY WILLIAM HERSCHEL. *This is a replica of the telescope with which Herschel discovered Uranus in 1781.*

URANUS AND TWO OF ITS SATELLITES. *Photograph by B. Warner and T. Saemundsson, University of London Observatory, 1960. Of the five satellites of Uranus, one (Miranda) is excessively faint; two (Ariel and Umbriel) are difficult objects, and the outer two (Titania and Oberon) are easier, though a moderate telescope is needed to show them. In this photograph, Titania and Oberon are shown. Surface detail on Uranus itself is almost impossible to record photographically.*

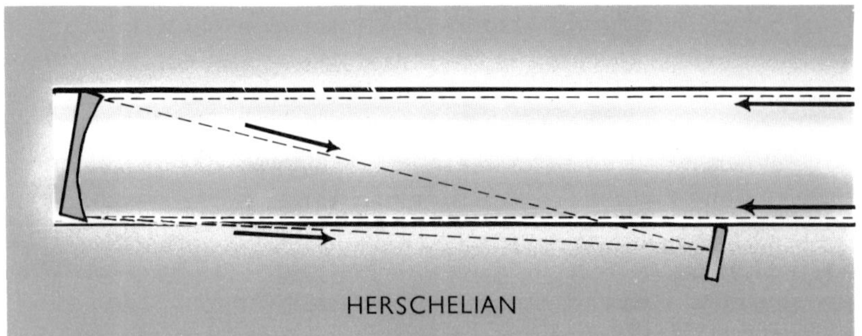

PRINCIPLE OF THE HERSCHELIAN REFLECTOR. *(Right) Herschel's method was to tilt the main mirror, thus doing away with the necessity of having a second mirror or 'flat'. Unfortunately the arrangement has numerous disadvantages, and is now seldom used.*

COMPARATIVE SIZES OF URANUS AND THE EARTH. *Uranus, with a diameter of 51,800 km, is a giant planet, though considerably smaller than either Jupiter or Saturn.*

THE OLD ROYAL OBSERVATORY, GREENWICH. *A view of one of the buildings. Photograph by Patrick Moore.*

Le Monnier never failed to quarrel with anyone whom he met, which again may or may not be correct! He died in 1799.

Uranus proved to be a giant planet; its diameter is over 50,000 kilometres, nearly half that of Saturn. Its mean distance from the Sun is 2,870,000,000 kilometres, and its 'year' is 84 times as long as ours. It is just visible with the naked eye, but no telescope will show much upon its pale, greenish disk, though we know that it has a gaseous surface and is probably mainly liquid inside, with a small solid core. We cannot pretend to know as much about it as we would like to do, and we must await the Voyager 2 fly-by of 1986, about which I will have more to say later. Meanwhile, Herschel discovered three of Uranus' satellites, Titania and Oberon in 1787 and Umbriel in 1802. Umbriel was then 'lost' for almost half a century before being recovered. Herschel also believed, mistakenly, that he had discovered several more satellites, as well as a ring round the planet. There is indeed a ring system, but it cannot be seen with ordinary telescopes, and Herschel's 'ring' was not genuine.

The discovery of Uranus altered Herschel's whole life. King George III was genuinely interested in science, and had personally observed the 1769 transit of Venus from his private observatory at Kew, which had been built specially for his benefit. He summoned Herschel to Windsor, and appointed him King's Astronomer at the modest salary of £200 per year. (Note that this was a unique appointment; Herschel was never Astronomer Royal – at that time the post was, as we have noted, held by Nevil Maskelyne.) Herschel saw that this was his chance to abandon music as a profession, and devote his whole time to astronomy. He accepted the King's offer; he left Bath, together with Caroline, moving first to Datchet and then to Observatory House in Slough, where he spent the rest of his life.

Even before leaving Bath, he had made up his mind to make a really large telescope, with a mirror no less than 91 centimetres in diameter. The downstairs room at No. 19 New King Street had been turned into a workshop, and here William and Alexander cast the blank disks which were to be turned into mirrors. A mixture of copper and tin was used, together with smaller amounts of other metals to make the finished mirror as bright as possible (large glass mirrors still lay in the future). There was an immediate disaster. The mixture was heated, and then poured into a mould; but the mould cracked. Metal ran out in a torrent across the floor, and the flagstones shattered. Caroline recorded that 'Both my brothers and the caster and his men were obliged to run out at opposite doors, for the stone flooring (which ought to have been taken up) flew about in all directions as high as the ceiling. My poor Brother fell exhausted by heat and exertion on to a heap of brickbats.' It must have been a dramatic moment, and for the time that was an end of mirror-making. Nothing more was attempted until William and Caroline were settled in at Observatory House.

This is the moment to say something more about Caroline. She was an excellent singer, and when reaching England intended to follow a musical career as well as keeping house for William, but she became a skilful astronomer in her own right; all the time that William was observing, Caroline stayed by his side, taking notes and keeping records. Without her, it would have been quite impossible for William to have accomplished as much as he did. When William married, in 1788, Caroline moved out of Observatory House, but remained nearby, returning every clear night to help in the observations. She even found time to discover six comets.

At Slough, work began on Herschel's largest telescope: a real giant, with a 122-centimetre mirror and a focal length of 12 metres*. Again there were problems. The first mirror distorted under its own weight, and it was August 1789 before the second mirror was ready. It was mounted in the tube which had been prepared for it, and at once Herschel used the telescope to discover two faint inner satellites of Saturn, Mimas and Enceladus.

One departure from the normal pattern was that Herschel abandoned the flat secondary mirror of the Newtonian system, merely tilting the main mirror and bringing the image to focus direct at the upper end of the tube. This seemed sensible, and obviously it saved a certain amount of light, but in practice it was never satisfactory, partly

*This is equivalent to 40 feet in Imperial measure, and the telescope is generally known as 'the 40-ft. reflector', just as the Palomar reflector is still called the '200-inch'.

FLAMSTEED HOUSE. *The old Royal Observatory in Greenwich Park, now renamed and used as a museum; one room is devoted to Herschel relics. The photograph, taken in 1961 by Patrick Moore, shows the Octagon Room, used by Flamsteed, and the celebrated time-ball.*

STARFIELD IN ORION *showing the gaseous nebula M42, photographed with a 30·5-cm reflector by H. R. Hatfield in 1971.*

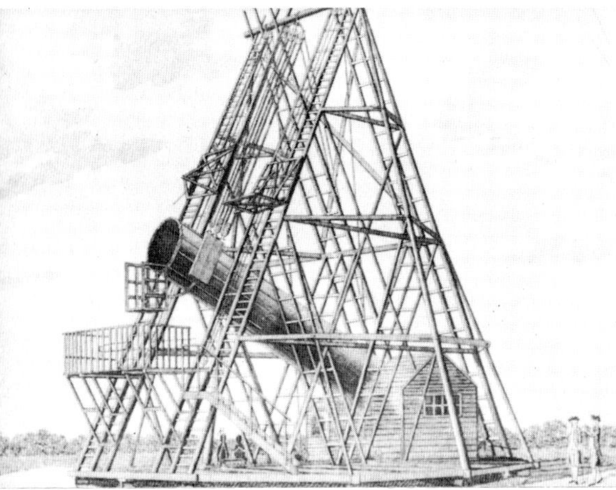

HERSCHEL'S 40-FOOT REFLECTOR, *with its scaffolding.*

ENGRAVING OF THE 40-FOOT REFLECTOR *in the garden of Observatory House, Slough.*

because of the difficulty of adjustment and partly because it introduced all kinds of undesirable distortions. The Herschelian system was fully tested, but it has long since become obsolete. (Incidentally, it seems that it was distortions in the 'front-view' system which led Herschel to believe that he had found a ring round Uranus.)

It must also be said, with regret, that the 12-metre reflector never lived up to its early promise. It was too cumbersome and awkward to use. The tube was mounted on a wooden framework, with the whole structure capable of being spun round on a circular revolving base; ladders fifteen metres long gave access to the observer's platform at the top of the tube, and the observer communicated with his assistants through speaking-pipes. Generally, of course, the main assistant was Caroline, sitting in a hut at the bottom of the structure. At least two workmen had to be in attendance to move the telescope around, and Herschel made much greater use of his smaller reflectors, the best of which had a 47-centimetre mirror and a focal length of six metres.

The giant telescope was not used after 1816. It was dismantled in 1839, and lay prone in the garden of Observatory House until it was destroyed by a falling tree. There is stayed until Observatory House was pulled down in 1960. The little that now remains of the tube is on view at the Old Royal Observatory at Greenwich, and the main mirror hangs up inside the museum.

Herschel made useful observations of the planets. For instance, he measured the rotation period of Mars to within an accuracy of a few seconds. But his main work was in connection with the stars, and in particular with the shape of the Galaxy.

Herschel believed that the apparent brightness of a star must be a reliable guide to its real distance from us, so that the most brilliant stars, such as Sirius, would be the closest. This is wrong, but Herschel had no means of estimating star-distances, so that his assumption was reasonable enough. The only way to find out the manner in which the stars were arranged was to count them. The regions containing the greatest numbers of stars would represent the greatest extensions of the Galaxy. It was clearly impossible to count all the stars within range of Herschel's telescopes, so he adopted the method of star-gauging, counting the stars in carefully selected regions.

There are many more stars in and near the Milky Way than in other areas, but Herschel soon made the interesting discovery that the percentage increase is greater for the fainter stars. For example, suppose that we take two areas, one near the Milky Way and the other some distance from it, and use a telescope which will show four times as many bright stars in the Milky Way zone as in the other. With a larger telescope, it will be found that the faint stars are ten times more numerous than the brighter ones. In other words, faint stars are unexpectedly plentiful near the Milky Way, and Herschel concluded that the Galaxy must be shaped rather like a double-convex lens. I have commented, rather unromantically, that the shape is not unlike that of two fried eggs clapped together back to back, though Herschel preferred to compare it with 'a cloven grindstone'.

In the main, Herschel was correct. He realized that the apparent crowding of stars in the Milky Way is nothing more than a line of sight effect, so that when we look along the main plane (**AB** in the diagram on page 98) we see many stars in almost the same direction. He was wrong in supposing that the Sun lies near the centre; actually it is at

GALAXY IN CASSIOPEIA, N.G.C. (= New General Catalogue) 147. *The photograph was taken in red light with the 508-cm Hale reflector at Palomar. This is an external system, and lies far beyond the boundaries of our own Galaxy.*

least 30,000 light-years out toward one edge, but he had taken the essential step. He also wondered whether some of the so-called nebulæ might be independent star-systems or galaxies, far beyond our own. This too is correct, though it was not proven until more than a century after Herschel's death.

He also managed to measure the movement of the Sun as compared with the stars. This was done by means of a well-known effect. If you drive along a road with trees to either side, the trees ahead of you will seem to 'open out' as you approach them, while the trees behind will 'close up'. The same sort of thing happens with the stars in the Galaxy as the Sun moves, taking the Earth and the other planets along with it. The shifts were remarkably small and difficult to measure, but in 1783 Herschel was able to announce that the Sun is moving towards a point in the constellation Hercules. The position which he gave is very close to the modern estimate.

He observed double stars, and discovered hundreds of new pairs. In 1802 he announced another great discovery. Many of the double stars are physically-associated or *binary* systems, with the two components moving together round their common centre of gravity much as the two bells of a dumbbell will do when twisted by their joining bar. This had been suggested earlier, but Herschel was the first to prove it by direct observation.

As the years passed by, Herschel continued his 'reviews of the heavens', discovering and listing clusters, nebulæ, double stars, variable stars and much else; he also found out that the Sun emits infra-red rays, which do not affect our eyes, thereby laying the foundations of what we now call invisible astronomy, though at the time he was

HERSCHEL'S PLAN OF THE GALAXY. *(Right) Herschel was the first man to propose a plan of the Galaxy or 'Milky Way' which proved to be reasonably accurate. Herschel's main error was in placing the Sun near the centre of the system. If an observer looks along the direction AB he will see many stars in much the same direction, giving rise to the Milky Way effect; if he looks along direction CD he will see fewer stars.*

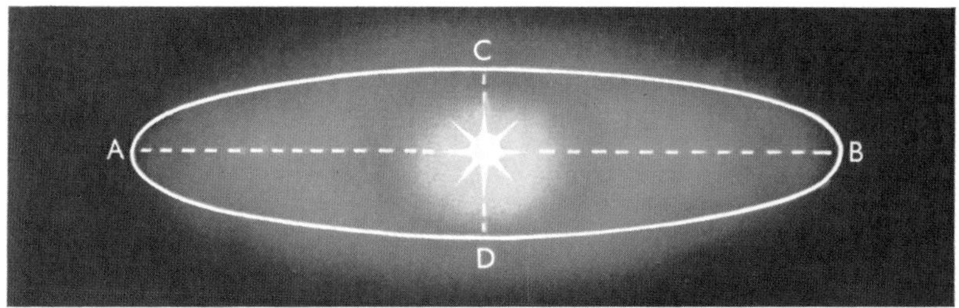

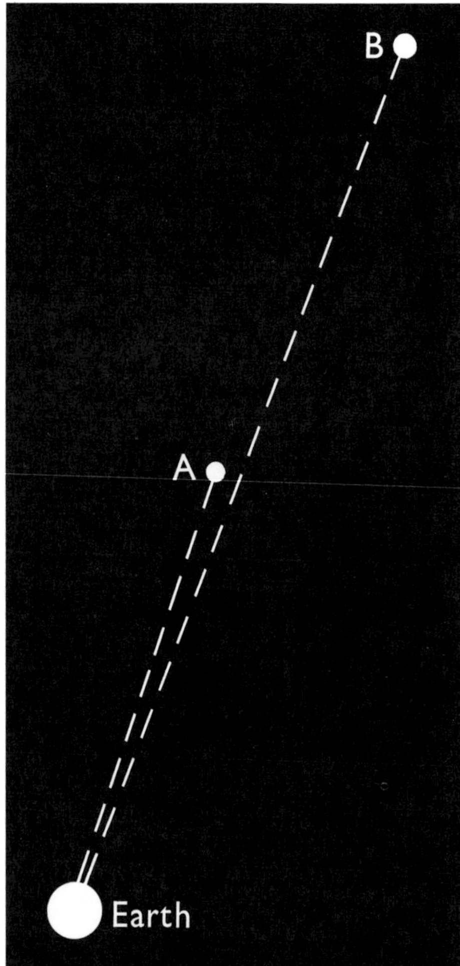

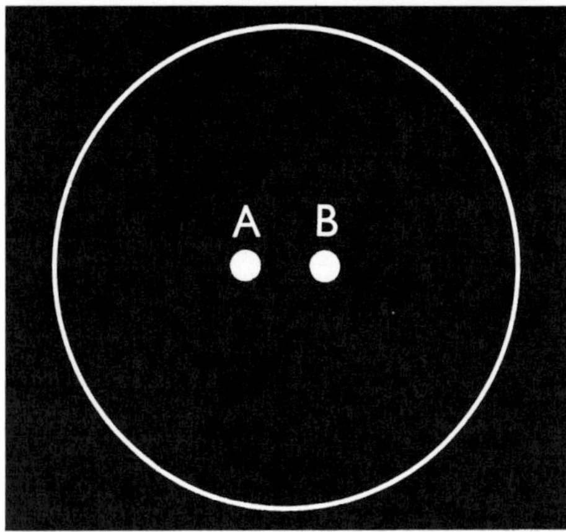

OPTICAL DOUBLE STAR. *As seen from the Earth, stars A and B lie in roughly the same direction, and telescopically they appear as in the circle, though in fact star A is much the closer to us and has no real association with B.*

naturally unaware of the tremendous importance of this discovery. Since he did not become a serious observer until he was nearly forty, it is amazing how much work he was able to do.

He received every honour that the scientific world could bestow. He was knighted in 1816, and in 1822 became the first effective president of the newly-formed Astronomical Society of London, now the Royal Astronomical Society. He presented his last paper when he was eighty years old, and was active almost to the time of his death on September 7, 1822. Caroline returned to Hanover (a decision which she subsequently regretted) and lived to the advanced age of ninety-eight.

William Herschel's son, John, carried on his father's work, extending the 'review of the heavens' to the southern hemisphere. But as a pure observer, William stands alone.

The site of Observatory House is now occupied by the Rank Xerox offices, at the side of a busy road in Slough, with a monument marking the exact spot where the 12-metre telescope once stood. As a personal note, I well remember my last sight of Observatory House, the day before demolition started. I stood on the grassy lawn where the 12-metre had been; today, as the traffic roars by, it is difficult to picture the scene as it must have been when William and Caroline worked patiently night after night.

Luckily, No. 19 New King Street in Bath still remains. For a while it too was neglected, but in 1978 the William Herschel Society was formed, and due mainly to the generosity of Dr and Mrs Leslie Hilliard the house has been acquired and restored. It has been turned into a Herschel museum, and is visited by many people who take an interest in the history of science. It was opened on March 13, 1981, exactly two hundred years after Herschel had stood in the garden to discover Uranus, and it was my great honour to deliver the centenary lecture.

Herschel was not infallible, and some of his ideas sound strange today. He believed, for instance, that the habitability of the Moon was 'an absolute certainty', and that there could even be intelligent beings living in a cool region below the surface of the Sun. But he was one of the greatest of all pioneers, and his life and work were aptly summed up by his contemporary, the Swiss astronomer Marc Pictet:

> 'This wonderful man, whose imagination is so brilliant, whose brains and physical powers are so well balanced, possesses besides these advantages the most agreeable manner and social qualities which would make him sought after if he had no other claim to public esteem. Visitors often take the unwarranted advantage of his courtesy and compliance, wasting his time and putting unnecessary and often ridiculous questions, but his patience is inexhaustible and he takes these inconveniences which are attached to celebrity in such good part that no one could guess how much they cost him.'

In the story of human achievement, William Herschel will never be forgotten.

16

Reflectors and Refractors

NEWTON had developed the reflecting form of telescope simply because he could see no way to make a refractor which would be free from the irritating false colour. The aerial telescopes of Huygens, Hevelius and Cassini were hopelessly clumsy, and unless better lenses could be made it was obvious that the refractor had little future.

So matters remained until 1729, when an English amateur named Chester More Hall, who lived near Harlow in Essex, made a series of interesting experiments and constructed the first compound or *achromatic* object-glass. Strictly speaking, this is not a single object-glass, but is made up of several lenses placed close together. The false colour due to one lens will thus tend to be cancelled out by the other. Actually More Hall's theory was basically wrong, but he did manage to build an achromatic refractor with a 6·4-centimetre object-glass and a focal length of only 51 centimetres. It was far from perfect, but it was a vast improvement on the old-type 'monsters'.

More Hall had no wish for fame, and he took no immediate steps to

PALOMAR OBSERVATORY. *(Left to right) The buildings shown are the dome of the 45·7-cm Schmidt camera; the residential quarters; the dome of the 122-cm Schmidt; the dome of the 500-cm Hale reflector; the garage, and the water tower and reserve tank.*

GREGORIAN TELESCOPE, *made by James Short in 1749.*

GREGORIAN TELESCOPE: *Another of Short's instruments.*

make his discovery known. At this time the reflectors made by men such as the Scottish optician James Short were regarded as much more promising. Most of Short's reflectors were Gregorians, and were remarkably good; many of them survive today.

Then, in 1758, John Dollond in London returned to the principle of the achromatic object-glass. Helped by his son Peter, he began to make lenses for sale, and before long refractors came back into favour. Lens-making is much more difficult than mirror-grinding, and the false colour trouble can never be entirely cured; at the same time, refractors have certain obvious advantages. They are easier to use, and are far less temperamental than reflectors.

Two great telescope-makers lived in the early part of the nineteenth century. One, of course, was Herschel, who made considerable sums of money by selling mirrors (King George III paid him 600 guineas each for four telescopes of focal length 3 metres). The other was a young German, Joseph von Fraunhofer. The main difference between the two was that while Herschel concentrated entirely upon reflectors, Fraunhofer was much more interested in refractors, and his ambition was to build a lens telescope better than anything that Herschel could produce.

Fraunhofer was born at Straubing, in Bavaria, in 1787. Both his parents died while he was very young, and he had little schooling. At the age of fourteen he was apprenticed to a Munich looking-glass maker, one Weichselberger. We often hear of cruel masters and starved, ill-treated apprentices, but in Fraunhofer's case the description fitted the facts, and the boy was desperately unhappy. Then came an incident which altered his whole career. The tumbledown lodgings in which he lived collapsed for no apparent reason, and young Fraunhofer was trapped in the ruins. The rescue operations were watched by the Elector of Bavaria, who happened to be driving past, and it pleased him to befriend the lad. He provided Fraunhofer with enough money to buy his release from Weichselberger, and to educate himself.

Later on the Elector would have had good reason to be pleased at the results of his good deed, since Fraunhofer became world-famous. He joined the Munich Optical Institute in 1806, and became its Director only seventeen years later. His lenses were by far the best produced up to that time, and when difficult problems arose he solved them step by step. He was also the true founder of spectroscopy. It was particularly tragic, therefore, that he died in 1826 at the early age of twenty-nine.

In 1817 Fraunhofer produced an object-glass of magnificent quality, 24 centimetres in diameter and of 4·3 metres focal length. It was bought for the Russian Government for the Observatory at Dorpat, in Estonia (which was then, as now, ruled by Russia). In 1824 the telescope was ready, and during the following years F.G.W. Struve, Director of the Observatory, discovered 2,200 new double stars with it. The 'great Dorpat refractor' has had a distinguished history.

In another way also the telescope opened up a new era; it was mechanically driven, so that it automatically followed the stars in their east-to-west movement across the sky.

The apparent daily motions of the Sun, Moon, planets and stars are due to the fact that the Earth is spinning on its axis from west to east. The movement is so slow that to the naked eye it is imperceptible over periods of a few minutes, but when a telescope is used the movement

becomes unpleasantly obvious. An eyepiece giving a magnification of, say, 200 will show only a very small area of the sky, and a star will seem to race across the field, so that the telescope has to be shifted all the time if the star is to be kept in view.

The trouble becomes worse with increased power, and it is not easy to push a telescope smoothly enough. The remedy is to mount the telescope *equatorially*, driving it mechanically so as to compensate for the Earth's rotation.

The idea of a driven telescope originated with Robert Hooke, but in his day no equatorial mountings had been built; even Herschel's 12-metre reflector was an *altazimuth*, so that it had to be moved up or down as well as east to west if it were to keep a star in view. The Dorpat refractor was set up on an equatorial mount, and was driven round by clockwork. At the time, this was a daring new development.

The next great advance lay not many years ahead. This was celestial photography, without which the modern astronomer would feel strangely helpless. But before going on with the story of instrumental developments, I must pause to say more about some of the other investigations made during the early nineteenth century – and in particular, the hunt for new members of the Sun's family.

183-CM REFLECTOR AT THE KARL SCHWARZCHILD OBSERVATORY, TAUTENBURG, EAST GERMANY. *(Left) This is an 'all purpose' reflector, using several optical systems; it can, for instance, be used as a Schmidt or a Cassegrain. It is the largest instrument of this kind yet built. Photograph by Patrick Moore, 1963.*

THE 102-CM REFLECTOR AT FLAGSTAFF. *(Centre) The tube is of the skeleton type, and the observatory has a slide-back roof instead of a dome. Photograph by Patrick Moore, 1964.*

THE ISAAC NEWTON 249-CM REFLECTOR, *(right) under construction at Newcastle. It is now in use at the Royal Greenwich Observatory, Herstmonceux. Photograph by Patrick Moore.*

17

The Celestial Police

HERSCHEL'S fame is so great that it tends to overshadow his contemporaries, just as, long before, Newton had overshadowed Hooke and Halley. But there were other skilful observers around the turn of the century, and in particular there was Johann Hieronymus Schröter.

Schröter was born at Erfurt, in Germany, in 1745. He was never a professional astronomer; he trained at Göttingen University as a lawyer, and in 1778 he was appointed chief magistrate of the little town of Lilienthal, near Bremen. In the following year he set up a private observatory, and began his long series of observations of the Moon and planets.

History has treated Schröter badly. It is quite true to say that he was not a good draughtsman; his drawings were clumsily made. It may also be true that his largest telescope, a 48-centimetre reflector made for him by a deaf Hamburg optical worker named Schräder, was not of the highest quality. Yet Schröter seldom made a serious mistake, and among his other telescopes were two made for him by no less a person than William Herschel, so that their excellence cannot be doubted. Schröter and Herschel never met, but they carried on a long correspondence which was very cordial – apart from one brief interlude, when Herschel expressed strong criticism of Schröter's belief that he had observed a high mountain on Venus. In this, of course, Herschel was justified, and fortunately Schröter's reply was calm and courteous, so that no bad feeling resulted.

Schröter was in fact the first man to record details on the cloud-covered surface of Venus, and he also observed what we now call the Ashen Light – the faint luminosity of the night side of the planet. The same sort of thing is seen with the Moon, and, as we have noted, Leonardo da Vinci explained it as reflected earthlight. But Venus has no satellite, and the Ashen Light is much less easy to account for. Many modern astronomers have dismissed it as a mere contrast effect, but it

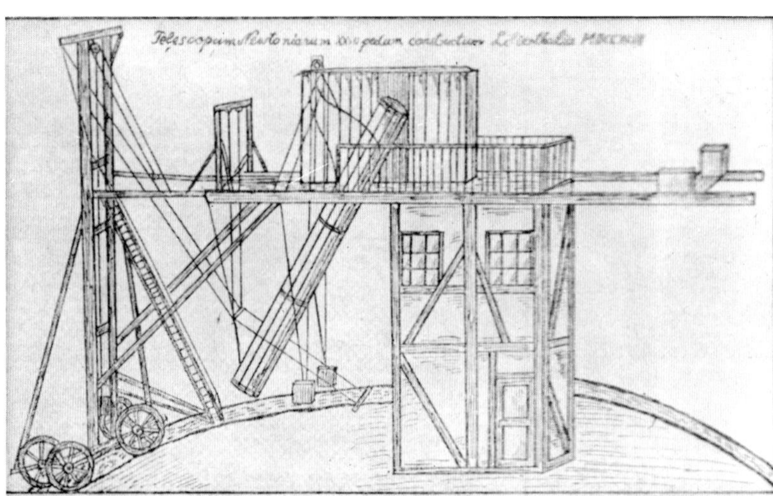

TELESCOPE USED BY SCHRÖTER. *Of Schröter's telescopes, it is possible that the largest (a 48·3-cm Schräder reflector) was of poor quality optically, but some of the others were undoubtedly good. One of them, made by Herschel, was extensively used for Schröter's work on the Moon and planets.*

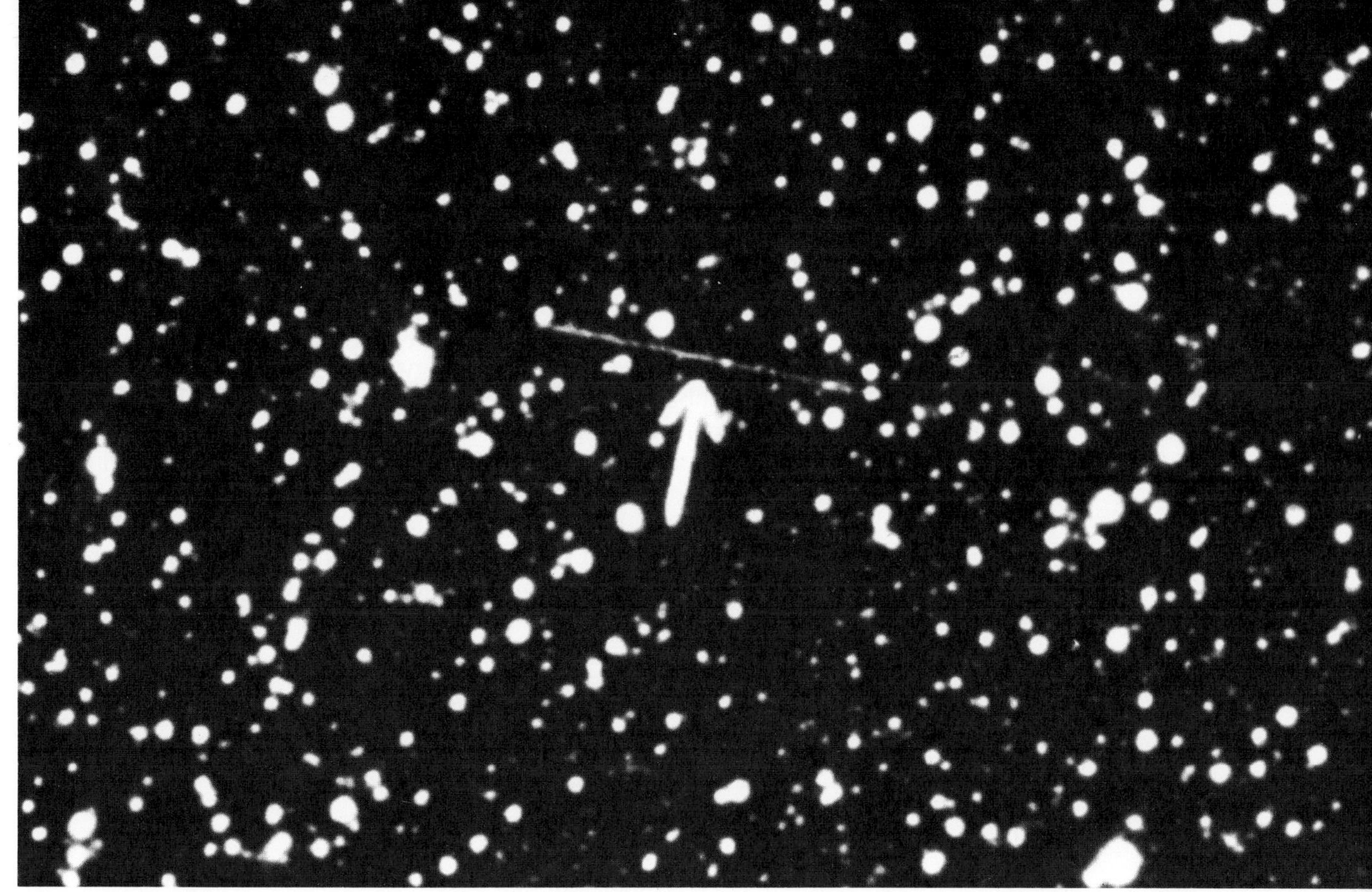

TRAIL OF ICARUS, 1949. *Icarus is an exceptionally interesting minor planet. It is not an 'Earth-grazer', but at its closest to the Sun its distance is less than that of Mercury. The movement among the stars is relatively rapid, and the trail of Icarus is indicated by the arrow; clearly the asteroid moved appreciably during the time of exposure. Photographed with the 122-cm Schmidt camera at Palomar.*

has been seen by almost every serious observer (and I must add that I have recorded it unmistakably myself). It is probably due to electrical phenomena in the upper atmosphere of Venus. At any rate, Schröter's keen eyes did not deceive him.

He also made many observations of the Moon, and drew the lunar mountains and craters in more detail then ever before. He is, in fact, the true founder of lunar study or *selenography*; he mapped most of the surface, and was the first to give a proper decription of the long, cracklike features which we now call clefts or rills, as well as making improved measurements of the heights of lunar peaks. He was in touch with most of the leading astronomers of the world; among his assistants was Friedrich Bessel, later to become famous as being the first man to measure the distance of a star. Lilienthal became very much of an astronomical centre. And it was here, in 1800, that the so-called Celestial Police first gathered. Their stated ambition was to find a new planet.

Even a casual glance at a map of the Solar System will show that the planets are divided into two groups. Mercury, Venus, the Earth and Mars are relatively small and close to the Sun. Then comes a wide gap, beyond which we have the four giants; Jupiter, Saturn, Uranus and Neptune, though of course Neptune was not known in Schröter's time. (Pluto, the ninth planet, is altogether different, and may not be worthy of planetary status, as we will see later.) It looked almost as though there should be a planet moving in the gap between the orbits of Mars and Jupiter. Kepler had suspected something of the sort, and had gone so far as to write, 'Between Mars and Jupiter I put a planet.'

In 1772 Johann Elert Bode came into the picture. He had been born in Hamburg in 1747, and had become Director of the Berlin Observatory; he published a star-catalogue (containing various new constellations, most of which have fortunately been forgotten) and was a prolific

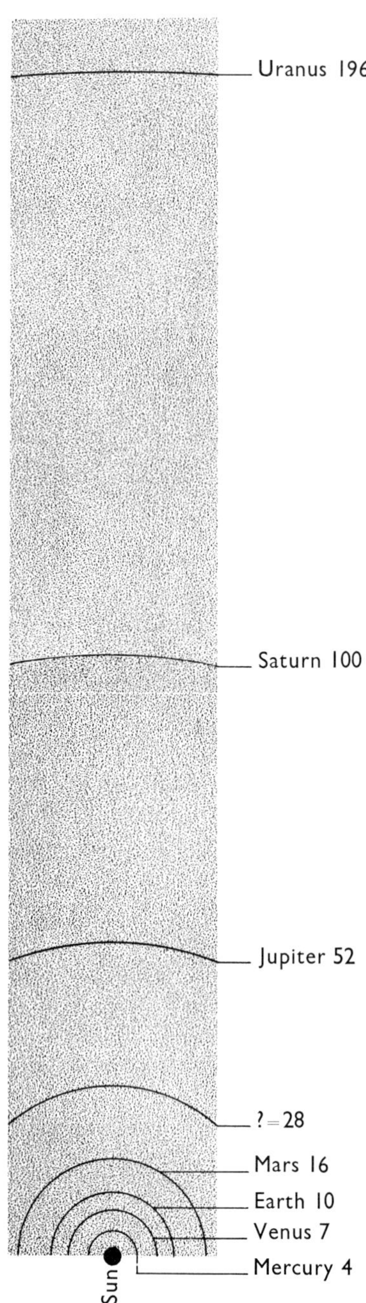

DISTANCES OF THE PLANETS FROM THE SUN ACCORDING TO BODE'S LAW. *Uranus, discovered in 1781, fitted excellently into the scheme. Note the 'gap' corresponding to number 28; it was this which led to an organized search, and the discovery of the first four Minor Planets*

writer who did much to popularize astronomy; he continued in active research until shortly before his death in 1826. It was Bode who suggested the name 'Uranus' for Herschel's planet. But on the whole he is probably best remembered in connection with the hunt for the missing planet beyond Mars.

Bode's Law, so called, should really be known as the Titius-Bode Law, because it was an otherwise obscure German – Titius of Wittenberg – who first drew attention to it; Bode merely popularized it and refined it. Whether or not it has any real significance (and I admit that I, for one, am highly sceptical), it is interesting. It is as follows:

Take the numbers 0, 3, 6, 12, 24, 48, 96 and 192, each of which (apart from the first) is double its predecessor. Now add 4 to each. Taking the Earth's distance from the Sun as 10, the remaining figures give the other planetary distances with reasonable accuracy, as the table shows.

Planet	**Distance from the Sun:**	
	According to Bode's Law	*Actual*
Mercury	4	3·9
Venus	7	7·2
Earth	10	10
Mars	16	15·2
–	28	–
Jupiter	52	52·0
Saturn	100	95·4
(Uranus	196	191·8)

I have put Uranus in brackets because it was still unknown in 1772, when Johann Elert Bode publicized the relationship. When it was found, nine years later, it fitted well into the sequence.

The only trouble was that there appeared to be no planet corresponding to Bode's number 28 – the gap between Mars and Jupiter.

If such a planet existed, it would certainly be faint, as otherwise it would have been found long before by astronomers busy compiling star catalogues – including Bode himself. Also, it was reasonable to suppose that its orbit would not take it too far from the plane of the ecliptic, in which case the planet would lie somewhere in the Zodiac.

In 1800, therefore, six astronomers assembled at Schröter's observatory at Lilienthal, and worked out an organized hunt for the missing planet. They were soon nicknamed the Celestial Police. Schröter became president; the secretary was a Hungarian baron, Franz Xavier von Zach, who had become director of the Seeberg Observatory at Gotha, and who played a great rôle in improving international cooperation in astronomy. The idea was to divide up the Zodiac, each 'policeman' being responsible for a definite section of it; all the stars down to a reasonable magnitude would be checked, to see if any one of them moved.

Ironically, they were forestalled. At the Palermo Observatory in Sicily, Guiseppe Piazzi had been compiling a new star catalogue. On 1 January, 1801 – the first day of the new century – he picked up a starlike point which behaved in a most unstarlike manner. It moved appreciably from night to night, and Piazzi thought that it might be a tailless

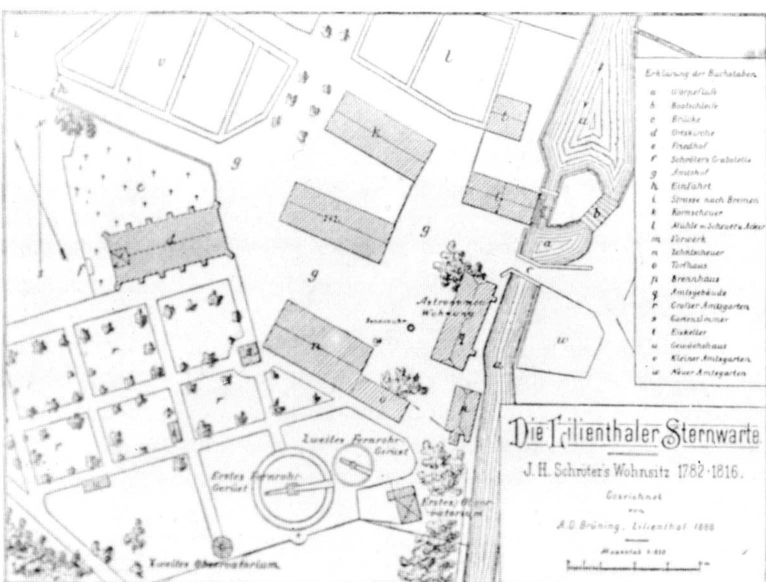

PLAN OF SCHRÖTER'S OBSERVATORY AT LILIENTHAL. *This observatory was for many years the 'astronomical centre' of Germany and, indeed, of much of Europe. It was from Lilienthal that the planet-hunt was organized, and it was here too that Schröter carried out almost all his pioneer work in lunar observation.*

comet. He was interested enough to write to von Zach, but his letter took some time to arrive, and by the time that von Zach received it the strange moving body had been lost in the evening twilight.

Fortunately, Piazzi had made careful observations of it. The German mathematician Karl Friedrich Gauss worked out an orbit and forecast its position, so that Heinrich Olbers, one of the 'police', was able to recover it exactly a year after Piazzi had first seen it. It did indeed prove to be a new planet, and it was named Ceres, in honour of the patron goddess of Sicily. Its distance on the Bode scale was 27·7, so that everything seemed to be highly satisfactory. The revolution period was found to be four years.

Yet there were nagging doubts. Ceres was too faint to be seen with the naked eye, and it was much smaller than any known planet. It was even smaller than the Moon, and we now know that its diameter is a mere 1003 kilometres. Olbers, among others, was not satisfied. He was an enthusiastic observer; by profession he was a medical doctor, and he had set up an observatory on the roof of his house in Bremen, where he paid special attention to comets. Together with the other 'police' he continued the search, and in 1802 he discovered a second small planet, Pallas, slightly further away than Ceres. In 1804 Karl Harding, one of Schröter's assistants at Lilienthal, discovered the third small planet, Juno; and in 1807 Olbers tracked down the fourth, Vesta.

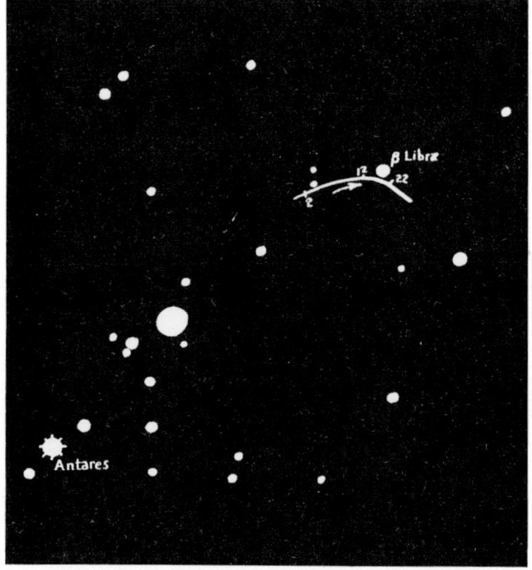

PATH OF CERES AMONG THE STARS. *The position of Ceres is shown for May 2, 12 and 22, 1959, close to the star Beta Libræ. At this time the planet Jupiter was near by, in the adjacent constellation of Scorpio, the Scorpion. Drawing by H. P. Wilkins.*

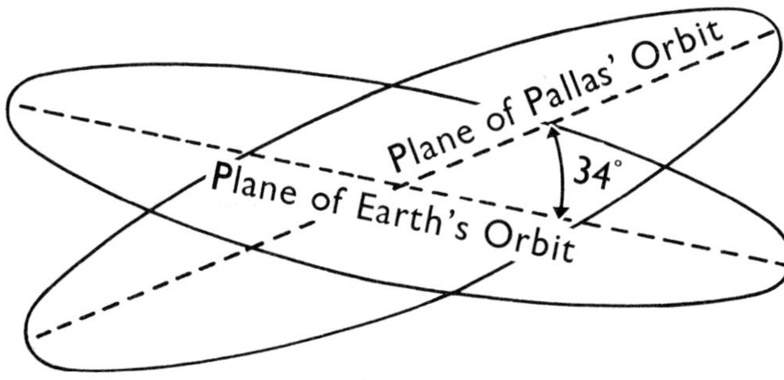

INCLINATION OF THE ORBIT OF PALLAS, *which amounts to over 34 degrees*

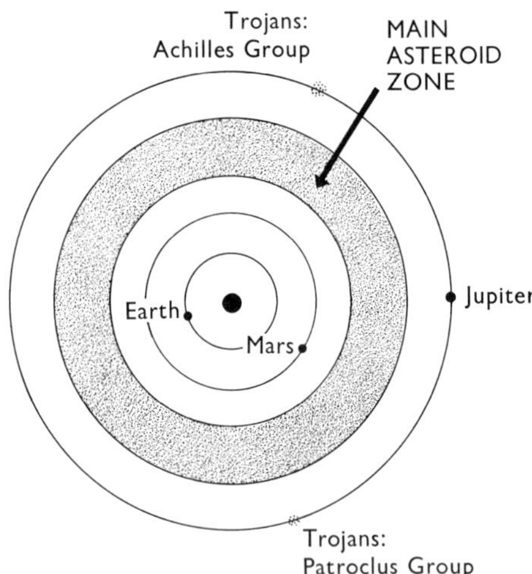

ORBITS OF THE TROJAN ASTEROIDS. *The Trojans revolve round the Sun at mean distances equal to that of Jupiter, so that in effect they move in Jupiter's orbit. One group of Trojans lies well 'ahead' of Jupiter and the other group well 'behind', so that there is no danger of the close encounters with Jupiter. The Trojans do not remain precisely 'fixed' in relation to Jupiter, but the diagram suffices to show the general situation.*

All this seemed very strange. Instead of one large planet in the Bode gap there were several small ones. Ceres remained the largest of them (though not the brightest; the closer and more reflective Vesta is the only one ever visible with the naked eye). Olbers suggested that the four new worlds might be the fragments of a larger planet that had met with disaster in the remote past, and had broken up. Collectively they became known as the minor planets, planetoids or asteroids. Were there others? The hunt was on; but no more asteroids came to light, and then came a tragedy which put an effective end to the organized search.

War had broken out once more between France and Germany. The invading French army, under Vandamme, captured Lilienthal and sacked it. Schröter's observatory was destroyed, together with all his unpublished observations; even his telescopes were plundered, because they had brass tubes, and the French soldiers believed them to be made of gold. The loss could never be made good. Schröter did organize the removal of the surviving equipment to Bremen, but he was too old to start again, and three years later, in 1816, he died. It was a sad end to a distinguished and highly successful career.

That, for the moment, meant a halt in the search for new members of the Solar System, but there is one curious fact that I have never been able to explain. As we have seen, Herschel had combed the heavens very thoroughly; how did he fail to discover at least one or two of the asteroids? Several, including Ceres, Vesta and Pallas, are above the eighth magnitude; Vesta is only marginally fainter than Uranus. At any rate, Herschel overlooked them, and the triumph of the Celestial Police was complete. (Piazzi, of course, was not a member of the team when he discovered Ceres, but he joined later.)

The next character in the asteroid story is Karl Ludwig Hencke, who was born at Driessen in Germany in 1793 and became postmaster there. In 1830 he decided to resume the search. Alone and unaided, he worked away for fifteen years – and at last, in 1845, he discovered Asteroid No. 5, now named Astræa. Two years later he found the sixth asteroid, Hebe. Interest was re-awakened, and more and more observers took up the problem. Since 1847 no year has passed without the discovery of several new asteroids, and by now thousands have had their orbits worked out. It has been estimated that the total membership of the swarm may exceed 40,000, but most of these asteroids are very small; a kilometre or two in diameter, or even less.

Of course, the early discoveries were visual, but in 1891 a new method was applied: that of photography. If the same area of the sky is photographed twice, with an interval of several nights, the stars will remain in the same relative positions, but an asteroid will move. In this way No. 323, Brucia, was discovered on December 20, 1891 by the German astronomer Max Wolf. In all, Wolf made 232 discoveries, though Karl Reinmuth, also of Germany, beat this record with no less than 246.

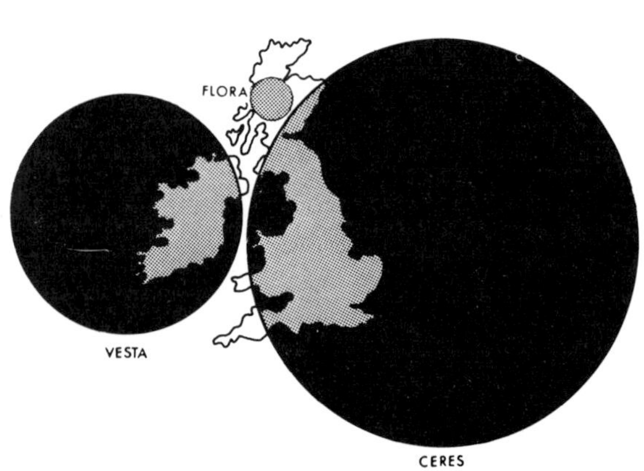

SIZES OF MINOR PLANETS COMPARED WITH THE BRITISH ISLES. *Ceres is the largest of the minor planets, and Vesta the brightest.*

It cannot be said that the asteroids have been universally popular! Photographic plates exposed for quite different reasons are often found to be crawling with asteroid trails, all of which have to be dealt with, and which waste an incredible amount of time. One irritated German went so far as to nickname them 'vermin of the skies', while another referred to the 'minor planet pest'. But today we think differently; the asteroids are interesting and important members of the

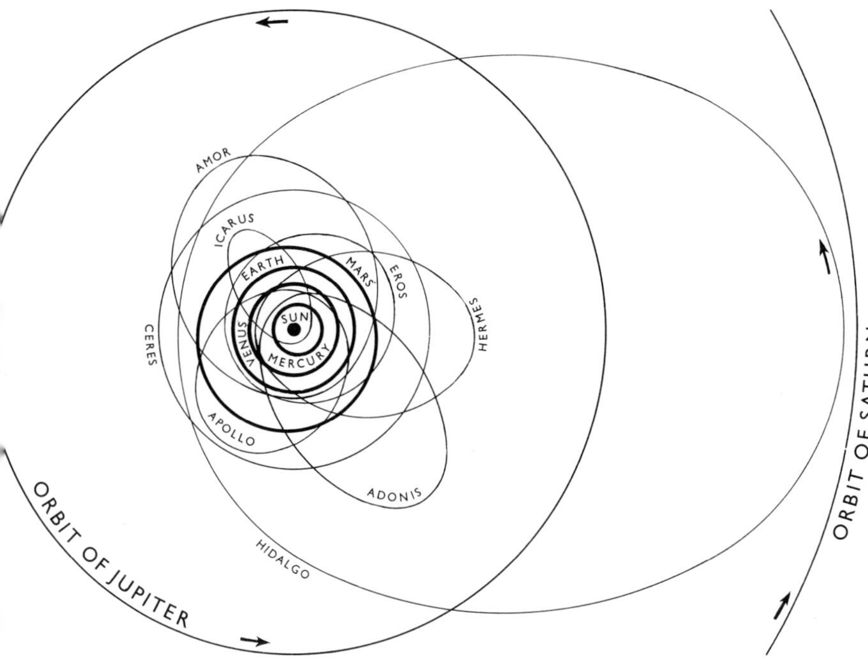

ORBITS OF SOME INTERESTING MINOR PLANETS. *Whereas most of the minor planets revolve round the Sun at mean distances between those of Mars and Jupiter, some have more eccentric orbits which take them away from the main swarm. Icarus, for instance, has a perihelion distance less than that of Mercury; Hidalgo travels out almost as far as the orbit of Saturn. However, most of these exceptional asteroids are very small, and so are extremely difficult to observe except when fairly close to the Earth.*

Solar System – particularly those which move in exceptional orbits, swinging them away from the main swarm between the paths of Mars and Jupiter.

Yet even if combined, the asteroids would still not make up one body as massive as the Moon, and Olbers' theory of a disrupted planet has fallen into disfavour. It is known that the main planets were formed out of a 'solar nebula' i.e. a cloud of material associated with the youthful Sun, between 4½ and 5 thousand million years ago. Very probably the strong gravitational influence of Jupiter prevented a large planet from being built up in the region of the asteroid belt, so that the end product was a swarm of small bodies instead of a single major one.

The original asteroids were given mythological names, but as the discoveries mounted the supply of gods and goddesses began to give out, and some of the later names are bizarre. We have, for instance, names of plants (987 Petunia, 973 Aralia); musical plays (1047 Geisha); foods (518 Halawe – the discoverer, R.S. Dugan, was particularly fond of the Arabian sweet halawe); shipping lines (724 Hapag – Hamburg Amerika Linie), and so on. Asteroid No. 1625 is even named after a computer: The NORC (Naval Ordnance Research Calculator, in Virginia). One asteroid was even auctioned! No. 250 was discovered by the Austrian astronomer Johann Palisa, assistant at the Vienna Observatory; he offered to sell his right of naming for a sum equivalent to £50, and the offer was accepted by Baron Albert von Rothschild, who named the asteroid Bettina, after his wife. No. 1000 is, appropriately enough, named Piazzia. (I cannot resist mentioning that Asteroid No. 2602, discovered in 1982 by Dr. Edward Bowell at the Lowell Observatory in Arizona, has been named afrer myself!)

An asteroid is given an official number and name only after its orbit has been worked out well enough for it to be kept under observation. Hundreds more asteroids have been insufficiently observed, and have been lost.

I will have more to say about the exceptional asteroids later. Meanwhile, we must not despise these dwarf worldlets. They have their place in the Sun's family, and it is surely unfair to dismiss them as 'vermin of the skies'.

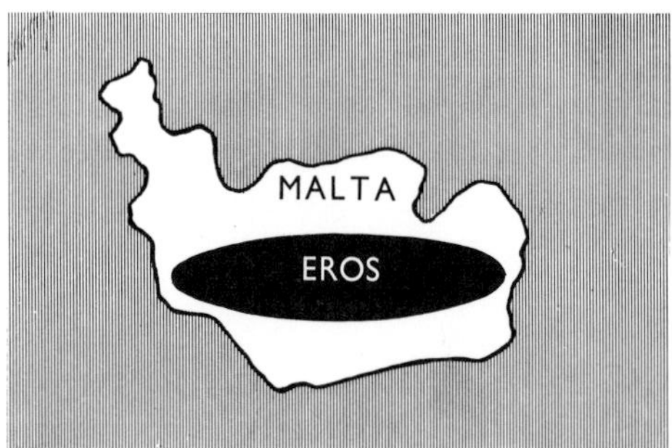

SIZE OF EROS COMPARED WITH MALTA. *Eros is one of the smaller asteroids, and is not even approximately spherical. Probably many of the other asteroids are of similar shape; but Eros is exceptionally easy to study, since its orbit occasionally brings it well within 33 million km of the Earth.*

EROS. *In 1975 the asteroid Eros made one of its closest approaches to the Earth. This photograph was taken on 9 February, 19.35 to 19.50 GMT, by Paul Doherty, with his 13·5-cm reflector. The bright star is Procyon, in Canis Minor; Eros is the speck closely left of it.*

18

Neptune

ONE of the difficulties about writing a history of astronomy is that the various themes overlap. For instance, astronomical spectroscopy really began early in the nineteenth century with the work of Foucault, and Bessel measured the first star-distance in 1838. But we have been discussing the planet-hunters, so let us jump forward to the next great discovery: that of Neptune, in 1846.

If Bode's Law were valid, any planet beyond Uranus would lie at a distance of 388 units on the scale. (Twice 192, the Bode distance for Uranus, gives 384; add 4, and we have 388.) If it were no larger than Uranus, it would be well below naked-eye range. At first there was no suggestion that a new planet might exist; but then astronomers became uncomfortably aware that something very peculiar was happening to Uranus. It was not keeping to its predicted path. It wandered away, and there seemed to be no obvious reason why.

Could the old pre-discovery observations (such as Flamsteed's of 1690) be inaccurate? The problem was taken up by the French mathematician Alexis Bouvard, who had begun his career as a shepherd-boy and had become a teacher. In 1784 Bouvard did the only possible thing: he rejected all the pre-discovery observations, and started again. Still Uranus refused to behave. Finally, in 1834, a Kent clergyman and amateur astronomer, the Rev. T.J. Hussey, made an important suggestion. Uranus was straying; therefore it was being pulled out of position by the gravitational force of an unknown, more remote body which was presumably a planet. Hussey wrote to the Plumian Professor of Astronomy at Cambridge, George Biddell Airy, and presumably hoped that his suggestion would be taken up.

SIR GEORGE BIDDELL AIRY

Airy plays an important part in the story, and I must digress briefly to say something about him. He was born at Alnwick in Northumberland in 1801, and soon made his mark as an excellent mathematician as well as a skilled designer of astronomical equipment. In 1835, some time after receiving Hussey's letter, he went to Greenwich as Astronomer Royal, and retained the post until 1881, only eleven years before his death. When he was appointed, the fortunes of Greenwich were at a low ebb. Halley had been succeeded as Astronomer Royal first by John Bradley, then (for only three years) by Nathaniel Bliss, and then by the energetic Nevil Maskelyne; but John Pond, who had become Astronomer Royal in 1811, had frankly neglected his duties, and in his later years seldom went to the Observatory at all. To be fair, it was not entirely Pond's fault, because he suffered from ill-health, but things went from bad to worse, and eventually Pond was politely asked to resign, which he did. Airy was faced with the task of re-building, and he was eminently well-equipped to do so. Under his régime Greenwich regained its old position of pre-eminence, and most of the credit was Airy's. On the other hand he was meticulous to a fault; it is said that he once spent a whole afternoon labelling empty packing cases 'Empty', and he insisted that his observers should be in their domes, ready to start work, even when rain was falling. There is a tale that even on inclement nights he would go round the Observatory, poking his head into the different domes and asking 'You *are* there, aren't you?' He also had great faith in his own importance, and once he had made up his mind nothing would persuade him to change it.

It was so with Hussey's suggestion. The idea was that if a planet existed it would show a small, unstellar disk, but before beginning a hunt it was essential to have some ideal of its position. It was a kind of cosmical detective problem. The victim's movements could be studied; the task was to work out the approximate position of the culprit. But Airy was unimpressed, and replied: 'I give it as my opinion, without hesitation, that we are not yet in such a state as to give the smallest hope of making out the nature of any external action on the planet.' Three years later he wrote to Bouvard: 'If it be the effect of any unseen body, it will be nearly impossible ever to find out its place.'

Others were not so pessimistic. Bessel, Director of the Königsberg Observatory (whom we will meet again later in connection with star-distances) believed that it would be possible to track down the proposed planet, but illness prevented him from tackling the problem. Then, in 1841, the next step was taken by a young Cambridge undergraduate, John Couch Adams, who came from Lidcot in Cornwall and had shown early signs of exceptional mathematical ability. To a a fellow undergraduate, George Drew, he confided: 'Uranus is a long way out of his course. I mean to find out why. I think I know.' He passed his degree – brilliantly – and then set to work. A few months of research convinced him that he was on the right track. He worked away, and in September 1845 he believed that he had found the answer. He knew where the planet ought to be.

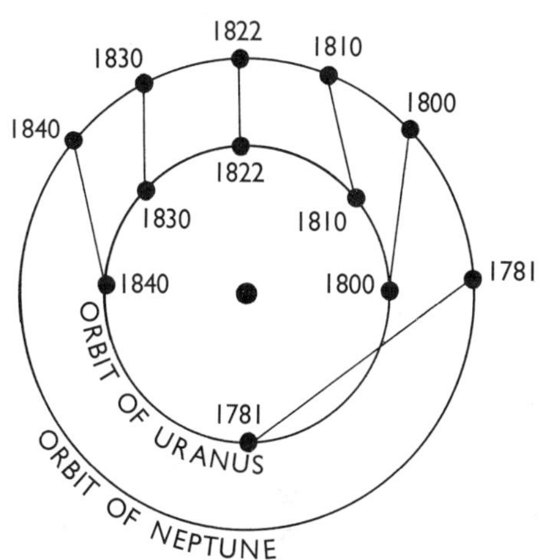

POSITIONS OF URANUS AND NEPTUNE, *1781-1840. Before 1822, Neptune tended to pull Uranus along; after 1822, to draw it back.*

He sent his results to James Challis, professor of astronomy at Cambridge, and Challis gave him a letter of introduction to Airy, now Astronomer Royal. That was the start of a chapter of accidents. Adams went to see Airy at Greenwich, only to find that Airy was away. He called again later; again Airy was out, so Adams left a card, which was taken to Airy's wife and then forgotten. When Adams returned, the butler told him that the Astronomer Royal was having dinner and could not be disturbed. Adams, feeling rebuffed, went away.

Adams then wrote to Airy. The reply was not encouraging, and Airy asked a question which showed that he had not really appreciated the nature of the problem. Adams did not reply, and things drifted aimlessly on. Meanwhile, there had been developments abroad.

The leading French astronomer of the time was François Arago, Director of the Paris Observatory. One of his assistants was a young mathematician, Urbain Jean Joseph Le Verrier, who had been born at St. Lo in Normandy in 1811 and was coming to the fore in celestial mechanics. Arago decided that Le Verrier was just the man to tackle the problem of Uranus, and in June1845 Le Verrier started work, blissfully unaware of Adams' existence.

By December 1845 Le Verrier had produced his first memoir on the subject. Airy, at Greenwich, read it and described it as 'a new and most important investigation' – yet still he took no action.

On June 1 1846 Le Verrier produced his second memoir, giving a position for the new planet. But just as Adams had had no help from observers in England, so Le Verrier was equally unlucky in France. Nobody began to search – not even Arago. Amazingly, the man stirred to action was George Biddell Airy! On July 9 he wrote to Challis, asking him to begin searching. Cambridge was the obvious place; it was equipped with a good telescope (the famous 'Northumberland' refractor), and there was no comparable equipment at Greenwich itself.

Challis was away. He did not return until July 18, and he then followed Airy's instructions, but not with any great energy. Moreover he had no good star-map of the region, and when he made observations he did not compare them promptly. He simply plodded on.

On September 23 Johann Encke, Director of the Berlin Observatory, received a letter from Le Verrier, who had lost patience with the French. Encke detailed one of his staff, Johann Galle, to search in the position indicated. During the discussion, Encke and Galle were joined by a young student, Heinrich D'Arrest, who asked permission to join in. As soon as darkness fell the two men set to work, using a 22·9-centimetre refractor (now, by the way, preserved in the Munich Museum). Unlike Challis, Galle and D'Arrest concentrated in the exact position which they had been given, examining the stars in the hope of finding something which showed a small disk. For some time they had no success, but then D'Arrest unearthed the relevant section of a new star chart which had just been received; Galle called out the magnitudes and positions of the stars, while D'Arrest checked them. Before long Galle came across an object of the eighth magnitude, and D'Arrest called out in excitement: 'That star is not on the map!' The search was over.

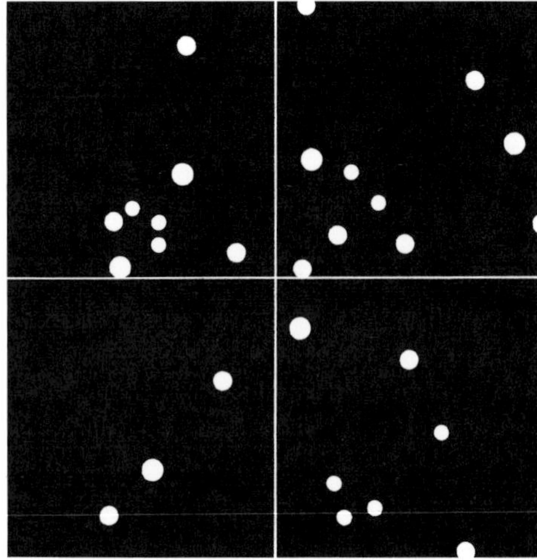

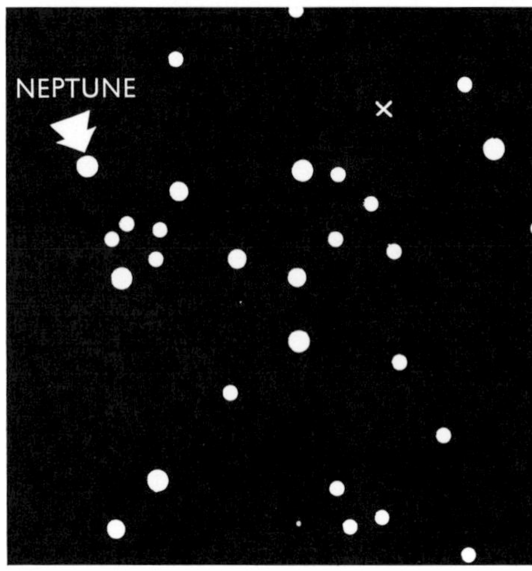

DISCOVERY OF NEPTUNE. *(Upper) Part of the star map used by Galle and D'Arrest. (Lower) The same map, with Neptune indicated by an arrow and Le Verrier's calculated position shown by a cross.*

Encke joined the two observers in the dome, and they followed the object until it set. By the next night they were confident that it was indeed the new planet, less than one degree from the position which Le Verrier had given. When Challis heard of the discovery, on October 1, he started checking the 3150 star positions he had recorded since starting work. He soon found that he had seen the planet twice in the first four days of his search – but he had failed to compare his observations.

Too late! The glory was Le Verrier's. When it was announced that Adams had arrived at the same result, considerably earlier, the French were furious. Arago even referred to Adams' 'clandestine' work and the 'flagrant injustice' of his claim. For a time the affair even threatened to become an international incident. Fortunately neither Adams nor Le Verrier took any part in the controversy, and when they met face to face, some time later, they struck up an immediate friendship which lasted for the rest of their lives.

In retrospect, it is fair to say that both Airy and Challis were guilty of negligence, and neither have I ever understood why Adams did not begin a personal search; he could have done so easily enough. But the facts remain, and today Le Verrier and Adams are recognized as co-discoverers of the planet we now call Neptune.

Neptune proved to be a giant, slightly smaller but more massive than Uranus, and of the same general type. Shortly after its discovery a well-known English amateur, William Lassell, detected the brightest of its three satellites, Triton. But one thing became clear at once: the Bode distance was only 300·7, instead of the predicted 388. Neptune's average distance from the Sun is 'only' 4,497,000,000 kilometres. This has led many people (including myself) to dismiss Bode's series as being mere coincidence. It is not strictly accurate for the closer planets; it breaks down completely for Neptune, and when a Law fails in so important a case it ceases to be a Law.

Le Verrier's fame became world-wide, and in 1853 he became Director of the Paris Observatory. Unfortunately he was touchy and irritable; it was once said of him that although he might not have been the most detestable man in France, he was certainly the most detested! In 1870 he was forced to resign his directorship, though he was reinstated two years later when his successor, Charles Delaunay, was drowned in a boating accident. Meanwhile, there had been another planet-hunt: this time an unsuccessful one.

Le Verrier had turned his attention to the movements of the innermost planet, Mercury. Here, too, he found slight differences between theory and observation. He worked out the position for an intra-Mercurial planet, and it was even given a name – Vulcan, after the mythological blacksmith of the gods.

Vulcan would be undetectable under normal conditions, since it would be lost in the Sun's glare. The only hope of seeing it would be to catch it as it passed in transit across the face of the Sun, as Venus and Mercury sometimes do. Consequently Le Verrier was delighted when, in 1859, he received a report from a French amateur, Lescarbault, that such a transit had actually been observed. Lescarbault lived in the small town of Orgères, and Le Verrier made haste to go and see him.

It must have been a strange interview. Lescarbault was very much of an amateur; he was the village carpenter as well as a doctor, and he used to record his observations upon wooden planks, planing them off when he had no further use for them, while his timekeeper was an old clock which lacked one of its hands. It is rather curious, then, that Le Verrier came away convinced that the observation had been correct, and he continued to believe in 'Vulcan' until he died in 1877.

But Vulcan has never been seen again, and it is significant that at the time of Lescarbault's observation a French astronomer named Liais, living in Brazil, had been watching the Sun and had seen nothing at all. No doubt Lescarbault was honest; equally certainly he was mistaken.

There was a flurry of interest in 1878, when two American astronomers, Swift and Watson, studied the area round the Sun at a time of total eclipse, and came to some startling conclusions; but their observations agreed neither with 'Vulcan' nor with each other, and it seems quite definite that they saw nothing more remarkable than ordinary stars. Finally, in our own century, Einstein's theory of relativity cleared up the irregularities in the motion of Mercury without the need for bringing in an extra planet. Many comets, and one known asteroid (Icarus) invade the torrid regions within the orbit of Mercury, but no large planet exists there.

The story of planet-hunting was not over. In the depths of the Solar System, Pluto has now been found, and even this may not be the outermost planet, but searches at these immense distances belong to the present century rather than to the time of Adams and Le Verrier.

THE TELESCOPE WITH WHICH GALLE AND D'ARREST DISCOVERED NEPTUNE. *The telescope was a refractor made by Fraunhofer in 1820.*

19

The 'Ferret of Comets'

I HAVE talked about the planet-hunters. Now let us turn to the comet-hunters, starting with the first and perhaps the greatest of them all: Charles Messier.

He was born in Lorraine in 1730, and at the age of twenty-one went to Paris to seek his fortune. He was interested in astronomy, and he had observed the bright comet of 1744 (De Chéseaux' Comet) which had at least six tails; but he had no qualifications – apart from neat, legible handwriting. Fortunately for him he was employed by the Astronomer of the Navy, Nicholas de l'Isle, who put him to work in keeping the Observatory records, and also gave him some lessons in elementary astronomy.

Then came an episode which changed his whole outlook. Halley's Comet was due at its first predicted return, and de l'Isle instructed Messier to search for it, using one of the small and rather elderly telescopes at the Observatory. Messier was only too glad to obey. On January 21, 1759 he duly located the returning comet, but two major disappointments awaited him. First, de l'Isle refused to let him announce the discovery, for reasons which remain obscure. Secondly unknown to Messier at the time, the comet had already been discovered by Palitzsch on Christmas Night, 1758.

Messier was not to be daunted. Comet-hunting had gripped him, and he began a long, fruitful period of research. After de l'Isle's retirement he had more freedom of action, and he made a whole series of discoveries – more than a dozen in all. In 1770 he was elected an Academician, and in the following year he became Astronomer to the Navy. Louis XV of France nicknamed him 'the Ferret of Comets'.

He was no mathematician, and was quite incapable of computing orbits. Much of this work was done for him by the French aristocrat Bochart de Saron, who met his end during the Terror of 1794; indeed, de Saron's last calculations were made in prison while he was awaiting execution, and these enabled Messier to recover a comet which he had found in the previous year. Messier was able to send de Saron news of this success only a day or two before the sentence was carried out.

Others who collaborated with Messier were Montaigne, a Limoges astronomer, and Pierre Méchain, who never failed to notify Messier immediately he discovered a new comet (in all, he found eight). Much less creditable was the career of the Chevalier Jean Auguste d'Angos, one of the Knights of Malta and a captain in the Navarre regiment of the French Army. D'Angos was one of science's charlatans. He dabbled in both astronomy and chemistry, with results which were uniformly disastrous. He was invited to set up an observatory in the Palace of Valetta, and one of his chemical experiments there gutted not only the Observatory, but most of the Palace as well. He claimed to have discovered three comets, but Messier was unable to confirm any of them, and later investigations showed the Chevalier's claim to be utterly bogus. He died in 1833, presumably regretted by none.

KOHOUTEK'S COMET, *January 24, 1974. This comet was a visual disappointment, as it never became brilliant; but it was scientifically valuable, and was the first comet to be studied from space (by the Skylab astronauts). Its period is approximately 75,000 years. Photograph, Royal Greenwich Observatory, Herstmonceux.*

Meanwhile Messier continued his work, but he was constantly held up by the faint, blurred objects which we call star-clusters and nebulæ. These frankly annoyed him. A dim nebula or cluster looks exactly like a comet, particularly with the low-powered telescopes which Messier used (his largest, apparently, had an aperture of less than 21 centimetres, and was probably of poor optical quality). Eventually he made up his mind to catalogue them as 'objects to avoid', and in 1781 he produced a list which is still in use. The original Messier catalogue contained 104 objects. They were of various kinds: open clusters, globular clusters, gaseous nebulæ, planetary nebulæ, galaxies and even one supernova remnant (the Crab Nebula). Messier made no attempt to divide them into different categories, for the excellent reason that he was profoundly uninterested in them. All he wanted to do was to weed them out while searching for his beloved comets.

By a stroke of irony, it is by his catalogue that Messier is remembered today. Thus the famous Orion Nebula is Messier 42, or M.42; the Andromeda Spiral is M.31; the Pleiades cluster is M.45; the Crab M.1, and so on.

Few of Messier's comets were of special interest – which was not his fault! An exception was the comet which he discovered on July 14, 1770. It was visible with the naked eye, and its apparent diameter was five times greater than that of the full moon. This was not because it was particularly large, but because it was exceptionally close; it passed within 2,000,000 kilometres of the Earth, which is still a cometary record. The orbit was worked out by the Finnish mathematician Anders Lexell, who gave it an orbital period of five and a half years. Unfortunately Lexell did not publish his results until 1778, by which time the whole situation had changed. Before its next return to perihelion, the comet had passed near the giant planet Jupiter, and its orbit had been so changed that it no longer passes within range of us; we can hardly hope to see it again. Though Messier had been the first to see it, it is still remembered as Lexell's Comet.

Messier died in 1817. By then another great comet-hunter had come to the fore: Jean Louis Pons, born in 1761. Pons went to the Marseilles Observatory in 1789, but not as observer or director; he was given the job of doorman and general caretaker. His interest in astronomy was so keen that the observers gave him all the help they could, and Pons went on to outshine them all. He was made assistant astronomer in 1813, and in 1819 left Marseilles to become Director of

MRKOS' COMET OF 1957, *photographed by E. A. Whitaker. This was the second reasonably bright comet of 1957; it was quite conspicuous to the naked eye for a few days in the autumn. It was discovered by the Czech observer Antonín Mrkos; the first observations from Britain were secured independently by a fifteen-year-old amateur, Clive Hare.*

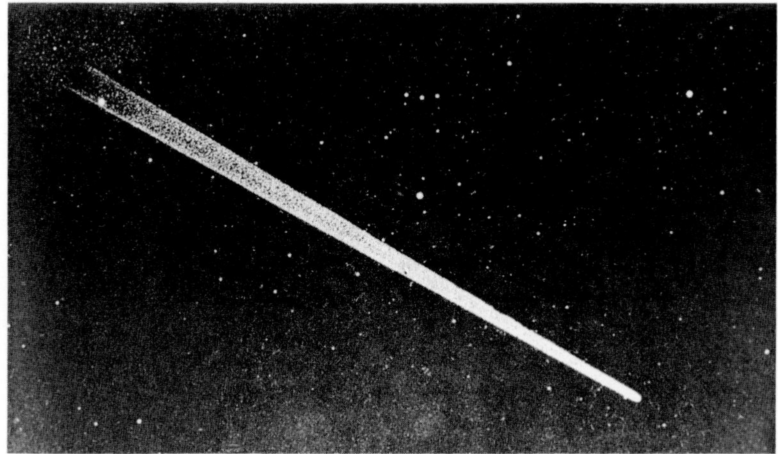

THE GREAT COMET OF 1811. *This woodcut shows one of the most spectacular comets ever observed.*

the new observatory at Marlia, near Lucca. He ended his career as Director of the Museum Observatory in Florence, where he died in 1831.

In its way, Pons' life-story is as romantic as any in science. There can be no other case of an observatory doorman rising to the rank of observatory director and earning an international reputation.

Pons discovered thirty-six comets. One of them, seen in 1818, proved to be of exceptional interest, because when its orbit was computed by Johann Encke, at that time assistant at the Göttingen Observatory, it was found to move round the Sun in a period of only 3.3 years. Working backwards, much as Halley had done, Encke decided that the comet was identical with those found in 1786 by Méchain, 1792 by Caroline Herschel, and 1805 by Pons himself (independently, by the French observer Thulis). In fact, Pons had discovered the same comet twice. Since then it has been seen regularly every 3·3 years, and it has been observed at over fifty returns. (Today, the powerful modern equipment can record it even when it is at its furthest distance from the Sun, at a distance of around 610,000,000 kilometres – well into the asteroid belt.) Fittingly, the comet has been named in Encke's honour.

This is one of a few cases in which a comet has been named after the mathematician who first computed its orbit (Halley's and Lexell's comets are other examples), but generally it is the discoverer who is commemorated; thus the famous comet of 1973, Kohoutek's, was named after Dr. Lubos Kohoutek of Hamburg Observatory, who first recorded it. There are also special numbering systems. Each time a comet is found, it is given a year and a letter; the first comet to be detected in 1981 was 1981a, the second 1981b, and so on. A year followed by a Roman numeral indicates that the comet has been listed according to its time of perihelion; the first comet to reach perihelion in 1981 was 1981 I, the second 1981 II, and so on.

Encke's was the first short-period comet, but nowadays over a hundred are known. All are faint, but their movements are so well worked out that we always know when and where to expect them. Yet they are of very small mass, and are strongly influenced by the pulls of the planets, particularly Jupiter; many short-period comets reach their aphelion points at about the mean distance of Jupiter, so making up

ALCOCK'S FIRST COMET OF 1959. *Drawing by G. E. D. Alcock.*

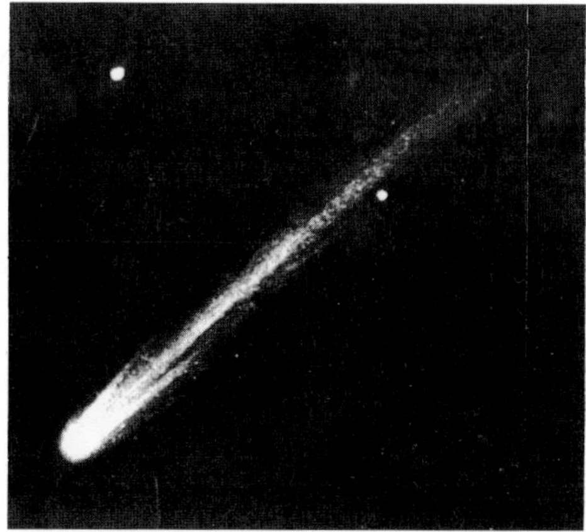

ALCOCK'S SECOND COMET OF 1959. *Drawing by G. E. D. Alcock.*

CHÉSEAUX' COMET OF 1744. *(Left) This was not one of the most brilliant of comets, but it was extremely spectacular, and was unique in developing a seven-fold tail. Ordinarily a great comet does not have more than one or two major tails, and nothing similar to Chéseaux' Comet has been seen since. The drawing given here is taken from an old sketch made at the time.*

BENNETT'S COMET, *(opposite) photographed by its discoverer, the South African amateur astronomer, J. Bennett.*

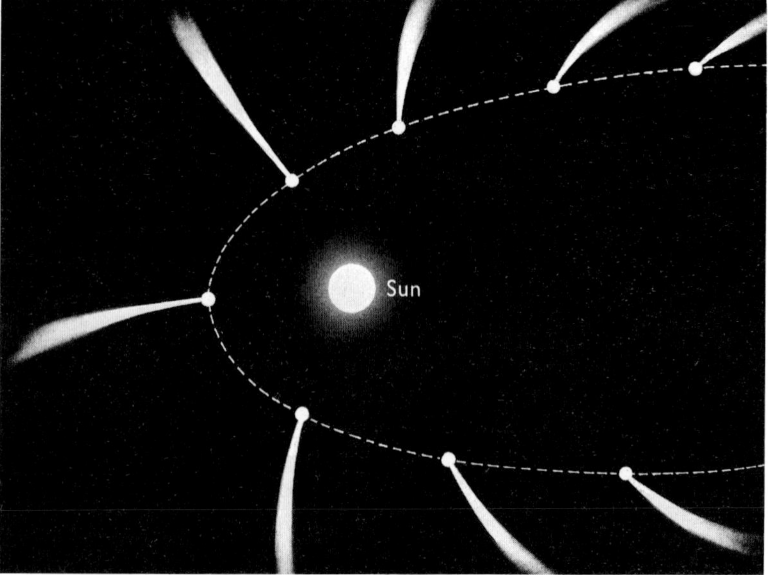

BEHAVIOUR OF A COMET'S TAIL. *(Right) When a comet is remote from the Sun its tail does not develop. As the comet nears perihelion, the tail may become very long, and it points more or less away from the Sun. When the comet has passed perihelion, it recedes from the Sun tail-first. A comet's tail always points approximately away from the Sun.*

what is termed Jupiter's comet family. Their orbits may be steeply inclined, and some of them move round the Sun in a wrong-way or retrograde direction. Comets have aptly been termed 'wanderers in space'.

Of the periodical comets, which return regularly to the neighbourhood of the Sun, only Halley's is bright. All the other brilliant comets seen over the ages have periods so long that they amount to hundreds, thousands or even millions of years, so that we cannot predict them. Such were the magnificent comets of 1811, 1843, 1882 and 1910 (the so-called Daylight Comet, which appeared a few weeks before Halley's). De Chéseaux' Comet of 1744 had several tails, as we have noted; Donati's of 1858 had two main tails, one of which was curved like a scimitar. A curved tail is made up of small 'dusty' particles, while a straight tail is composed of gas.

Even a large comet will be tailless when a long way from the Sun. As it sweeps inward, the ices in its head or coma begin to evaporate; a tail is formed, and may become very long indeed – that of the Great Comet of 1843 attained a record length of 320,000,000 kilometres, more than twice as great as the distance between the Earth and the Sun.

One might assume that a comet travels head-first, with its tail streaming out behind it, but this is not always true. When approaching the Sun the comet does move more or less head-first, but after it has passed perihelion, and has begun to move outwards, it is the tail which takes the lead. The old explanation for this curious behaviour was that it was due to light-pressure. Light certainly does exert a small force, and in 1900 the Russian physicist Theodor Bredikhine suggested that it was this which drove comets' tails outwards, but nowadays it is known that this is not the main cause. The Sun sends out streams of electrified particles, making up the *solar wind*, and it is these which act upon the 'dust-grains' in the comet's tail, each of which is a tiny fraction of a millimetre across. If a comet loses some of its mass by evaporation each time it passes through perihelion, it must be short-lived on the cosmical scale. Several short-period comets which used to orbit the Sun during the nineteenth century have now disappeared, and this brings me on to the close link between comets and meteor streams.

COMET IKEYA-SEKI, 1965, OCTOBER, *photographed by D. Andrews at the Boyden Observatory in Africa.*

20

Celestial Fireworks

IN 1846 the Danish astronomer Theodor Brorsen discovered a periodical comet. It was not bright enough to be seen with the naked eye, but it was a conspicuous telescopic object. The period proved to be 5½ years, and it was seen at almost every return until 1879, when it 'went missing' and has never been seen again. It is not the only casualty. Westphal's Comet, which had a much longer period (62 years) faded out during its return in 1913, while in 1926 Ensor's Comet became diffuse as it approached perihelion and then vanished completely. Of course, one must not be too hasty. Holmes' Comet, first seen in 1892, was lost for more than half a century after its return in 1906, and was re-discovered only in 1964. But comets do have short lives, and the most startling story of all is that of Biela's Comet.

It was found in 1826, almost simultaneously by Wilhelm von Biela, an officer in the Austrian Army, and Jean Gambart, assistant at the Marseilles Observatory. It proved to have a period of 6¾ years. It was seen again in 1832; missed in 1838 because it was badly placed, and picked up once more in 1845. It then created an astronomical sensation by dividing into two parts, so that the appearance was likened to two comets sailing through space in convoy. In 1852 the twins came back on schedule, rather more widely separated. In 1858 conditions were again poor, but in 1866 they should have been good, and astronomers waited eagerly to see what would happen.

The answer was – nothing. The calculations were right, but in spite of the most intensive searches the comet did not appear; it had vanished as completely as the hunter of the Snark. Something had happened to it.

The next return was due in 1872. Again the comet was absent, but this time a spectacular meteor shower was seen, coming from the point where the comet ought to have been. Clearly the comet itself was dead, and the meteors marked its funeral pyre!

Actually the full explanation is not so straightforward as this, but it is certainly true that there is a close link between comets and meteors. The fate of Biela's Comet proved it, and for years afterwards meteors were still recorded every November whenever the comet should have been. Even today we see a few Bieliid meteors, though the shower has become very feeble.

As a comet moves through space, it leaves a trail of débris behind it, and this débris is gradually spread out all along the comet's orbit. This explains why comets are so short-lived, though obviously those which come back to perihelion only at relatively long intervals last longer than those of short period.

When the Earth passes through a shoal of meteors, the result is a shower of shooting-stars. In fact, a shooting-star is due to a tiny particle dashing into the upper air at a speed of anything up to seventy-two kilometres per second, and burning away (what we actually see is not the tiny particle itself, but the effects which it produces in its last

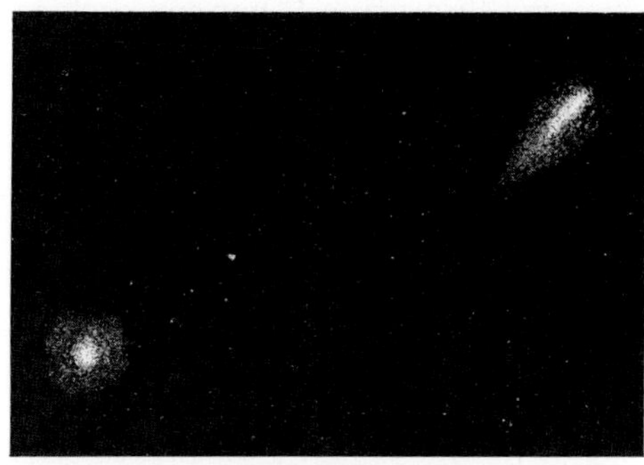

DRAWING OF BIELA'S COMET, *made in 1845. At this time the comet had separated into two distinct parts – the first time that such a phenomenon had been observed. In view of later events, it seems that this marked the beginning of the complete disintegration of the comet.*

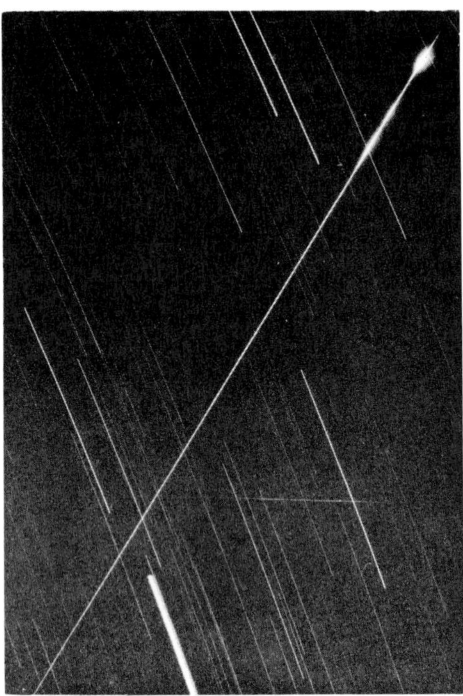

PERSEID METEOR, *photographed by H. B. Ridley on August 15, 1958.*

LEONID METEORS, *photographed from Arizona during the great show of November 17, 1966. Photograph by D. R. McLean.*

moments). The shower effect is simply one of perspective, just as the parallel lanes of a motorway will seem to meet at a point in the distance. The meteors in any particular shower are moving through space in parallel paths, so that they seem to issue from a definite point or *radiant*. The radiant of the Bieliid shower is in the constellation of Andromeda.

Not all meteors are members of showers. The non-shower or *sporadic* meteors may appear from any direction at any moment, and normally amount to about eighty per cent of the total, but it is the showers which are the most spectacular, and several of the annual displays can be linked with known comets.

For instance there are the Perseids, which last from about July 25 to August 17 each year, with a maximum around August 12. The comet in this case is a fairly dim one – Swift-Tuttle, discovered in 1862, with a period of about 120 years; in 1862 it was just visible with the naked eye. The fact that the meteors are spread out all round the orbit, producing an annual shower, shows that the meteor stream must be an old one, and it is very reliable. In 1980 the Z.H.R. or Zenithal Hourly Rate reached 120 at peak activity. (The Z.H.R. is the number of meteors which would be seen with the naked eye by any observer under ideal conditions, with the radiant at the zenith or overhead point. In practice these ideal conditions are almost never attained, but the Z.H.R. number is a good general guide.) Stare up into a dark, clear sky for a few minutes at any time during the first fortnight of August, and you will be unlucky not to see at least one or two shooting-stars. Yet the Perseids of 1981 and 1982 were no more than normally plentiful. Comet Swift-Tuttle may have come and gone unseen.

The Leonids of November 17 each year are quite different. This time the meteors are not spread uniformly around the orbit of the parent comet, Tempel-Swift. The comet itself has a period of 33⅓ years, taking it from the path of the Earth out to beyond Uranus, and we see a brilliant display only when we meet the main swarm. Brilliant displays were recorded in 903 (the 'Year of the Stars') and 1366, for example. More recently there have been excellent showers in 1799, 1833 and 1866, when it was said that the meteors 'rained down like snowflakes'. After that, the orbit was perturbed by the action of the giant planets, and the expected showers of 1899 and 1933 did not materialize. The Leonids were back in strength in 1966, as I well remember; I had appeared on BBC television and asked for volunteer

PERSEID METEOR AND SATELLITE TRAIL, *(right) August 12/13, 1974; K. Kennedy, Dundee. The satellite trail is the fainter of the two.*

observers during the night of maximum, but over England very few Leonids were seen. The shower was at its most intense during daylight over Europe, and lasted for only a few hours, but viewers in other parts of the world were treated to a dazzling display of cosmic pyrotechnics. The fact that the shower was so brief shows that the meteors are still bunched up. Of course, a few Leonids are seen every year around November 17, but the next major display is likely to be deferred until 1999, though one never knows.

Many other comet-meteor associations have been established, but sometimes a shower may have no known cometary parent. Such is that of early January, known as the Quadrantid shower because the radiant lies in one of Bode's rejected constellations: Quadrans, the Quadrant, now included in the Great Bear.

In the 1830s a Swiss astronomer named Wartmann noticed that many meteors were coming from the Quadrans area at about January 3 each year, and the shower was given in the first general catalogue of meteor radiants, drawn up in 1839 by Jacques Quetelet, Director of the Brussels Observatory. The annual display lasts for only a few hours, and the orbit of the meteor stream is tilted at a steep angle to that of the Earth: seventy-one degrees. The most interesting point about the Quadrantids is that the shower may be a temporary one so far as we are concerned. It was not noted before the 1830s, and the orbit is being slowly but steadily altered by the gravitational action of Jupiter, so that in five hundred years from now the swarm will miss the Earth completely – and we will have no more Quadrantids. It has been suggested that the parent comet used to have a period of 5½ years, but faded away between 1300 and 1700 years ago.

How high are meteors? This may seem a difficult problem to answer, but it was solved by the method of triangulation, first put into practice as long ago as 1798 by two German students named Brandes and Benzenberg. If a meteor is recorded by two different observers at two different sites, it will seem to pass against a different background of stars, and this means that the altitude can be worked out – provided, of course, that the distance between the two observers is known. Most meteors burn away by the time they have penetrated to eighty kilometres above ground level. Still smaller particles, known as micrometeorites, cannot produce luminous effects, and they too finish their earthward journey in the form of very fine 'dust'.

There are also larger pieces of cosmical débris, which can survive the complete drop to the ground and land more or less intact. They are then termed meteorites, and may produce craters; for instance the famous Arizona Crater, not so very far from the town of Flagstaff, was certainly produced by an object which hit the desert about 22,000 years ago, and other similar craters are known, notably Wolf Creek Crater in Australia. But for many years the idea of 'stones from the sky' was regarded as absurd. The fall which caused a change in official opinion was that of 1803.

On April 26, at one o'clock in the afternoon, the inhabitants of the little French village of L'Aigle were disturbed by a strange sound. It was not unlike a violent roll of thunder; yet the skies were cloudless, and there was no sign of a storm. As the villagers rushed out of their houses in alarm, they caught sight of an immense ball of fire racing across the sky, and as it vanished there came a series of explosions which could be heard over a radius of eighty kilometres. A few minutes

WEST'S COMET OF 1976, *photographed on March 9, 1976, 0500 GMT; P. Doherty, Stoke-on-Trent. West's Comet was a fine naked-eye object for several successive mornings.*

WOLF CREEK CRATER *in Australia, which is almost certainly meteoritic. Photograph by G. J. H. McCall.*

later a great number of stones fell to the ground, each of which landed at a speed great enough for it to bury itself in the earth.

The report reached Paris, and a leading French astronomer, Jean-Baptiste Biot, was sent to investigate. He examined the strange stones, and came to the conclusion that they really had arrived from space. Once this had been established, other meteorites were identified. One was the large stone which had fallen at Ensisheim, in Alsace, in 1492, and is now on display at Ensisheim Church. The famous Sacred Stone in the Holy City of Mecca is also a meteorite.

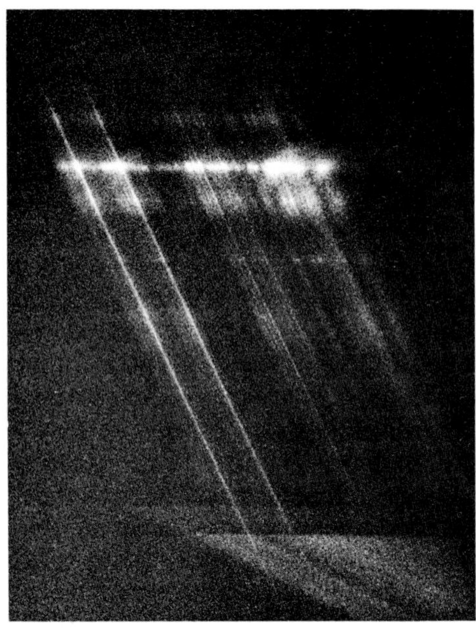

SPECTRUM OF A TAURID METEOR, *photographed by H. B. Ridley on October 29, 1954. This was the first meteor spectrum obtained in the United Kingdom.*

Meteorites range in size from pebbles up to huge blocks weighing tons. The present holder of the heavyweight record is still lying where it fell, in prehistoric times, at Hoba West, near Grootfontein in Southern Africa. I doubt whether anyone will try to run away with it, since it weighs well over sixty tons. The largest meteorite on display is the Ahnighito, found in Greenland by Robert Peary in 1897; it is now at the Hayden Planetarium in New York.

The most famous fall of modern times was that of June 30, 1908, in the Tunguska region of Siberia. As seen from the town of Kansk, 600 kilometres away, the descending object was said to outshine the Sun, and detonations were heard over a distance of 1000 kilometres. Pine-trees were blown flat over a wide area, and it was just as well that the region was uninhabited. The first expedition to the site, led by Leonid Kulik, did not arrive until 1927; no meteoritic fragments were found, and it is now generally believed that the object was the icy nucleus of a small comet, which presumably evaporated during the final descent and impact. A second Siberian fall occurred on February 12, 1948, in the Sikhote-Alin area; it was widely observed, and over a hundred craters were located.

I must mention a much less spectacular fall – that at Barwell, in

THE HOBA WEST METEORITE IN AFRICA. *The weight exceeds 61 tonnes.*

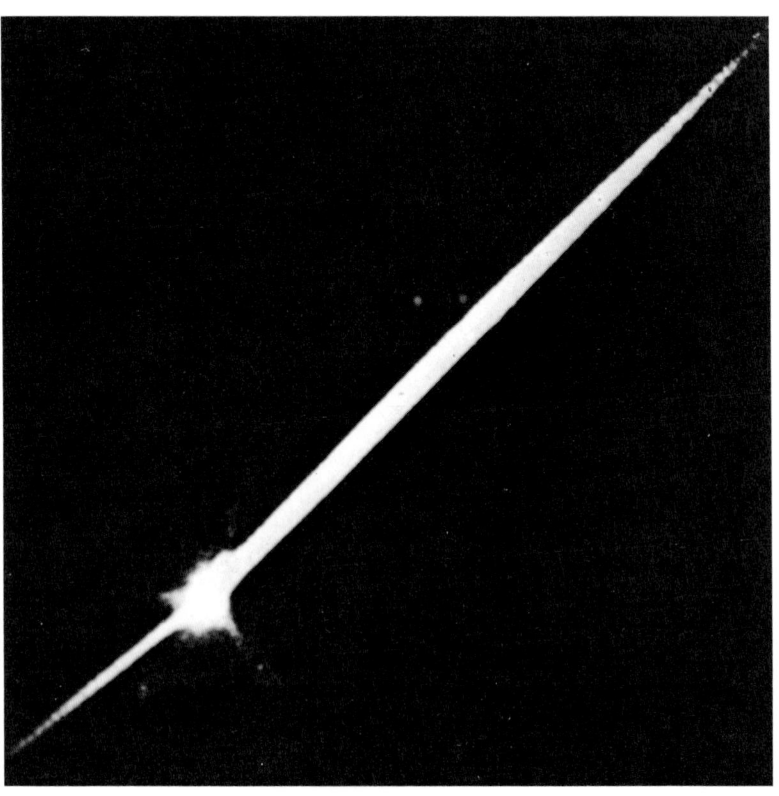

EXPLODING ANDROMEDID METEOR, *(right) photographed by H. Butler in 1895.*

Leicestershire, on Christmas Eve 1965; the meteorite shot over England, and scattered various fragments – the original diameter may have been a metre or two. The last British meteorite, to date, was that of April 25 1969. Fragments were found at Bovedy in Northern Ireland, but probably the main mass fell in the sea. It too flashed over England – and I missed it by a matter of seconds! I had just gone indoors after a period of observing.

There is no known case of anyone having been killed by a meteorite. The only casualty has been a dog, which was struck by a meteorite which came down in Egypt. But it must be stressed that a meteorite is not simply a large meteor. The two classes of objects are completely different. Meteorites are not associated with comets; they are made up either of stone or iron, or a mixture of the two, and they are much more closely associated with asteroids. Actually, there may be no difference between a large meteorite and a small asteroid.

Before taking our temporary leave of the Solar System, let me make brief mention of the strange and often beautiful glows in the sky which have been seen throughout history, and which were well known by the end of the eighteenth century. Auroræ, or polar lights, are due to electrified particles sent out from the Sun; they cascade down into the upper atmosphere, and produce splendid displays of coloured lights – aurora borealis in the northern hemisphere, aurora australis in the southern. The particles are drawn to the magnetic poles of the Earth, so that auroræ are best seen in high latitudes (though it is on record that an aurora was once seen from Singapore, only a few degrees from the equator). The first display of northern lights was described by Pierre Gassendi from France in 1621, but of course auroræ had been seen much earlier, and from places such as Iceland and North Norway they are seen most of the time during darkness. It is also recorded that on one occasion the Roman emperor Claudius sent his fire-engines to quench what he took to be a fire in Ostia, the port of Rome, only to find that he had been misled by a brilliant red aurora. The first record of the southern lights was made by Captain Cook in 1773. In general, auroræ extend from 100 to 1000 kilometres above ground level.

PIECE OF BARWELL METEORITE, 1965. *Photograph by Patrick Moore, who discovered this fragment in a field near Barwell village.*

THE ZODIACAL LIGHT, *photographed at Chocaltaya, Bolivia, at an altitude of 5182 m, with the Sun 18 degrees below the horizon. The photograph was taken by D. E. Blackwell and M. F. Ingham. A single lens camera was used (aperture f/18, focal length 12·7-cm) with an exposure of 10 minutes, which accounts for the star trails on the plate.*

Then there is the Zodiacal Light, first reported by Giovanni Cassini in 1683; Cassini correctly suggested that it must be caused by sunlight reflected from interplanetary 'dust' spread along the main plane of the Solar System. It is cone-shaped, and extends away from the Sun, so that it is usually observable for only a short period either after the Sun has set or before it rises. The diameters of the Zodiacal Light particles are around 0·1 to 0·2 microns; since a micron is one-millionth of a millimetre, it is clear that the particles are very tiny indeed. A much more elusive glow, exactly opposite the Sun in the sky, is due to the same cause. It is known as the Gegenschein or Counterglow, and was first described by Esprit Pézénas in 1731. It is very faint, and few people will see it without making a deliberate search.

It is true to say that up to the beginning of the eighteenth century, astronomers were concerned mainly with our near neighbours, the bodies of the Solar System, using the stars as a convenient background. There were exceptions; Herschel, of course, was one. But as the years passed by, more and more attention was given to the stars themselves. We are coming on to the era of spectroscopy and photography, as well as to the construction of telescopes powerful enough to extend Man's surveys millions of light-years into space.

21

Birr Castle

NOW and then, the history of science leads to an episode which, in modern parlance, may be termed a 'one-off'. Such was the story of Birr Castle, which provides a link between the old and the new phases of astronomy. Though the main part of it did not begin until the 1840s, I think it will be best to describe it here, because it is in a class of its own.

The main character was the third Earl of Rosse, an Irish landowner, who had been born in 1800 and was educated first at Trinity College and then at Oxford. (Before succeeding to the title, he was known as Lord Oxmantown.) In 1822 – the year of William Herschel's death – he graduated with first-class honours in mathematics, but he felt bound to play an active part in public life, and in 1823 he entered Parliament. As a parliamentarian he was highly regarded, but his political career ended in 1834; he realized that he had to make a choice, and science was his main interest.

The family seat was Birr Castle in what is now County Offaly, not many miles from Athlone in Central Ireland. The climate is far from good, and as a site for an observatory it was singularly unpromising, but Lord Rosse had no choice; Birr was his home, and he had too many responsibilities to leave it even if he had wished to do so. He could have bought a telescope, but he preferred to make one for himself. He made some experiments with refractors, but never very seriously; he concentrated upon reflectors, and he taught himself the art of mirror-making. By 1838 he had a 91-centimetre reflector in full working order. It was then the largest telescope in the world, and it was built upon the conventional Newtonian pattern, with a main mirror and a flat secon-

BIRR CASTLE

OBSERVING WITH THE 183-CM ROSSE REFLECTOR. *(Right) From a painting now in Birr Castle. It shows the great telescope in its heyday (around 1850) with the third Earl of Rosse preparing to begin a night's observing.*

dary. Lord Rosse considered the front-view Herschelian arrangement, but rejected it, in which he was wise.

The 91-centimetre was a success, and useful observations were made with it, but Lord Rosse was not satisfied. Up to that time Herschel's giant reflector had been the largest ever attempted; Lord Rosse felt that he could do better. He set to work, and planned a telescope which would have a mirror no less than 183 centimetres across.

THE 183-CM ROSSE REFLECTOR AT BIRR CASTLE. *An engraving of this great telescope, for many years the largest in the world.*

Obviously the main mirror was the most important part of the instrument, and at that time it was quite impossible to use glass; techniques were hopelessly inadequate, and so the first step was to cast a huge metal disk from which the mirror would be made. Lord Rosse had no skilled helpers, but he trained labourers from his estate, and built a furnace to melt the mixture of copper and tin. By 1842 all was ready. We have an eyewitness acount of the scene, written by Dr. Romney Robinson, the famous and somewhat forbidding Director of the Armagh Observatory in Northern Ireland (who, incidentally, caused the main Belfast-Dublin railway to be diverted by several kilometres, because he feared that passing trains would shake his telescopes). Robinson wrote:

> 'The sublime beauty can never be forgotten by those who were so fortunate as to be present. Above, the sky, crowded with stars and illuminated by a most brilliant moon, seemed to look down auspiciously on their work. Below, the furnaces poured out huge columns of nearly monochromatic yellow flame, and the ignited crucibles during their passage through the air were fountains of red light, producing on the towers of the castle and the foliage of the trees such accidents of colour and shade as might almost transport fancy to the planets of a contrasted double star.'

The casting was slowly cooled, and then put on to a special machine invented by Lord Rosse for preliminary grinding. But it would have been too much to hope for prompt success, and in some manner or other the mirror was broken. Immediate preparations were made for a second casting. This time all went well, and at last the mirror itself was ready.

Mounting such a huge telescope presented novel problems. The 91-centimetre at Birr had been set up on a mounting based upon Herschel's design, but to attempt to mount the 183-centimetre in the same way would have been sheer folly. The mass was too great; and when using high magnifications, with small fields of view, a really steady mounting is essential. A lesser man would have tried to build something fully rotatable. Lord Rosse did not. The tube of the telescope was nearly eighteen metres long, and the only possible course was to mount it between two massive stone walls, pivoting it at the bottom of the tube. The main disadvantage was that the telescope could reach only a strip of the sky in a north-south direction, and a star could be followed for only an hour or so as it passed over the meridian; both before and afterwards it would be hidden by the walls.

Lord Rosse was sensible enough to accept this admittedly serious limitation, and the strange-looking 'observatory' was built. It was unlike any made either before or since. The telescope tube itself was 2·4 metres across, made from 2·5-centimetre thick deal staves hooped with iron clamp-rings; it was moved up and down by a strong cable chain fixed to the top of the telescope, and from side to side by a wheel turned by an assistant.

It was said that anyone who set out to use the telescope had to be not only an astronomer, but also an expert mountaineer! True, the system of stairs and galleries must have been somewhat terrifying, but Romney Robinson wrote that 'though it is rather startling to a person who finds himself suspended over a chasm sixty feet deep, without more than a speculative acquaintance with the properties of trussed beams, all is perfectly safe'. There was no accident during the whole of the sixty years that the telescope was in operation. Of course the whole arrangement was clumsy, but at the time it was the best that could be done.

THE WHIRLPOOL GALAXY, M.51 IN CANES VENATICI. *The spiral nature of some external galaxies was first revealed by the Rosse 183-cm reflector. Photograph taken with the 508-cm Hale reflector at Palomar.*

The first practical tests were made in 1845, and at once it became clear that the telescope was every bit as effective as Lord Rosse had hoped. It was turned towards some of the star-clusters and nebulæ that Messier and others had listed, and before long the first really important discovery was made.

As we have seen, Messier's catalogue contained objects of all kinds, ranging from open clusters such as the Pleiades to gaseous nebulæ such as the Sword of Orion. One object, his No. 51 in the little constellation of the Hunting Dogs, near the Great Bear, had been classed as a nebula: but when Lord Rosse looked at it with the 183-centimetre, he saw that it was spiral in shape, like a Catherine wheel. This was completely unexpected. From 1847, when regular observing began (following the end of the worst of Ireland's troubles of that period), more and more spirals came into view, though other Messier objects looked more like patches of gas or straightforward star-clusters.

Lord Rosse could not hope to measure the distances of the spirals, but he knew that they must be very remote, and he believed – correctly – that they were rotating. Like Herschel, he wondered whether some of them could be far beyond our Galaxy, and this too was later found to be the case. The spirals are so far away that their light takes millions or even hundreds or thousands of millions of years to reach us. At the time when Lord Rosse began his work, only the 183-centimetre 'Leviathan' was powerful enough to show the spiral forms. Any astronomer who wanted to see them had to come to Birr Castle.

Many other objects came under study, including Messier's No. 1; it was Lord Rosse who gave it the nickname of the 'Crab' by which it has been known ever since. Neither were the Moon and planets neglected. Work went on all the time, though inevitably many nights were lost because of cloud.

Lord Rosse himself became a skilled observer, and fortunately he believed in sharing his knowledge with others. All who came to Birr were welcomed, and it has been said that nobody who asked Lord Rosse for help and advice ever went away unsatisfied. Assistant astronomers were engaged; one of them, Robert Ball, afterwards became Astronomer Royal for Ireland, while another, John Louis Emil Dreyer of Denmark, was responsible for drawing up the important New General Catalogue of clusters and nebulæ ('N.G.C.') which officially superseded Messier's, though the M numbers are always used together with those of the N.G.C.

It has sometimes been suggested that the 183-centimetre reflector never came up to expectations. This is demonstrably untrue. It could never be used for photography; it could not easily be guided by mechanical means; its view of the sky was limited, and it was cumbersome to use – yet it did all that its maker had hoped, and Lord Rosse achieved his ambition of seeing further into space than had ever been possible before. The discovery of the spirals would alone have made the project worthwhile.

Lord Rosse died in 1867, and was succeeded by his son, who also became a well-known astronomer. The fourth Earl was interested mainly in measuring the tiny quantity of heat sent to us from the Moon, and in this he was successful. He built special equipment, and his results were later found to be very accurate. Work with the 183-centimetre Leviathan went on until the end of the century, but as time went by the great reflector was overtaken by the modern-type instruments which were so much easier to handle, and had glass lenses or mirrors. The Leviathan's career came to an end with the death of the fourth Earl in 1908. The mirror was taken away to the Science Museum in London, and the main mounting was dismantled in 1926.

Go to Birr Castle today, and you will still see the massive stone walls in position. The tube of the reflector has been set up between them, and a museum has been erected on the site. There are pieces of equipment, drawings, diagrams and historical exhibits of all kinds. The achievement has been aptly summed up by Sir Bernard Lovell:

> 'The observations of Lord Rosse mark a vital stage between those of Herschel and the discoveries of the modern age, for he found that some of the nebulæ present a spiral structure. The drawings made by Lord Rosse of these distant objects remain today as a remarkable tribute to his observational skill. Later in the century photographic techniques were to be applied to the observations with large telescopes, but a comparison of his drawings with the modern photographs reveals the astonishing precision of his work. For example, his drawing of the Whirlpool Galaxy, M.51 in Canes Venatici, made in 1848, appears to be almost identical with the photograph taken with the 200-inch telescope on Palomar in California a century later.'

With no professional help, Lord Rosse built a telescope far larger and more powerful than any of its predecessors, and used it to make fundamental discoveries. Such a thing has never happened before in the history of science; neither can it ever happen again.

SPIRAL GALAXY, *drawn by Lord Rosse.*

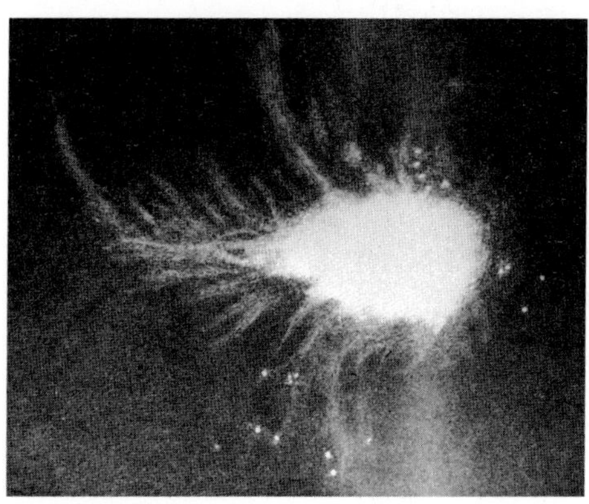

THE CRAB NEBULA *as drawn by Lord Rosse – who also gave it its nickname.*

22

Stars of the Far South

ASTRONOMY began in the northern hemisphere. This does not mean that no southern stars could be seen; of course they could; for instance, Sirius lies well to the south of the celestial equator. But some of the really glorious constellations, notably the Southern Cross, can never be seen from Europe or most of the United States, and inevitably the stars of the far south were badly neglected. Halley's expedition to St. Helena was in the nature of a preliminary reconnaissance, and it was not until the eighteenth century that really detailed observations were made. The chosen site was South Africa.

Actually the story began in 1685, when a Jesuit priest, Father Guy Tachard, stopped off at the Cape of Good Hope on his way to Siam. He set up a temporary observatory, mainly for navigational purposes, and noted that the brightest star of the Cross, Acrux, is double; but the first true astronomer to make use of the South African conditions was Nicolas Louis de La Caille, later the Abbé de La Caille, who was born near Rheims in 1713 and became an expert astronomer as well as a surveyor. He arrived at the Cape in 1751, and remained for two years, drawing up a catalogue of 10,000 stars. This is all the more remarkable because his telescope was a refractor with an object-glass less than two centimetres across.

The origins of Australian astronomy date back to 1788, when William Dawes landed with his First Fleet and set up a temporary observatory. In 1821 Sir Thomas Brisbane, the Governor of New South Wales, established a timber observatory at Paramatta which operated for some time, and later on came major observatories in Sydney, Perth and Melbourne, but all in all it is fair to say that the most important of the earlier observatories were South African.

In 1820 the English authorities decided to set up a fully-fledged observatory at the Cape, where conditions were expected to be good, and the whole of the southern sky could be seen. The first Cape Astronomer was the Rev. Fearon Fallows, a native of Cumberland who was known to be an excellent mathematician. Fallows accepted the post, but the task facing him upon his arrival in 1821 was a daunting one.

As a site for the new observatory, he selected an area between six and seven kilometres from Cape Town. The area was very unlike its aspect today. Red dust swirled along the rough track-road to the town; the land was marshy; there were hippopotami in the rivers and snakes everywhere. Living conditions were primitive, and Fallows had few assistants – one of whom, incidentally, was his wife; she made her mark in 1830, when she discovered a comet in the south polar constellation of Octans, the Octant. Fallows did his best, and by the time he died, in 1831, the Observatory was at least in working order, but it can hardly have been an inviting place.

Fallows' successor was a Scot, Thomas Henderson. He detested the whole situation from the start; he nicknamed the area the 'Dismal

CRUX AUSTRALIS, *(opposite) the Southern Cross, photographed by Jack McBain (Bulawayo, Rhodesia) in 1975.*

Swamp', and he stayed for only two years, though while at the Cape he made an important series of measurements which led him on to work out the distance of the bright star Alpha Centauri. This story belongs to a later chapter, but Henderson's contributions must not be forgotten. No portrait of him exists, and quite recently I found out why. A distant descendant of his told me that Henderson had a pronounced squint, and stubbornly refused to let any artist paint his picture!

The next Cape Astronomer was Thomas Maclear, an Irishman. He qualified as a doctor, but turned to astronomy at an early age, and stayed at the Cape from 1834 to 1870. He arrived to find Henderson's 'Dismal Swamp'; when he eventually retired the Cape Observatory had become respected all over the world. Yet there can be little doubt that the true pioneer of the southern skies was John Herschel, son of William.

John Herschel was born at Slough in 1792. He graduated from Cambridge, and gave his aging father tremendous help in the latter part of Sir William's career as well as making important observations on his own account. He was knighted in 1831, and spent much of his time in completing and extending his father's survey of the northern sky. But this was not enough. William Herschel never saw the Southern Cross, the Centaur or the Clouds of Magellan; they remain below the European horizon. John set out to complete the survey by taking in the far south as well, and in January 1834 he arrived at the Cape, bringing with him the famous 46-centimetre reflector of 6 metres focal length (now to be seen at Flamsteed House in the Old Royal Observatory at Greenwich).

John Herschel was not officially connected with the Cape Observatory, but he was on the best possible terms with Maclear, who had reached the Cape ten days before him. Herschel set up his own observatory at Feldhausen, ten kilometres from Cape Town and by June he was ready to start work.

Before going any further, it is worth saying a little about the way in which the sky is seen from various parts of the Earth. There are many people who still believe that the Southern Cross cannot be seen without travelling across the equator. The facts are different. Remember, the declination of a star is its angular distance north or south of the celestial equator, just as latitude on the Earth is the angular distance north or south of the terrestrial equator. To find the limiting visibility of a star (or any other celestial object), simply take its declination away from 90. Thus Canopus, the second brightest star in the sky, has a declination of −53°, or 53 degrees south. 90 − 53 = 37. Therefore, Canopus rises as seen from any point on the Earth which is south of latitude 37 degrees north. It can be seen from Alexandria (latitude 31°N) but not from Athens (38°N), which was one of the reasons given by Aristotle in demonstrating that the world is round rather than flat.

The Southern Cross, the Magellanic Clouds and other superb objects are further south than Canopus, but they are excellently placed from South Africa (or from Australia), and John Herschel was breaking new ground. He spent four years at the Cape; he catalogued the southern stars, clusters and nebulæ; he discovered many new double stars; he made observations of Halley's Comet at its 1835 return, and was actually the last to see it until it came back once more at the return of 1910. When he came home, it took him many years to sort and arrange his Cape observations, and it was he, more than anyone else,

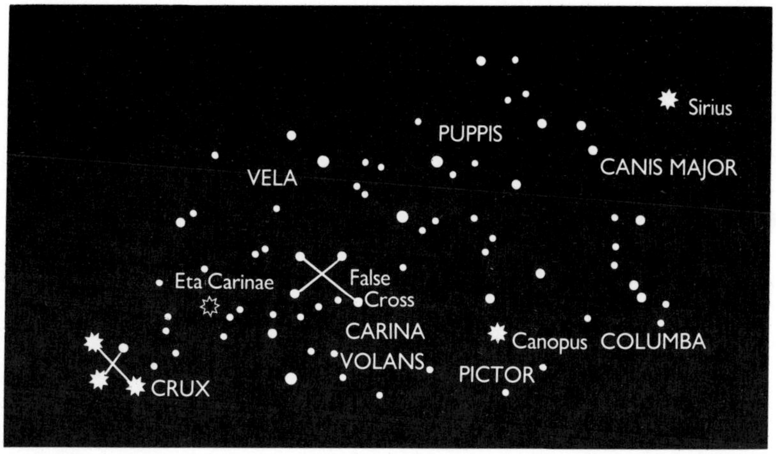

POSITION OF THE ERRATIC VARIABLE, ETA CARINAE.

who put southern-hemisphere astronomy on a really firm footing.

One star which caught his attention was Eta Carinæ, in the constellation of Carina (the Keel of the Ship), not far from Canopus. Eta Carinæ had long been known to be variable, and had been listed by most of the old observers such as La Caille, but it had never been brilliant. During Herschel's spell at the Cape it brightened up until it surpassed every star in the sky apart from Sirius. At that time it seems to have been the most luminous star in the whole of the Galaxy, and may have been about 6,000,000 times more powerful than the Sun, so that even from its immense distance of nearly 7000 light-years it was striking. Subsequently it faded slowly, and for more than a century now it has been too faint to be seen with the naked eye, though binoculars will show it. Through a telescope it appears as a kind of 'red blob', surrounded by nebulosity, quite unlike a normal star.

Eta Carinæ is in a class of its own, and even today we are by no means sure of its nature. One recent suggestion is that it is preparing to explode in a supernova outburst, though this may not happen for many thousands or even several millions of years. In any case, it is always worth watching, and at some time in the future it will probably put on a really spectacular display.

Nowadays nothing is left of John Herschel's observatory. The site is occupied by the Grove Primary School at Claremont, a suburb of Cape Town, but at least an obelisk has been erected in the exact place where the reflector once stood. It was from here that the first overall survey of the southern stars was carried out, due to the energy and skill of Sir John Herschel – a worthy son of a famous father.

23

How Far are the Stars?

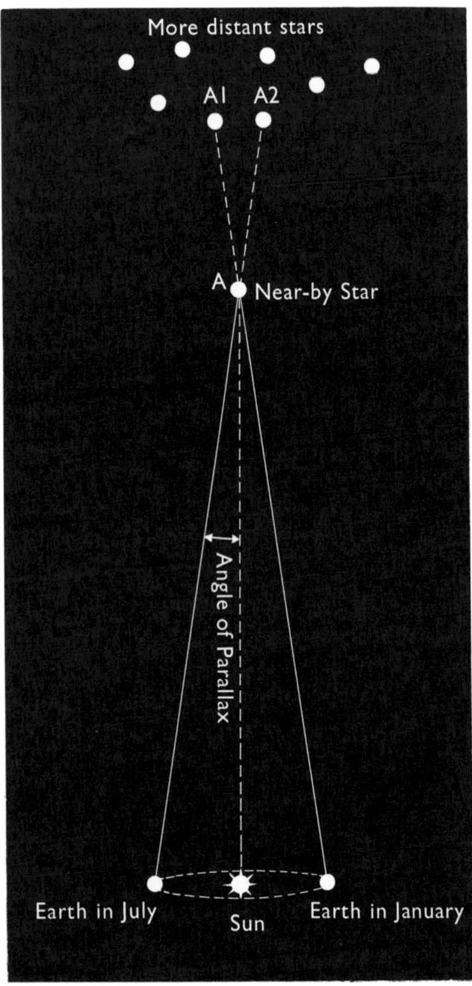

THE PARALLAX METHOD OF MEASURING THE DISTANCE OF A STAR. *The relatively close star, A, seems to shift in position over the course of a year; in January it is seen at A1, in July at A2. However, it must be remembered that the diagram has had to be drawn completely out of scale. The parallax shifts are extremely small even for the closest stars, and it is not surprising that even Herschel failed to detect them. This method of trigonometrical parallax is practicable only for stars in the neighbourhood of the Sun, since at greater distances of, say. a few hundred light-years, the shifts become so small that they are swamped by unavoidable errors in observation.*

WORKING out the scale of the Solar System had been difficult enough, but finding out the distances of the stars was much harder still. The first man to tackle the problem really seriously was the Rev. James Bradley, who succeeded Halley as Astronomer Royal in 1742 and held office until his death twenty years later.

Bradley, born in 1693, was educated at Oxford, and became Vicar of Bridstow in Monmouthshire. Then, in 1721, he went back to Oxford as Professor of Astronomy, and remained there until moving to Greenwich. His first aim was to improve existing star catalogues, and he did so very thoroughly; he gave the positions of over 60,000 stars so accurately that his catalogue is still valuable today.

Yet Bradley's greatest discovery was made in 1728, before he became Astronomer Royal. The story is an interesting one, and shows how careful an observer Bradley was. It was connected with his attempts to measure star-distances by the method of *parallax*.

Suppose you hold a ruler at arm's length, and line it up with a distant object such as a tree, using one eye only. Now, without moving your head or the ruler, use your other eye. The ruler will no longer be lined up with the tree; it will apparently have shifted, because it is being viewed from a slightly different direction. The angle by which it seems to have moved is a measure of its parallax. If you know the distance between your eyes, and also the angle of shift, you can easily work out the distance of the ruler from the "baseline", or the line joining your eyes.

Surveyors who want to measure the distance of some inaccessible object, such as a mountain-top, use parallax in precisely this way. A longer baseline is needed, but the principles are the same.

Bradley adapted this principle to the stars. In the diagram, **A** is a relatively close star, seen against a background of more distant stars; the positions of the Earth are shown at opposite sides of its orbit – that is to say, at a six-months' interval. When the Earth is at its January position, the star will appear at **A1**. Six months later the Earth is in its July position, and the star will have shifted to position **A2**. We therefore know the parallax, and we also know the length of the baseline; it is twice the radius of the Earth's orbit, or 300,000,000 kilometres. From this, the distance of star **A** can be calculated.

Obviously the shifts are very small; no star shows a parallax of as much as one second of arc, but Bradley began his work with high hopes. Using a special instrument owned by James Molyneux and erected in Kew, he set out to measure the position of the star Gamma Draconis, which passes directly overhead as seen from this altitude, and was therefore very conveniently placed for accurate measurement.

He found shifts indeed, but they were extremely puzzling, and did not appear to be due to parallax. Gamma Draconis moved around in a tiny circle, returning to its original position after one year. Bradley checked with other stars, and found the same effect.

POSITION OF GAMMA DRACONIS. *This was the star studied by Bradley in an attempt to determine its parallax. Gamma Draconis was selected because it passes directly overhead at the latitude of Greenwich, and was thus convenient for observation with the special instruments used by Bradley. The parallax was not detected, but Bradley's research led to his discovery of the aberration of light.*

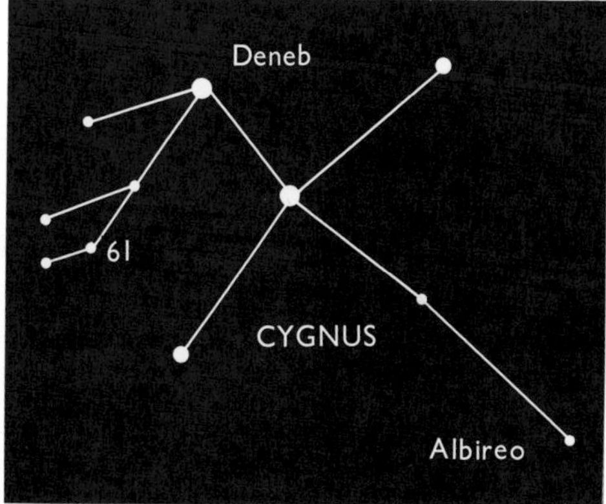

POSITION OF 61 CYGNI. *The star numbered 61 in Cygnus by Flamsteed was the first object outside the Solar System to have its distance measured; the work was carried out by Bessel, and published in 1838. 61 Cygni is of the 5th magnitude, and is therefore visible to the naked eye, but it is far from conspicuous. It is a relatively wide binary, and has an exceptionally large individual or proper motion, which was why Bessel concentrated his attention on it.*

According to a famous tale which is probably true, Bradley hit upon the solution one day when he was out sailing on the Thames. He noticed that when the direction of the boat was altered, the vane on the mast-head shifted slightly, even though the wind remained the same as before. At once he realized that this was the principle which accounted for the behaviour of Gamma Draconis. Light, as Rømer had found, does not travel instantaneously, but at a velocity of 300,000 kilometres per second. The Earth also is moving; its mean orbital speed round the Sun is 29·8 kilometres per second, so that its direction in space is changing all the time. Let the Earth represent the boat, and the incoming light represent the wind, and the situation becomes clear; there is always an apparent displacement of a star toward the direction in which the Earth is moving at that moment. This is *aberration*.

Bradley had made a discovery of vital importance – in fact it was the first observational proof of the Earth's motion; but he had not succeeded in measuring the distance of Gamma Draconis or any other star, and Herschel was equally unsuccessful. However, it was during his efforts to measure parallaxes that Herschel discovered the binary nature of many double stars, as had been tentatively suggested earlier by an English amateur astronomer, John Michell.

The next attempts were made in the 1830s by three astronomers; Henderson at the Cape, Friedrich Bessel (formerly Schröter's assistant) at Königsberg, and F.G.W. Struve, at the Dorpat Observatory in Russia. They worked quite independently of each other, and selected different stars.

Bessel decided upon the fifth-magnitude star 61 Cygni, in the Swan. It is a binary, with the two components far enough apart to be separated with a small telescope, but Bessel was interested in it mainly because of its relatively large proper motion. Relative to the background stars, it moves by over five seconds of arc per year. This means that it takes 350 years to shift by an amount equal to the apparent

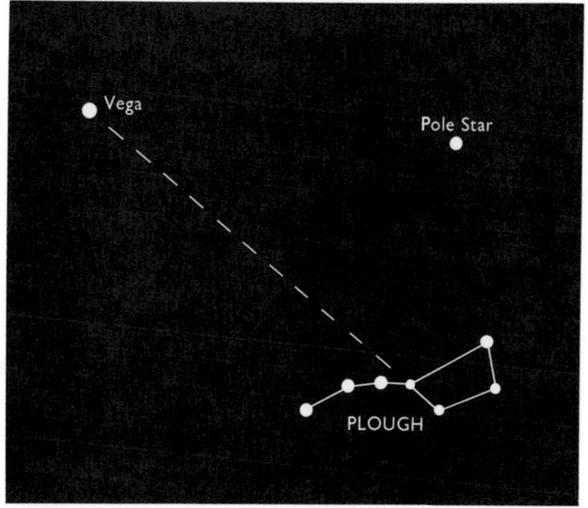

POSITION OF VEGA. *O. Struve made efforts to measure the distance of Vega at about the same time as Bessel and Henderson were carrying out their researches. Struve's result was less accurate, mainly because Vega is relatively remote. It is the fifth brightest star in the sky, and is easy to find, both because of its brilliancy and because of its decidedly bluish colour; the Great Bear may be used as a direction-finder to it. Vega is the chief star of the small but interesting constellation of Lyra, the Lyre. It is fifty times as luminous as the Sun.*

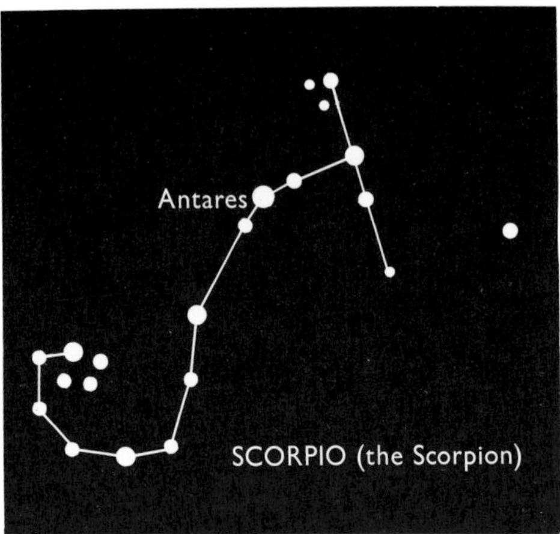

THE SCORPION. *The leading star of Scorpio, the Scorpion, is Antares, known as 'the rival of Mars' because of its strong reddish colour. Antares is a typical Red Giant star, of relatively low surface temperature but immense size. Its distance from us is about 360 light-years, and it is accompanied by a faint companion which is decidedly greenish. Scorpio is a brilliant constellation, but is never seen to advantage in Great Britain or the northern United States because of its southern position; from London and New York, the 'tail' never rises at all.*

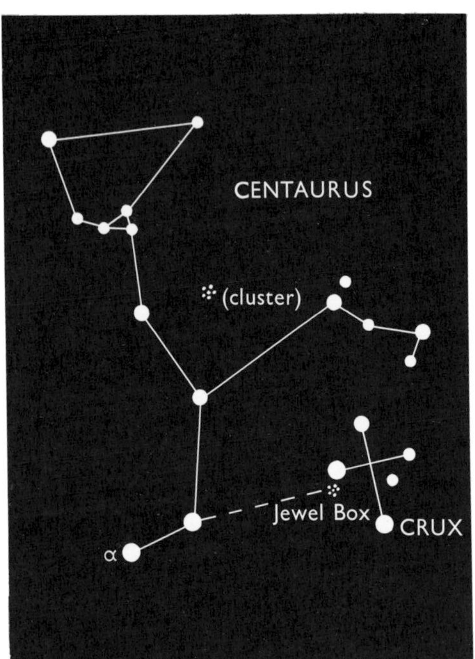

POSITION OF ALPHA CENTAURI. *The bright southern star Alpha Centauri was studied for parallax shift by Henderson, at the Cape of Good Hope. Measurable shifts were found, and Henderson was able to give a reliable figure for the star's distance, which amounts to rather more than 4 light-years. Alpha Centauri is thus the nearest of the brilliant stars. It is a splendid binary, and a fainter member of its system, Proxima, is the nearest known star to the Sun.*

diameter of the Moon, but even so it is unusually fast-moving and Bessel correctly decided that it must be one of the nearest stars in the sky.

He began work in 1837, and only a year later he was able to announce that both components of 61 Cygni showed a parallax of 0·29 seconds of arc, giving a real distance of just over eleven light-years – approximately 95 million million kilometres. His measurements were excellent, and the values which he gave were almost the same as those accepted today. Yet the parallax shift is much less than the apparent diameter of a coat-button seen from fifteen kilometres!

Henderson's measurements at the Cape had been made earlier, but had not been worked out, and so Bessel must rightly claim priority. Actually Henderson's task was easier than Bessel's; Alpha Centauri, his chosen star, is only a little over four light-years off, and shows a parallax of 0·76 seconds of arc. Alpha Centauri is a glorious binary, and there is a third member of the group, Proxima, which is slightly closer to us and remains the nearest star known. Neither is it likely that any are closer; they would certainly have been discovered by now. Vega, the star selected by Struve, is twenty-seven light-years away, with a correspondingly smaller parallax; understandably Struve's results were less accurate, but they were at least of the right order.

The parallax method can be used only for relatively near stars. Beyond about 150 light-years the shifts have become so small that they are hard to measure accurately, and by 600 light-years they are swamped by unavoidable errors in observation, so that less direct methods have to be used. But it was Bessel, Henderson and Struve who made the great step forward, and led us on to an understanding of how vast the universe really is.

24

Craters, Canals and Rings

BEFORE the end of the nineteenth century, refractors had well and truly come into their own. True, the Rosse 'Leviathan' was still the largest telescope in the world, but it belonged by a bygone age, and – good though it was – it could not be compared with some of the later instruments. This, too, was a period when great emphasis was still being placed upon visual observations of the Moon and planets. I say 'visual' because for planets, at any rate, the human eye could see more than could be shown on any photograph.

It had become possible to cast and shape large lenses, and astronomers took full advantage of it. In 1874 James Lick, a wealthy manufacturer of musical instruments, came into the story by giving a handsome present to the University of California, to provide 'a telescope superior to and more powerful than any telescope yet made' and to instal it in a suitable observatory. The University accepted the gift with thanks, and fourteen years later the great Lick 91·4-centimetre refractor was ready for use. It was the second of the generation of large refractors – the first, at Nice in France, had been completed in 1880; it had a 76·2-centimetre object-glass. The Lick refractor was followed by others, and even today it remains the second largest telescope of its type. Pride of place goes to the 101·6-centimetre refractor at the Yerkes Observatory, completed in 1897. This also was financed by an American businessman, in this case Charles T. Yerkes, who provided the money after approaches from George Ellery Hale (about whom I will have more to say later). Other powerful refractors were set up at Meudon, near Paris (83-centimetre objective; 1893), Potsdam in what is now East Germany (81·3-centimetre; 1899), and Allegheny Observatory, Pittsburgh (76·2-centimetre; 1914). Significantly, the Allegheny refractor was the last of its kind. After that, the emphasis shifted to reflecting telescopes.

I must pause here to say something about a telescope which failed. In 1894 a Monsieur Deloncle made an interesting proposition to a group of French astronomers. He proposed to master-mind a refractor with a lens 125 centimetres in diameter, and show it at the Paris Exposition of 1900. To reduce the false colour trouble, the focal length was to be very long, well over fifty metres. The tube, weighing 21 tons, was to be horizontal, the light from the object under observation being sent to the stationary object-glass by a rotating mirror. This principle is sound enough in the modern sense, and is widely used, but it is clearly unsuitable for refractors. The telescope was actually built and exhibited, but it was never used for research; after the Exposition ended it was dismantled, and I believe that the lens is still stored somewhere in the cellars of the Paris Observatory. Yet in its way it was useful, because it demonstrated that there is a limit to the size of a refractor. Lenses have to be supported round their edges, and if too heavy they sag under their own weight, which produces unacceptable distortions in the image. It is very doubtful whether the Yerkes

ALPETRAGIUS. *One of the great lunar craters, photographed from Orbiter.*

THE VOLCANIC CRATER OF HVERFJALL, IN ICELAND *(opposite, above). This vulcanoid, close to Lake Mývatn, bears a strong superficial resemblance to a lunar crater; the central peak is a prominent feature. Photograph by Brian Gulley, 1960.*

THE ARIZONA METEOR CRATER *(opposite, below), photographed from the air by Patrick Moore. This is undoubtedly due to the impact of a large body which hit the desert in prehistoric times, and is almost 1·6 km in diameter. No doubt many of the smaller craters on the Moon were produced in this way, but whether the larger lunar craters are volcanic or meteoric is still a matter for debate.*

101·6-centimetre will ever be surpassed in size. Meanwhile, let me turn to what these telescopes told us about our near neighbours, the members of the Solar System.

The Moon, of course, is a comparatively easy body to study, partly because of its nearness and partly because its lack of atmosphere means that its surface is never veiled. Schröter had shown the way. He never produced a complete map (or at least, if he did it was destroyed with the looting of his observatory), and the chart by Tobias Mayer, completed in 1776, remained the best for over sixty years. Yet Schröter laid the foundations of selenography, and later observers were able to build upon his work.

Meanwhile, a larger-scale map of the Moon was badly needed. Wilhelm Lohrmann, born in Dresden in 1796 and trained as a land

THE 83-CM REFRACTOR AT MEUDON OBSERVATORY. *Meudon is a station of the Paris Observatory. The telescope is shown here with Dr. Audouin Dollfus. Photograph by Patrick Moore, 1975.*

THE LICK OBSERVATORY. *The following domes are shown* (left to right): *the 66-cm Crossley reflector, the 56-cm Tauchmann reflector, the 91·4-cm reflector, the 30·5-cm refractor, the 50·8-cm Carnegie Astrograph, and the 305-cm reflector. Lick Observatory photograph.*

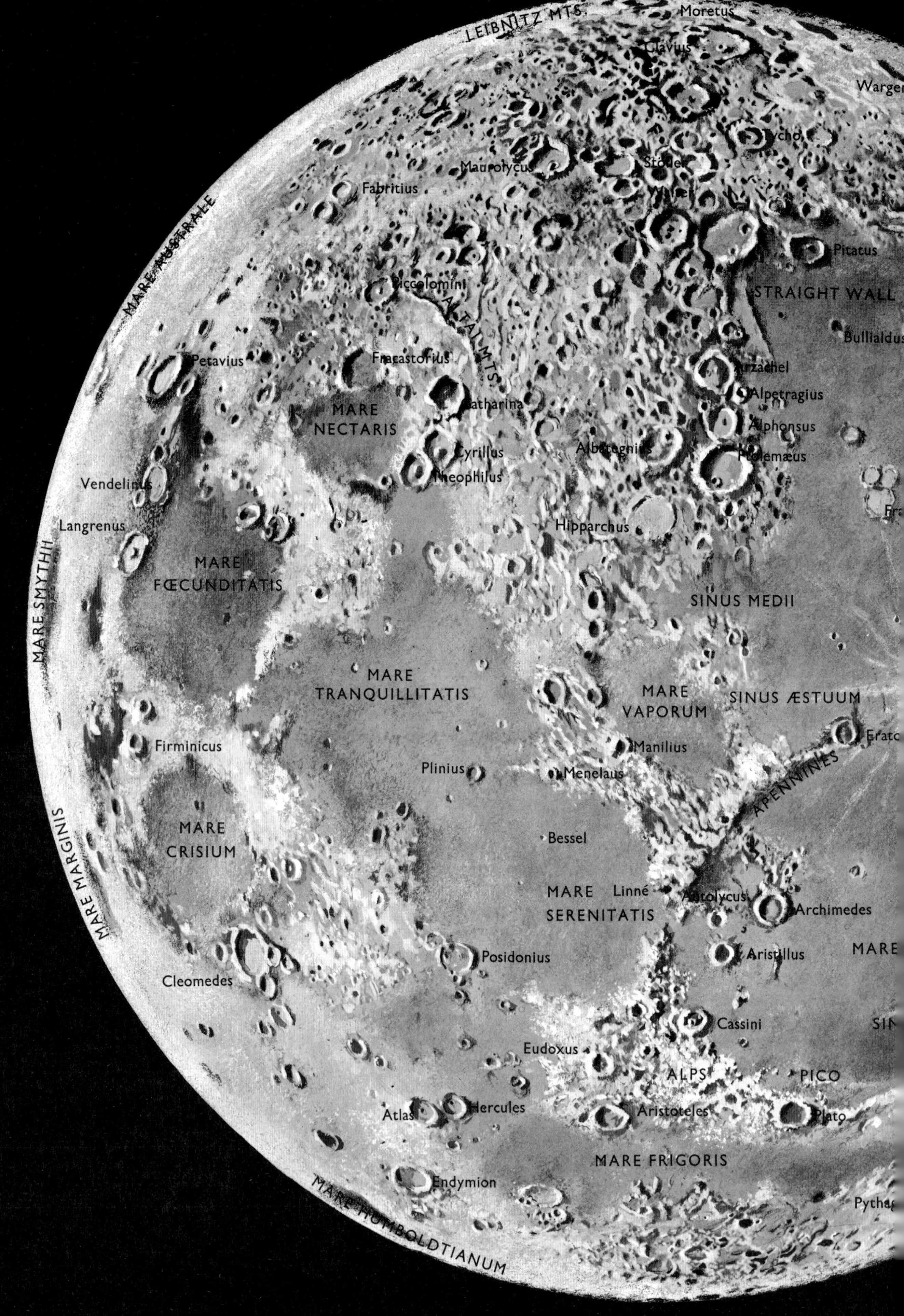

LEIBNITZ MTS
Moretus
Clavius
Tycho
Maurolycus
Fabritius
MARE AUSTRALE
Pitatus
STRAIGHT WALL
Piccolomini
ALTAI MTS
Bullialdus
Petavius
Fracastorius
Alpetragius
Catharina
MARE
NECTARIS
Alphonsus
Cyrillus
Vendelinus
Theophilus
Langrenus
Hipparchus
MARE
FŒCUNDITATIS
MARE SMYTHII
SINUS MEDII
MARE
TRANQUILLITATIS
MARE
VAPORUM
SINUS ÆSTUUM
Firminicus
Manilius
Plinius
Menelaus
APENNINES
MARE MARGINIS
MARE
CRISIUM
Bessel
MARE
SERENITATIS
Linné
Archimedes
Posidonius
Aristillus
Cleomedes
Cassini
Eudoxus
ALPS
PICO
Atlas
Hercules
Aristoteles
Plato
MARE FRIGORIS
Endymion
MARE HUMBOLDTIANUM

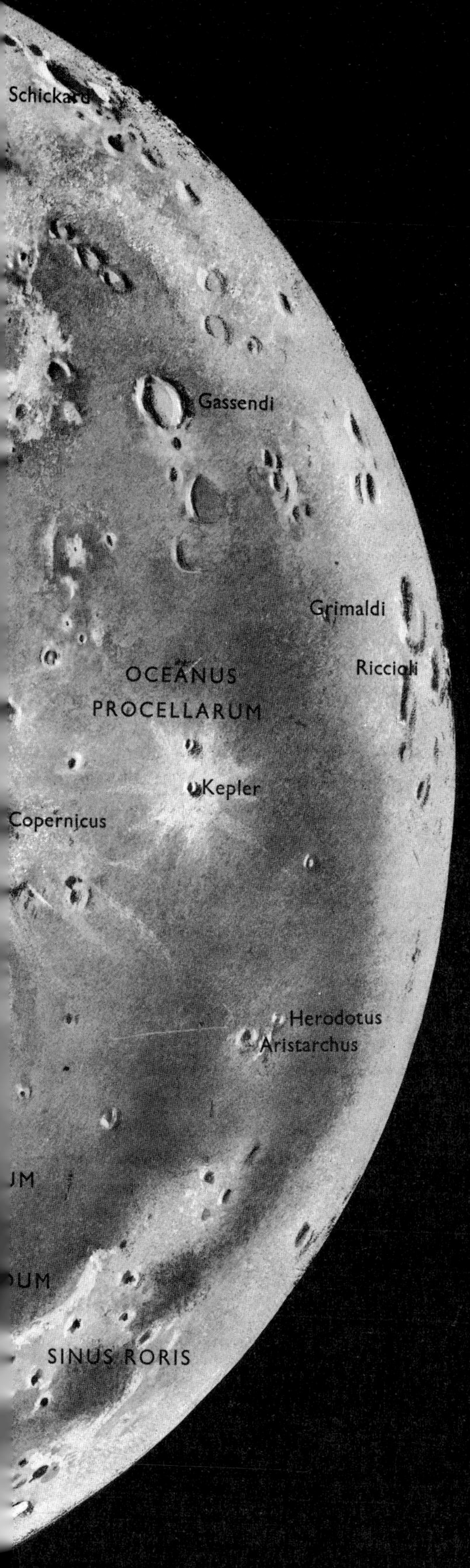

Map of the Moon

This chart has been based on two half-moon photographs. The shadows are therefore not consistent, but the method enables most of the important craters to be shown in recognizable form. The features near the limb are not so clearly marked, since both the photographs concerned were taken under fairly high light for these areas, but various special features, such as the dark-floored craters Grimaldi and Riccioli, can be made out. In the map, the most prominent features are named, with the craters lettered in script and other objects (such as mountains) in capitals. For the 'seas', the Latin names are used, so that for instance the Sea of Showers becomes the 'Mare Imbrium'. This is by far the best method, since the Latin names are international

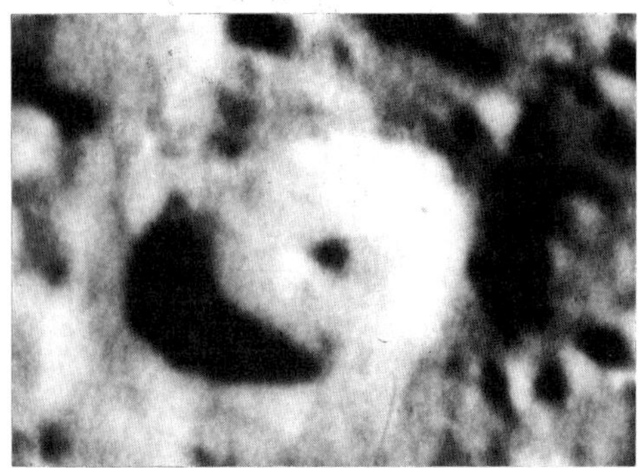

THE LUNAR CRATER TYCHO. *The central peak is a prominent feature. Lick Observatory photograph.*

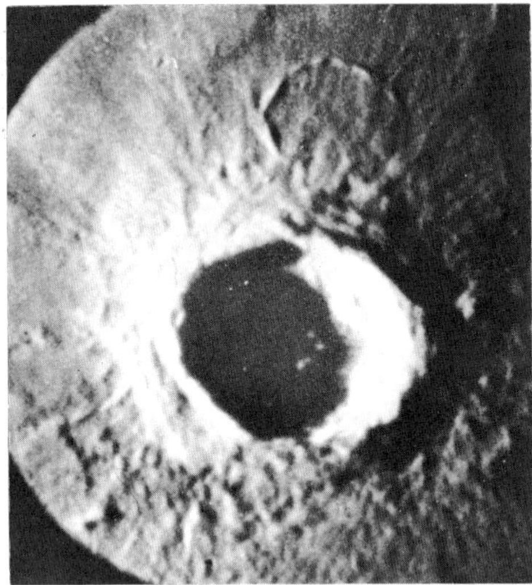

THE LUNAR CRATER ARISTILLUS. *A prominent crater, photographed at the Pic du Midi Observatory.*

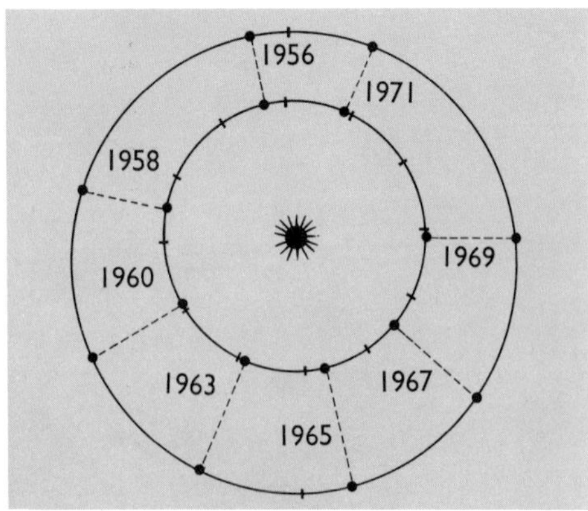

OPPOSITIONS OF MARS, 1956-71. *It will be seen that the most unfavourable oppositions are those of 1963 and 1965.*

surveyor, made up his mind to draw one; he completed four sections of it, but unfortunately his eyesight failed, and he had to give up. He died in 1840.

At about the same time Johann von Mädler, a Berlin teacher, was giving private lessons in astronomy to a wealthy banker named Wilhelm Beer, brother of the famous composer Meyerbeer. Beer was an apt pupil, and before long the two men joined forces. Beer equipped his observatory with a fine 9·5-centimetre refractor made by Fraunhofer, and together they began work on a lunar chart. It appeared in 1837-8, together with a book, *Der Mond*, which contained a description of all the important surface features.

The map was remarkably good, and a triumph of patient, skilful observation. It was regarded as the 'last word' on the subject, but the results of its publication were unexpected. For some decades afterwards most observers neglected the Moon altogether, and turned their attention elsewhere.

The reason for this was simple. Beer and Mädler believed the Moon to be a dead world: barren, lifeless, almost or quite without atmosphere, and completely inert. In their opinion, nothing had happened on the Moon for millions of years, or would do so in the foreseeable future. They had charted the lunar surface as well as possible, and other observers felt that there was nothing left to be done.

This became the official view – but not perhaps the popular one, as the celebrated 'Moon Hoax' showed. And before going on, I must say something about it, because it caused a great deal of interest at the time.

Between 1834 and 1838 Sir John Herschel, as we have noted, was busy at the Cape, studying the southern stars. Richard Locke, a graceless reporter on the New York *Sun*, had a bright idea. Herschel was on the other side of the world; communications in those days were slow and uncertain; who was there to check any statements that he made?

Locke saw his chance, and took it. On August 25, 1835 the *Sun* came out with a startling headline: 'Great Astronomical Discoveries made by Sir John Herschel at the Cape of Good Hope', and an account of how Herschel had built a new telescope powerful enough to show the Moon in amazing detail. Locke's article was so cleverly worded that it sounded almost plausible. It was said that Herschel had used his telescope to form an image, and had then reinforced this image by means of a light-source in the observatory itself: 'a transfusion of artificial light through the focal point of vision'!

The way was open, and the *Sun* kept up the good work for the next week. The lunar scenery was varied and colourful; 'a lofty chain of obelisk-shaped or very slender pyramids... of a faint violet hue, and very resplendent.' Living creatures included 'a strange amphibious creature of spherical form, which rolled with great velocity across the pebbly beach.' There were also bat-men 'four feet in height... covered with short and glossy copper-coloured hair... faces of a yellowish flesh-colour, a slight improvement upon that of the large orang-outang.' On one occasion Herschel forgot to cover up the main lens, so that when the Sun shone upon it the lens acted as a burning-glass and set fire to the observatory.

The articles met with a mixed reception. The *New York Times* declared that the discoveries were 'both probable and plausible', while the *New Yorker* stated that the discoveries had 'created a new era in astronomy and science generally'; a women's club in Massachusetts wrote to Herschel asking for advice upon how to convert the bat-men to Christianity. Others were more cautious. The hoax was exposed by a rival paper within a few days, and the *Sun* itself confessed on September 16, but Locke's joke had been magnificently successful. Before we laugh, let us remember that his articles were no less unlikely than some of the books and papers written about flying saucers within the last few years!

With this aside, let me return to true science.

Shortly after the appearance of *Der Mond*, Mädler left Berlin to become Director of the new observatory at Dorpat in Russia, and never undertook much more lunar work; Beer, too, was content with casual observing. Their map remained unrivalled for many years, and in the mid-nineteenth century the only really serious lunar worker was Julius Schmidt, a German who spent much of his life in Greece and became Director of the Athens Observatory. Schmidt studied the Moon consistently, and in 1866 made the spectacular announcement that a small, well-marked crater named Linné, on the grey plain of the Sea of Serenity (Mare Serenitatis), had disappeared, to be replaced by a tiny pit in the middle of a white patch.

Tremendous interest was generated, and astronomers began to turn their telescopes back towards the Moon. Actually, there seems little doubt that no change had occurred in Linné or anywhere else. Mädler, who had drawn the feature as a craterlet in the 1830s, observed it again after Schmidt's announcement and commented that it looked exactly the same as before, which seems conclusive. However, the Linné episode was beneficial, and since then the Moon has been under regular observation, at first mainly by amateurs and more recently by professional astronomers as well. In England, a good lunar map was published in 1876 by Edmund Neison, who had a strange career. His real name was Nevil, but he wrote under a pseudonym. In 1882 he went to South Africa to become director of the newly-founded Natal Observatory at Durban, but met with no success there; funds were always short, and when the observatory was closed down in 1910 Neison came home to live in retirement in Eastbourne.

In 1878 Schmidt produced an excellent and elaborate lunar chart, based on Lohrmann's, and from 1895 the Lunar Section of the British Astronomical Association took up the work. Its first director was Thomas Gwyn Elger, who himself produced an excellent outline map. But before going any further, it is worth pausing to say rather more about the nature of the lunar surface.

Any casual glance will show that the Moon is quite unlike the Earth, and the reason for this is not far to seek: the Moon has no atmosphere. This means that it must also lack water, and the wide grey areas which are known as seas or *maria* are not watery; there is no trace of moisture anywhere on the Moon. For well over a century the harsh nature of the lunar environment has been known.

The *maria* are much smoother than the bright uplands, and many of them are more or less circular, though as seen from Earth they appear elliptical by foreshortening. The regular *maria* are mountain-bordered; thus the lunar Apennines and Alps make up part of the boundary of the huge Mare Imbrium or Sea of Showers. There are practically no terrestrial-type ranges, but there are vast numbers of isolated peaks and clumps of elevations.

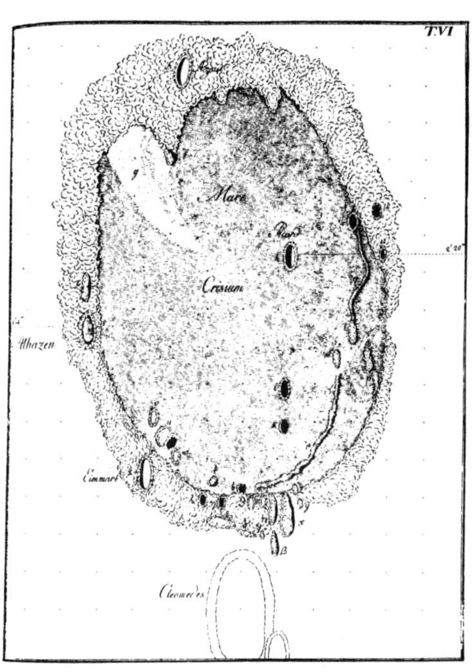

THE MARE CRISIUM, AS DRAWN BY SCHRÖTER. *This is a typical drawing by Schröter, who was the first of the great lunar observers. It is said, and with justification, that Schröter was a clumsy draughtsman; nevertheless he seldom made a serious mistake, and the details are always clearly shown, as may be seen by comparing this drawing with a modern photograph. It shows the Mare Crisium or Sea of Crises, near the western limb of the Moon – one of the smaller but most conspicuous of the dark plains.*

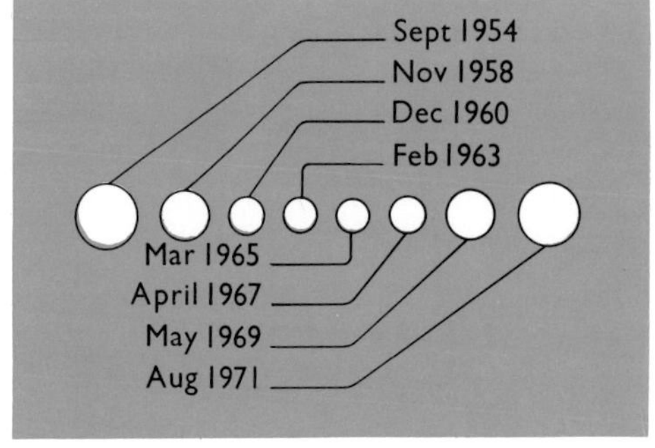

APPARENT OPPOSITION SIZE OF MARS. *The maximum size for each opposition between 1954 and 1971 is shown to the same scale.*

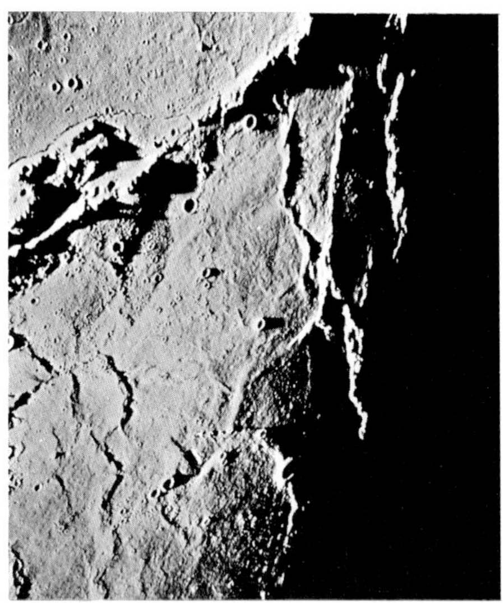

THE APENNINES, *photographed from Apollo 15 before it landed.*

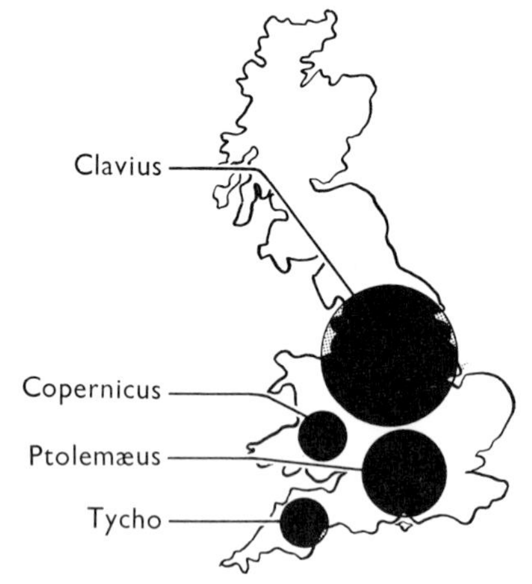

SIZES OF SOME LUNAR CRATERS, *compared with the British Isles.*

The craters dominate the entire scene. They range from huge enclosures well over 200 kilometres in diameter down to tiny pits too small to be seen from Earth. No part of the Moon is free from them; they cluster thickly in the uplands, and are also to be found on the *maria*, inside larger craters and even on the tops of mountains. They break into each other and deform each other, so that obviously they are of different ages, and some of them are almost obliterated. Such is Stadius, near the magnificent 90-mile ray crater Copernicus. Stadius has walls which can be little more than ten metres high anywhere, and the formation looks as though it has been overwhelmed by the *mare* lava.

Craters are of various types. (I am here using the term 'crater' to include all the walled formations, though in many cases 'walled plain' would be more accurate.) Some have high, terraced walls and lofty central peaks or groups of central peaks; some are low-walled, with floors which are comparatively smooth. Aristarchus, 37 kilometres across, is so reflective that it is easily seen by earthshine alone, and there seems no doubt that even an observer so experienced as William Herschel mistook it for an erupting volcano on several occasions. On the other hand, the 96-kilometre Plato has a floor so dark that Hevelius, in 1645, named it 'the Greater Black Lake'.

However, well-developed craters have several features in common. They are essentially regular in outline, and their floors are sunken with respect to the outer surface. If there is a central elevation, it never rises as high as the outer walls. The wall slopes are relatively gentle, and if drawn in profile a crater is more like a shallow saucer than a steep-sided mine-shaft. An observer standing in the middle of a large crater such as Plato would be unable to see the walls, as they would be below his horizon.

A lunar crater is at its most spectacular when seen near the terminator, or boundary between the sunlit and night hemispheres. It is then shadow-filled and is most imposing, whereas when the Sun has risen high over it, and the shadows have shortened, it may be hard to identify at all. This is one reason why the best telescopic views of the Moon are obtained during the crescent, half or gibbous stages. At or near full the shadows almost disappear, and the overall effect is one of confusion, particularly as some of the craters, such as Tycho in the southern uplands and Copernicus and Kepler closer to the lunar equator, are the focal points of systems of bright streaks or rays which overlie all other features and tend to mask them. The rays are surface deposits; they cast no shadows, and are not well seen except under high illumination.

Of the less obtrusive features, the clefts or rills are of special interest. They look like cracks in the surface, and some of them, such as the 240-kilometre rill near the little crater Ariadæus, are visible with small telescopes; some large craters, such as Gassendi, have whole systems of rills on their floors. Note also the magnificent winding valley close to Aristarchus and its companion crater Herodotus, known as Schröter's Valley (though unfortunately the crater named after Schröter is nowhere near). Another interesting feature is the Hyginus Rill, which is basically made up of a chain of small craters which have run together – a lunar 'string of beads'.

Low swellings, known as domes, are seen in selected areas; many of them have symmetrical summit craters, so that they must have been

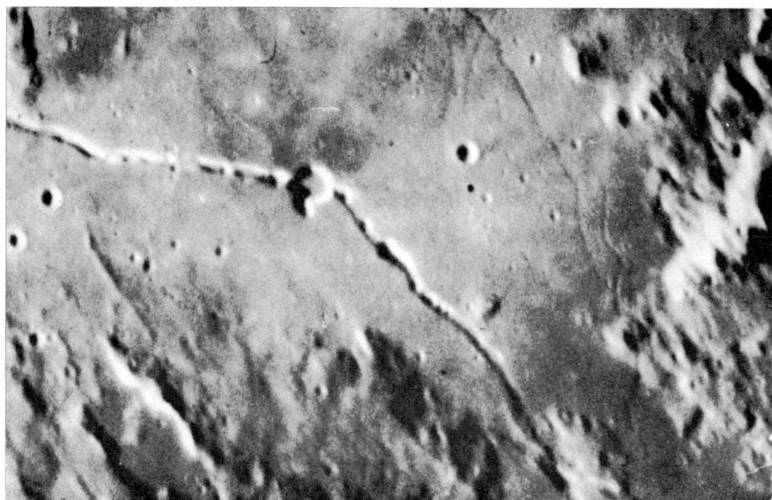

THE HYGINUS CLEFT, *(left) photographed at the Pic du Midi Observatory. This is a prominent feature, visible in a very small telescope. Under low magnification it appears as a true crack or cleft, but close examination with a higher power shows that it is, in part at least, a crater-chain and not a genuine cleft at all. In this photograph, many of the craterlike enlargements are shown. Hyginus itself lies in the middle of the picture.*

TYCHO. *Tangled rocks outside the wall of the crater, photographed from the Surveyor craft which landed there.*

produced by internal action rather than by meteoritic bombardment. One crater, Wargentin – 88 kilometres across – is filled with lava almost to the brim, so that it has become a plateau. Note also the so-called Straight Wall, which is not straight, and is not a wall. It is a fault in the surface, near the little crater Birt in the Mare Nubium (Sea of Clouds). It is 125 kilometres long, and the surface drops to the west* by almost 250 metres. Before full moon it casts a shadow, and appears as a black line; after full moon the sunward face is illuminated, and the feature appears bright. It used to be thought that the cliff-face was steep, but in fact the angle of incline is no more than forty degrees. In the future it will no doubt become a lunar tourist centre!

There is plenty to see on the Moon, and after the Linné episode there were plenty of observers to study it. In 1904 W.H. Pickering in Jamaica produced a new kind of photographic atlas, showing each area of the surface under several different angles of illumination; and from then until the start of the Space Age amateur observers were very active. It was only when the rockets began to fly, and it became clear that the Moon was within reach, that professional astronomers joined in.

Of the planets, Mars came in for special attention, and this brings me on to the story of the canals, which led to an argument which went on for almost a century. It began in 1877, and was not finally settled until 1965.

Drawings of Mars had been made by many of the early observers. Indeed, the first sketch to show recognizable detail was made by Christiaan Huygens as long ago as 1659. On it, Huygens showed the V-shaped marking which we now call the Syrtis Major. Herschel, too,

*There has been long-standing confusion about lunar east and west. Before the 1960s, lunar charts and photographs were usually printed with south at the top, as seen in a normal astronomical telescope, which gives an inverted image; west was to the left – that is to say, the Mare Crisium was near the west limb, and the huge dark-floored walled plain Grimaldi near the east limb. The International Astronomical Union, the controlling body of world astronomy, then decreed that in future north should be at the top, and east and west reversed – i.e. Mare Crisium at the *east* limb. At the meeting at which this was decided, I led a battle against the change, but I was decisively defeated! Let me make it clear that in this book I am following the I.A.U. convention; Mare Crisium to the *east*, Grimaldi to the *west*.

MARS AND THE PLEIADES. *September 16, 1952, 1·05 hours. Photograph by Ramon Lane.*

MARS FROM MARINER 7, 1969. *These pictures were taken at an interval of 47 minutes during which time Mars rotated through an angle of 12 degrees. Prominent in the picture are the bright, ring-shaped Nix Olympica and the complex bright streaks of the Tharais-Candor region.*

drew Mars; he measured the rotation period very accurately as being only slightly more than half an hour longer than our own, and he also traced the seasonal growth and shrinking of the white caps covering the Martian poles. Reasonably enough, he believed that the whiteness was due to ice or snow. The caps had been discovered by Cassini in 1666, and are easy to see with a small telescope when Mars is well placed.

Originally it was thought that the dark patches on the surface were seas, and that the reddish-ochre tracts were continents. During the first half of the nineteenth century Schröter made some useful drawings of Mars, though admittedly he misinterpreted them (he attributed them to masses of clouds). Next Beer and Mädler, during the 1830s, produced a preliminary map of the surface.

Much depended upon the exact nature and extent of the Martian atmosphere. Remember, Mars is much smaller than the Earth; its diameter is a mere 6787 kilometres, and its escape velocity is no more than 5 kilometres per second, so that it could hardly be expected to hold on to a dense, moist atmosphere similar to ours. In 1860 Emmanuel Liais, in Brazil, suggested that since Mars appeared to be a distinctly dry place, the dark patches were not likely to be seas. Instead, he attributed them to areas of primitive-type vegetation.

Not everyone agreed. In 1871 Richard Proctor wrote that 'there must be rivers on Mars... there must be volcanic eruptions and earthquakes... there must be mountains and hills, valleys and ravines, watersheds and watercourses.' Proctor also published a map of Mars in which he gave the various regions romantic names, such as Beer Continent, Arago Strait, Herschel Strait, Mädler Continent, Fontana Land and so on. It all sounded very inviting, but could there really be large sheets of water upon such a world? Nobody knew.

Then, in 1877, came an astronomical bombshell.

One of the leading Italian astronomers of the time was Giovanni Schiaparelli, who had made for himself a great reputation; for instance, he was one of the first to establish a definite connection between comets and meteor showers. He had become Director of the Brera Observatory in Milan, and used the fine 23-centimetre refractor there. In 1877 Mars was exceptionally well placed, and was almost at its minimum distance from us (56,000,000 kilometres), and Schiaparelli made the most of the opportunity. He produced a new map, and renamed the surface features, so that, for instance, Proctor's Mädler Continent became 'Chryse' (the region, incidentally, in which the Viking 1 probe came down nearly a century later). But Schiaparelli also recorded some straight, artificial-looking lines, criss-crossing the ochre areas. He called them *canali*, or channels. Inevitably the word was translated into English as canals – and canals they remained.

The most remarkable thing about the canal network as described by Schiaparelli was that it seemed to follow a definite pattern. There was nothing haphazard about it. Either the canals followed great-circle tracks, or else they were gently curved. Whether curved or not, they ran from dark area to dark area; there was not a single case of a canal breaking off abruptly in the middle of a 'desert'. Altogether, Schiaparelli drew forty canals during the 1877 opposition, and in 1879, when Mars was again on view, he reported that some of the canals had the strange ability of becoming abruptly double. The original single track was replaced by two streaks, as perfectly parallel as the two tracks of a railway line.

PERCIVAL LOWELL, *Yerkes Observatory photograph.*

Streaky features had been reported before, but nothing of this nature, and at first astronomers as a whole were sceptical, because nobody apart from Schiaparelli could see the canals at all. Then, in 1886, the canals were confirmed by the French astronomers Perrotin and Thollon, using the large refractor at Nice Observatory. After that, canals became all the rage. Most observers saw them, sometimes with surprisingly small telescopes, and maps of Mars began to look very peculiar indeed.

The next development was due to Percival Lowell, who came of an old New England family. He was born in Boston in 1855, and after spending some years as a diplomat in the Far East he decided to devote the rest of his life to astronomy in general and Mars in particular. He was wealthy, and so he was able to set up a major observatory at Flagstaff in Arizona, where the conditions were expected to be (and are) good.

He equipped it with a 61-centimetre refractor, and from 1894 until his death in 1916 he and his assistants carried out intensive studies of Mars. All in all, Lowell must be the most famous figure in the whole of Martian history.

Schiaparelli had been cautious about the nature of the canals. Lowell was not. He was firmly convinced that they were artificial, built by intelligent engineers to make up a planet-wide irrigation system. He knew that Mars was short of water, and reasoned that the main supplies must be locked up in the frozen polar caps; therefore the Martians had to conserve every scrap of water – and hence the canals. He did not suggest that the canals as seen from Earth were sheets of open water, because evaporation would have been an insuperable problem. Rather he believed a canal to be a narrow channel, possibly piped, flanked to either side by strips of fertile land.

What could the Martians be like? Lowell did not know; he did not even speculate. He wrote that 'to speak of Martian beings is not to speak of Martian men'. But at least he regarded their existence as indisputable, and meanwhile other observations were coming in. W.H. Pickering found that the canals crossed the dark regions as well as the deserts; also, many observers described a seasonal 'wave of darkening', spreading from the pole towards the equator during Martian spring and early summer, as though vegetation were being revived after its winter hibernation.

This sounds bizarre today, but it was not so very unreasonable in Lowell's time. After all, in some ways Mars is not so very unlike the Earth. Its seasons are of the same general type, though much longer (the Martian year is equal to 687 Earth-days or 669 Martian days, and the axial tilt is almost the same as ours). And let it be said at once that if Lowell's drawings had been accurate, then Mars would have been inhabited. The canal network could not possibly be natural.

Yet doubts arose at once. Other observers, using telescopes just as powerful as Lowell's, could not see the canal network at all. One such observer was Eugenios Antoniadi, Greek by birth, who spent most of his life in France and had the use of the 83-centimetre Meudon refractor, still the largest of its type in Europe and the third largest in the world. Antoniadi saw vague streaky features, but he had no patience with Lowell's canals, and stated baldly that they did not exist. He maintained that they were due to nothing more than tricks of the eye.

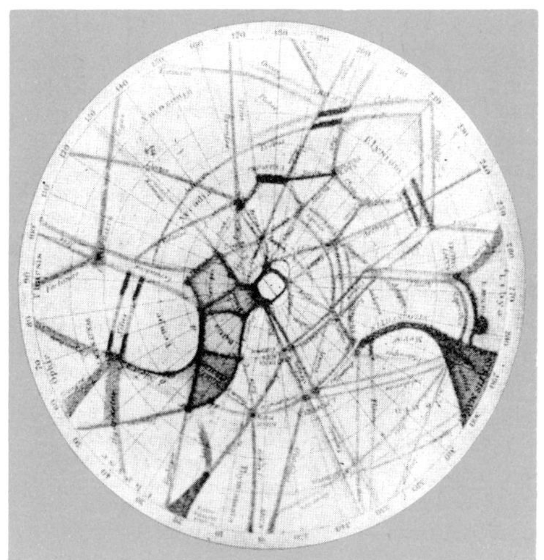

THE SOUTH POLAR AREA OF MARS, 1888, *according to Schiaparelli; many canals are shown.*

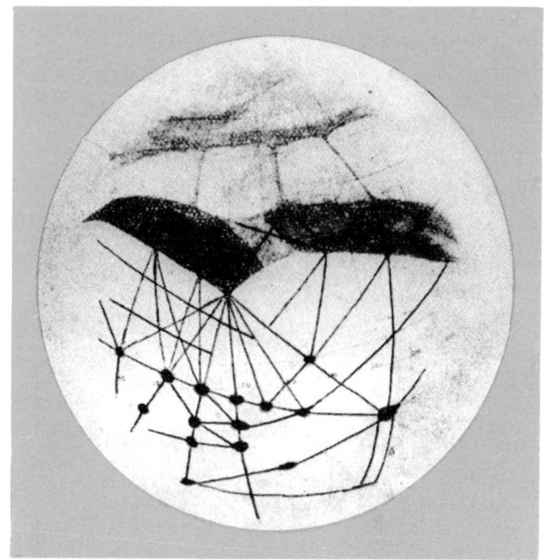

DRAWING OF MARS BY LOWELL. *Here we see the canal network in its most developed form.*

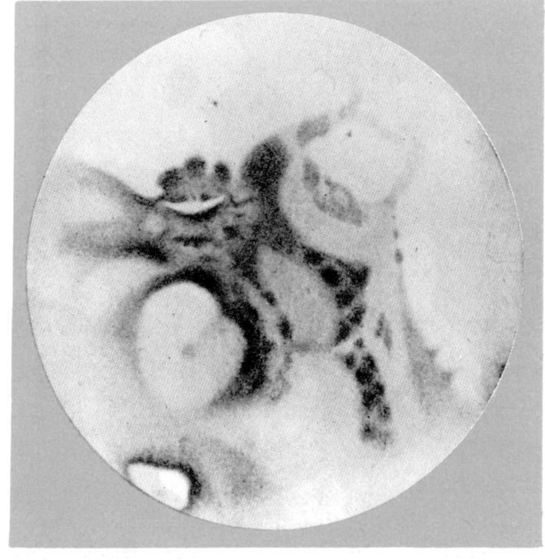

MARS, 1909, DRAWN BY E.M. ANTONIADI. *The telescope used was the Meudon 83-cm refractor. The difference between Antoniadi's representation and those of Schiaparelli and Lowell is very obvious.*

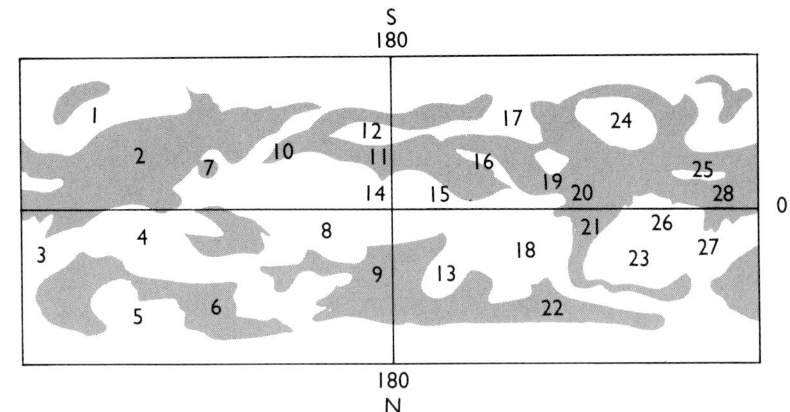

GREEN'S NOMENCLATURE FOR MARS. Superseded by Schiaparelli's nomenclature in 1877. Most of the features are identifiable; thus Green's Lockyer Land is the modern Hellas; while the Kaiser Sea or Hourglass Sea is the present Syrtis Major.

Key:

1 Jacob Land
2 De la Rye Ocean
3 Burton Bay
4 Mädler Continent
5 Rosse Land
6 Airy Sea
7 Terby Sea
8 Secchi Continent
9 Oudemans Sea
10 Pratt Bay
11 Maraldi Sea
12 Webb Land
13 Fontana Land
14 Trouvelot Bay
15 Huggins Bay
16 Burckhardt Land
17 Cassini Land
18 Herschel I. Continent
19 Flammarion Sea
20 Main Sea
21 Kaiser Sea (or Hourglass Sea)
22 Delambre Sea
23 Beer Continent
24 Lockyer Land
25 Phillips Island
26 Schmidt Bay
27 Dawes' Forked Bay
28 Herschel II. Strait.

Antoniadi's views carried a great deal of weight, and as the years passed by fewer and fewer astronomers felt able to accept Lowell's theories. The *coup de grâce* was given to them in July 1965, when the first successful Martian probe, Mariner 4, by-passed the planet and sent back information from close range. Since then we have had Mariner 9 and the Vikings, to be described later, and the results are conclusive. There are no canals on Mars. They do not even have a basis of reality, as I demonstrated when I took a modern chart and superimposed Lowell's canal network on to it.

What was the cause of this strange episode? Again we come back to unconscious prejudice. Canals had been reported; therefore, canals must exist. And yet there was nothing there.

It is said that men are remembered for their mistakes rather than for their triumphs, and this is certainly the case with Lowell. He carried out a great deal of valuable work, and it is unfair to dismiss him merely for his admittedly wild theories about Mars. The Lowell Observatory at Flagstaff remains today as one of the world's major astronomical centres. Lowell himself carried out precise measurements of the movements of the two tiny satellites of Mars, Phobos and Deimos, which had been discovered by Asaph Hall in the famous year of 1877.

The other inner planets presented problems of a different sort.

ROCKY MARTIAN PANORAMAS FROM VIKING 2, *September 1976. Some of the first pictures taken by Viking 2, they show the site on a northern plain of Mars known as Utopia Planitia. The surface is strewn with rocks, some porous and sponge-like, some dense and fine-grained. There are also small drifts of very fine sand similar to that seen at the Viking 1 landing site some 7400 km to the southwest. The slope of the horizon in both pictures is an illusion created by the 8-degree tilt of the spacecraft.*

Venus generally showed a more or less featureless disk, and it was clear that what we were seeing was not the actual surface but merely the top of a layer of cloud. What lay underneath was a matter for conjecture.

There was, too, the interesting case of the ghost satellite of the planet. It was reported in 1686 by Cassini, at Paris, who claimed that it had a diameter about one-quarter that of Venus itself. Further accounts of it were given regularly during the next seventy years, notably in 1740, when it was described by the famous telescope-maker James Short. Then, in 1761, Montaigne of Limoges recorded it several times, and for a brief period its existence was regarded as established. An orbit was worked out by the German mathematician Lambert, who found that the distance from Venus was 417,000 kilometres, and the period 11 days 5 hours.

The satellite was reported again in 1764 by three observers, two in Denmark and one in France, but these were the final accounts of it. It has never been seen since, and we are now sure that Venus is moonless. The old records were certainly due to telescopic 'ghosts', caused by instrumental defects.

There was also the Ashen Light, or faint luminosity of the planet's unlit hemisphere. Franz von Paula Gruithuisen, a rather imaginative German astronomer, suggested that it might be due to vast forest fires lit on the surface of Venus by the local inhabitants to celebrate the election of a new Government! This, I fear, is distinctly improbable, and nowadays it seems likely that the Light is caused by electrical effects in the upper atmosphere of Venus.

Mercury is always so close to the Sun that it is hard to study properly. It is never visible against a dark sky; it is seen with the naked eye only when low in the west just after sunset, or low in the east just before sunrise. Schiaparelli – who else? – was the first to attempt a map of the surface, and, sensibly, he elected to study Mercury in the daytime, when it was high above the horizon. Of course the Sun was high too, but that could not be helped. Schiaparelli drew various dark patches on the surface, and came to the conclusion that the axial rotation period must be the same as the period of revolution round the Sun: 88 Earth-days. In this case Mercury would keep the same face turned sunwards all the time. Part of the surface would be in permanent daylight, and another part in everlasting night. There would be only a narrow 'twilight zone', between these two extremes, from which the Sun would bob up and down over the horizon.

Antoniadi, using the Meudon refractor in the 1920s, agreed with this idea. It was, of course, obvious that Mercury could retain very little in the way of an atmosphere; it is only 4880 kilometres in diameter, with a low escape velocity (2·6 kilometres per second). It was generally thought that the surface might not be so unlike that of the Moon. Antoniadi's map, published in 1934, was regarded as the best authority, and it was not his fault that it finally turned out to be very wide of the mark.

Jupiter was much more rewarding, if only because its surface was always in a state of change. Obviously it could not be solid, and in view of its high escape velocity (60 kilometres per second) it might be expected to hold on to all its original gases, including hydrogen and helium, which are so light that they have long since escaped from the Earth. The main features were the streaky 'cloud belts' crossing the planet, virtually parallel with the Jovian equator. Several belts could

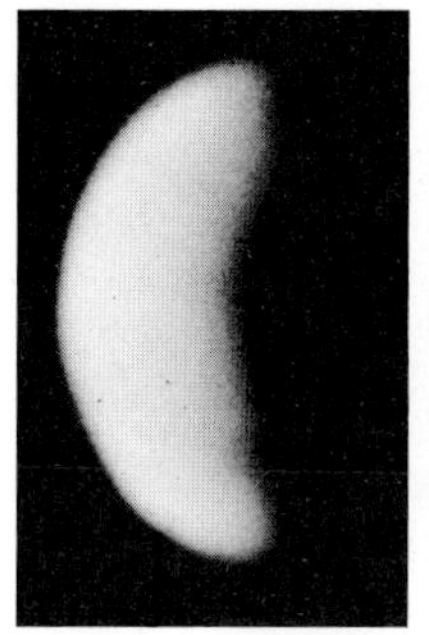

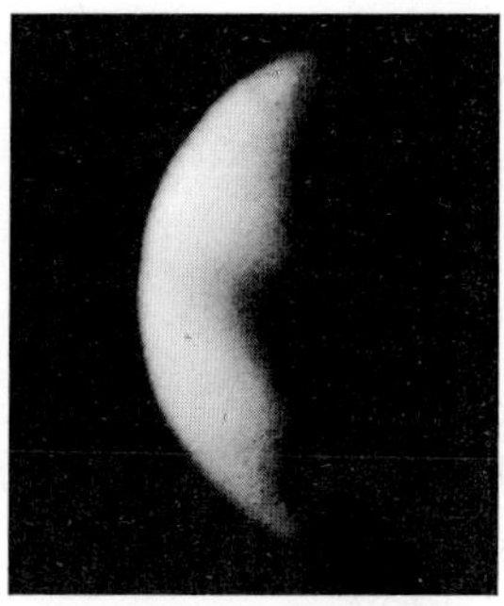

TWO PHOTOGRAPHS OF VENUS, *taken with the 254-cm reflector at Mount Wilson.*

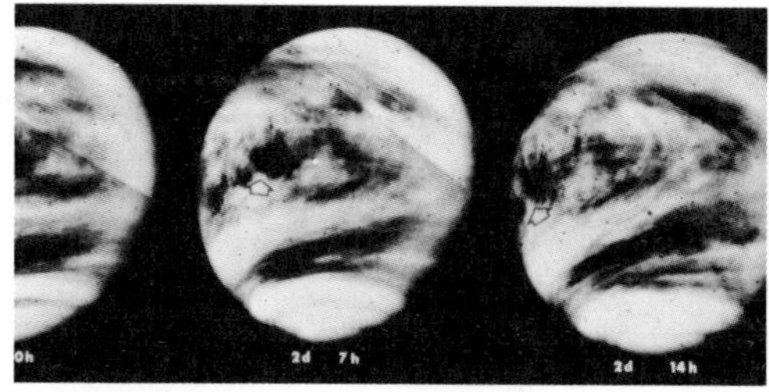

ROTATION OF VENUS. *The axial rotation of the planet is demonstrated in these photographs from Mariner 10; they gave full confirmation of the 4-day period of the upper clouds.*

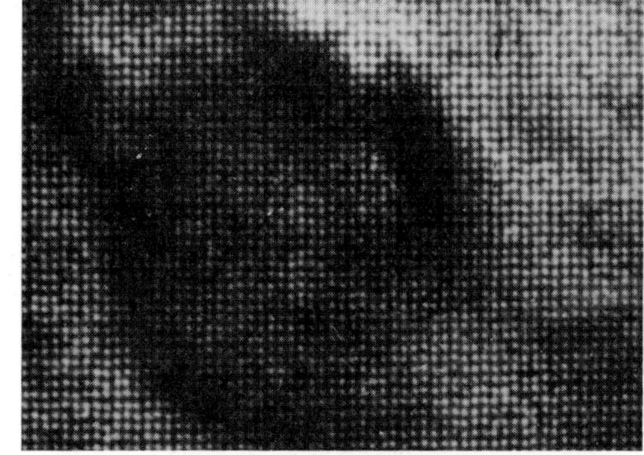

CRATER ON VENUS, *as mapped by R. M. Goldstein and his colleagues at the Jet Propulsion Laboratory in California. The area is just north of Venus' equator. The crater is about 160 km in diameter.*

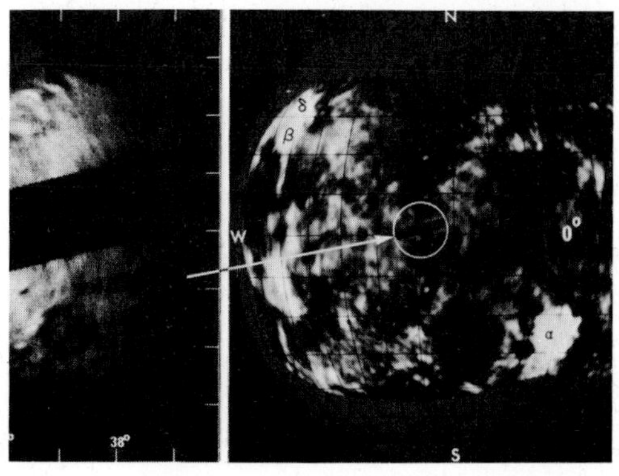

RADAR CHARTS OF VENUS, *showing craters. The black strip is due to the equipment used.*

SATURN IN 1949. *(Above) From an observation made by Patrick Moore on June 2, 1949, 22 h., 15·2-cm refractor × 400. The rings were almost closed. Four satellites are shown; from left to right, Dione, Enceladus, Tethys and Rhea. When the ring-system is at this angle to the Earth, surface features on the globe of Saturn are best placed for observation, since when the ring-system appears more open, an appreciable part of the globe is masked. However, the belts are always very much less conspicuous than those of Jupiter. Drawing by D. A. Hardy.*

VARIOUS ASPECTS OF SATURN'S RINGS. *(Right) Drawings by D. A. Hardy.*

SATURN, *(below) photographed with the 152-cm reflector at Mount Wilson.*

VENUS. *From observations by Patrick Moore. (Opposite, above) July 17, 1959, 16h. 55m., 61-cm reflector × 350. (Opposite, below) January 3, 1958, 16h. 30m., 31·7-cm reflector × 250. The Ashen Light is shown, but has been somewhat exaggerated for the sake of clarity. Drawing by D. A. Hardy.*

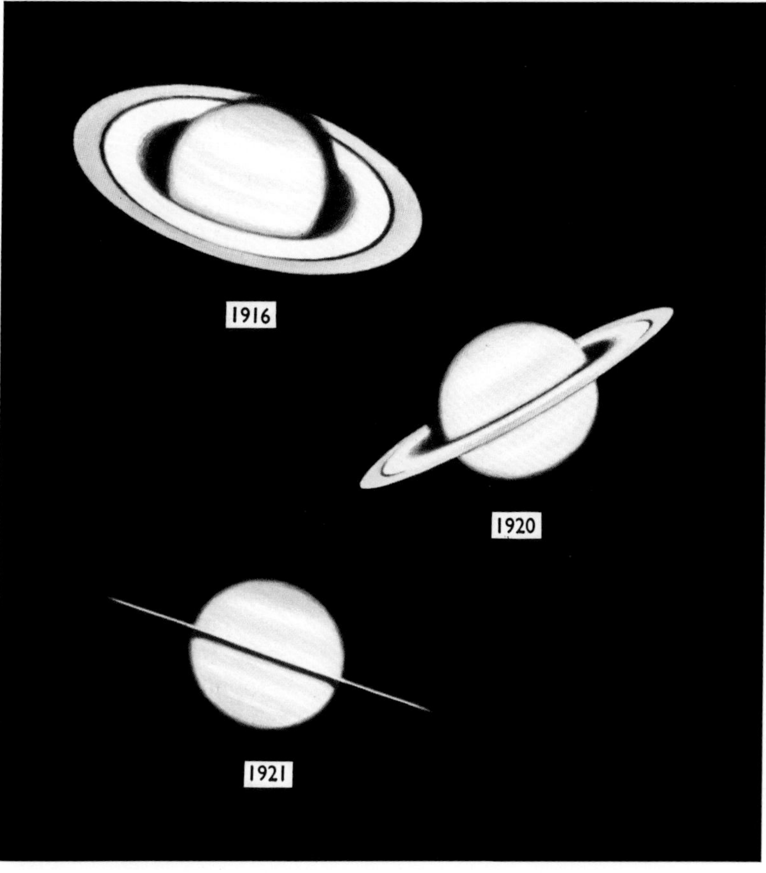

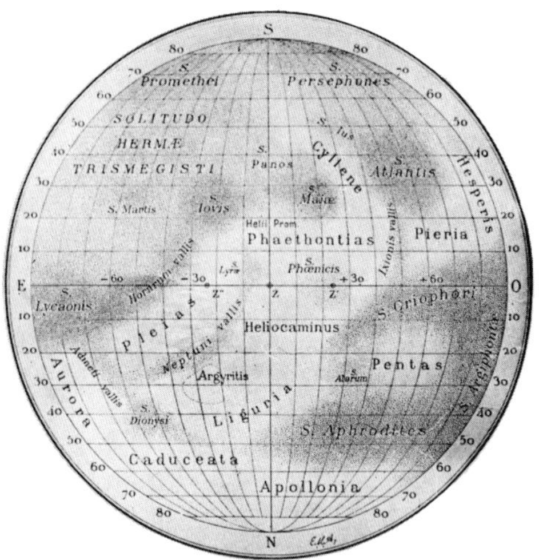

MAP OF MERCURY *by E. M. Antoniadi drawn from observations made with the Meudon 83-cm refractor. Antoniadi's nomenclature is now in general use.*

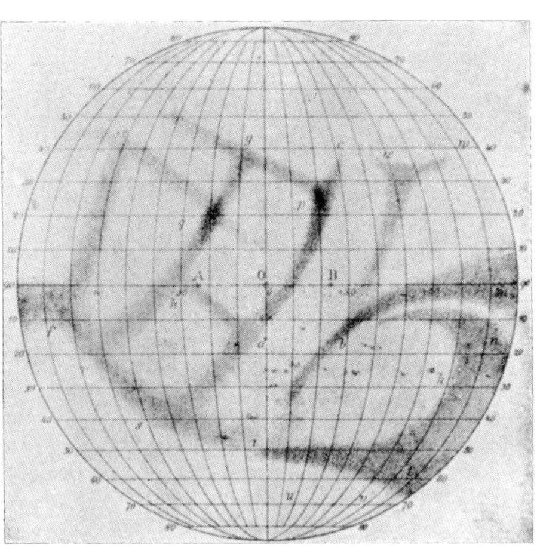

MAP OF MERCURY *by G. V. Schiaparelli*

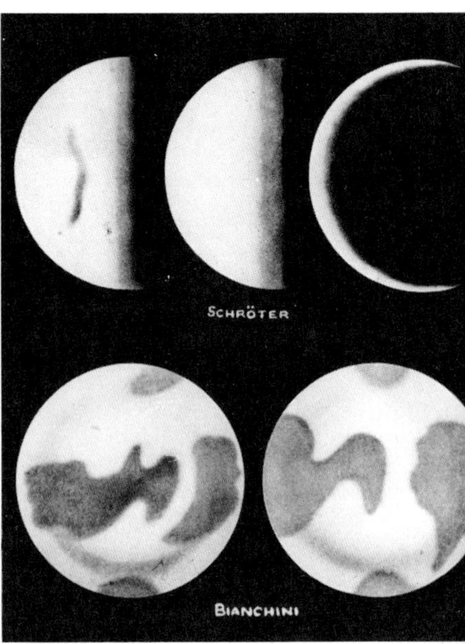

OLD DRAWINGS OF VENUS *by Bianchini and Schröter. The markings recorded by Bianchini, at least, are certainly illusory.*

always be seen, of which those to either side of the equator were particularly prominent. There was also the Great Red Spot, which had been seen intermittently for many years, and which suddenly became striking in 1878, when it was described as brick-red and oval in shape. It was then about 48,000 kilometres long by 11,000 broad, so that its surface area was greater than that of the Earth.

Observations showed that Jupiter's rotation period is very short. On the equator, a 'day' is less than 9 hours 51 minutes long. The period in higher latitudes is a few minutes longer, so that Jupiter like the Sun, does not spin in the way that a solid body would do. It is the quick rotation which causes the obvious flattening of the globe. The equatorial diameter is 142,200 kilometres, the polar diameter only 134,700 kilometres; the flattening is obvious when Jupiter is seen through a small telescope.

But was Jupiter a miniature sun, sending out appreciable light and heat? Many astronomers believed so; it was also suggested that the Red Spot could be an erupting volcano. It was not until the 1920s that theory proved that though Jupiter must have a hot core, its outer gases are bitterly cold, and there is nothing sunlike about it.

The four Galilean satellites, Io, Europa, Ganymede and Callisto, had been known since 1610. They are large globes; only Europa is smaller than our Moon, and Ganymede is larger than Mercury. In 1892 the keen-sighted Edward Emerson Barnard, using the Lick 91-centimetre refractor, discovered a fifth satellite, Amalthea, closer-in than any of the Galileans. It was the last planetary satellite to be found by visual observation; all the later discoveries have been made either by photography or from space-probes.

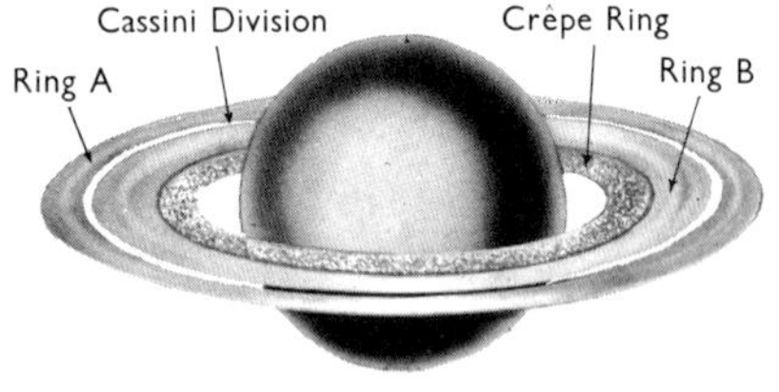

DIAGRAM OF SATURN'S RING SYSTEM *showing the Cassini Division, Rings A and B, and the Crêpe Ring (Ring C).*

Saturn is much more remote than Jupiter. Its mean distance from the Sun is 1,427,000,000 kilometres, as against 778,000,000 for Jupiter, and its 'year' is 29½ times as long as ours, as against 11¾ for Jupiter. Like its larger brother, Saturn is flattened at the poles; the equatorial diameter is 119,300 kilometres. The belts are much less conspicuous than Jupiter's, though a modest telescope will show at least two. Spots are rare, and there is nothing nearly so striking as the Great Red Spot on Jupiter.

Like Jupiter, Saturn has a gaseous surface, and it also was once thought to be a minor sun. In 1882 Proctor wrote a popular book, *Saturn and its System*, in which he stated that 'we seem to be forced to the conclusion that they [Jupiter and Saturn] are in a state of intense heat... over a region hundreds of thousands of square miles in extent, the glowing surface of the planet must be torn by subplanetary forces. Vast masses of intensely hot vapour must be poured forth from beneath.' This, remember, was written only a hundred years ago. We have learned a great deal since then!

The glory of Saturn lies in its ring system, which is visible with a small telescope when the rings are 'open'. When edge-on, as they were in 1980, they almost vanish; but at their best, they make Saturn the most glorious object in the sky.

The ring-system is of vast extent; from side to side the main part of it measures almost 275,000 kilometres. There are two bright rings, **A** and **B**, separated by a gap known as Cassini's Division in honour of its discoverer. The outermost bright ring **(A)** is 16,000 kilometres wide, and Cassini's Division has a width of 2,700 kilometres. Then comes the brightest ring **(B)**, 26,000 kilometres wide. Johann Encke, once Director of the Berlin Observatory and the man who instructed Galle and D'Arrest to hunt for Neptune, discovered a much narrower division in Ring **A**. The inner ring – the Crêpe or Dusky Ring – has a curious history, which takes us back to the early days of astronomy in America.

In 1835 William Cranch Bond was watching Halley's Comet from his private observatory. At that time there were no large telescopes in the United States, and it occurred to Bond and others that the time had come to remedy matters. Citizens in Boston and elsewhere subscribed enough money to buy a large instrument, and in 1847 a 38-centimetre refractor, with a wooden tube, was set up at Cambridge, Massachusetts. That was the start of the now world-famous Harvard College Observatory.

On July 16, 1850 Bond's son was observing Saturn with the 38-centimetre when he noticed a third ring, closer to the planet than the known ones, and much fainter. Indeed, it seemed to be almost transparent. This discovery, as interesting as it was unexpected, was made independently at about the same time by an English amateur, the Rev. William Rutter Dawes.

The Crêpe Ring is not a particularly difficult object to see, and it is strange that nobody recognized it before Bond and Dawes did so, though Galle from Berlin had recorded vague traces of it.

It was once thought that the rings must be solid or liquid sheets, but in 1848 this idea was disproved by the French mathematician Édouard Roche. Within a certain limit, now known as Roche's Limit, such a ring would be promptly broken up by the powerful gravitational pull of the planet, and the rings lie inside Saturn's danger zone. Roche therefore proposed that the rings must be composed of numerous small particles, each moving round Saturn in the manner of a dwarf moon. Final proof was given later by James Keeler, at Lick Observatory, who showed that the inner rings move round Saturn faster than the outer ones – in strict obedience to Kepler's Laws.

Today we know that the ring system is amazingly complex, but these revelations had to await the fly-by missions of the Voyager probes in 1980 and 1981.

SATURN IN 1933, *showing the famous white spot discovered by W. T. Hay. This is Hay's original drawing.*

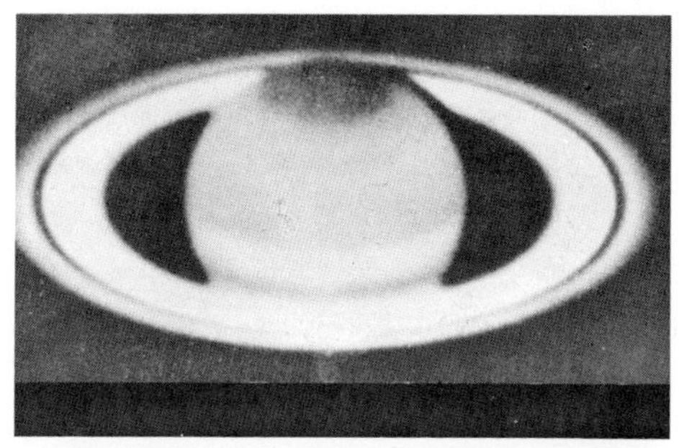

SATURN, *photographed at the Lowell Observatory.*

SATURN AND TITAN, *February 11, 1974; Commander H. R. Hatfield, 30·5-cm reflector. Titan is planet-sized, and has an atmosphere whose ground pressure is ten times that on the surface of Mars. Titan is here seen to the right of Saturn.*

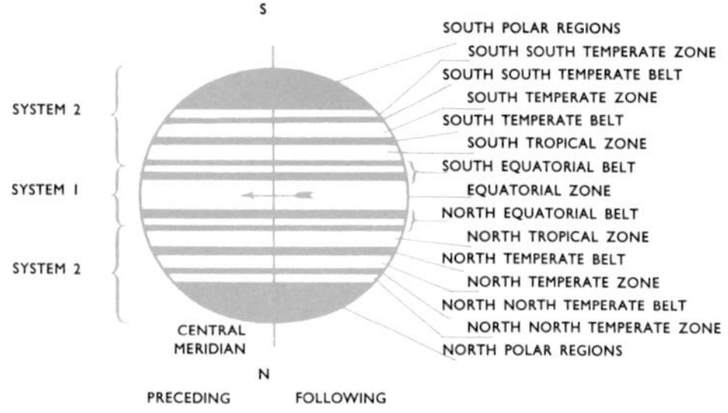

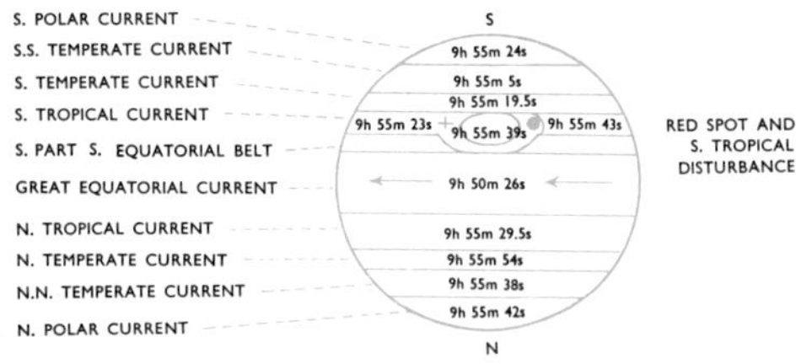

ROTATION PERIODS OF THE VARIOUS ZONES OF JUPITER. *The 9h. 50m. 26sec. zone is termed System I; the rest of the planet, System II.*

Of Saturn's satellites, the first – Titan, discovered by Huygens in 1655 – is much larger than our Moon, and is visible with a small telescope. Of the rest, Cassini's four (Iapetus, Rhea, Dione and Tethys) are between 1000 and 1500 kilometres in diameter. The two satellites discovered by Herschel when he first used the 12-metre reflector (Mimas and Enceladus) are smaller, and Hyperion, found by Bond in 1848, is smaller still; in 1981 Voyager 2 showed that it is irregular in shape, and it has been compared with a cosmic hamburger! The outermost satellite, Phœbe, was discovered by W.H. Pickering in 1898. It is very faint, and moves round Saturn in a wrong-way or retrograde direction, so that it may well be a captured asteroid. Yet another satellite was announced by Pickering in 1904, and named Themis; but it has never been seen again, and probably does not exist. All the other members of the Saturnian family have been discovered since 1978.

The great refractors, then, had provided a wealth of knowledge about the planets, even though they could show nothing on the tiny disks of Uranus or Neptune, and Pluto did not come to light until much later. The refractors were used, too, for stellar research. But their domination was coming to an end; the twentieth century belonged to the reflector, and the focus of research shifted from Europe to the United States.

SATURN, *photographed with the 508-cm Hale reflector at Palomar.*

JUPITER, *February 5, 1967; photograph by Commander H. R. Hatfield, 30·5-cm reflector.*

25

Great Telescopes

WE have reached the twentieth century. From here on, it is no longer possible to tell the history of astronomy in strictly chronological order. There is too much overlap. So I feel that the best course is to take each branch separately, tracing it through to modern times and beginning, naturally enough, with the story of the telescope. The central figure is that of George Ellery Hale.

Hale was born in Chicago in 1868. Astronomy fascinated him very early on; he began observing while he was still at school, after which he studied at Harvard and then Berlin. At the age of twenty-three he became famous for his invention of the spectroheliograph, which makes it possible to photograph the Sun in the light of one element only (for instance, hydrogen or calcium). At that period he was interested mainly in the Sun, and in 1908 he discovered that sunspots are centres of powerful magnetic fields. The importance of this discovery in solar physics can hardly be over-estimated.

His first observatories were privately financed by his father, a prosperous Chicago businessman, but they were too small to satisfy

THE NEW DOME OF THE INT *(Isaac Newton Telescope) at La Palma. Photograph by Patrick Moore, 1982.*

the young astronomer. In every way it was desirable to have a much larger telescope, and at first he looked at the possibilities of a big refractor. At that time the largest refractor in the world was the 91-centimetre at Lick Observatory. Hale wanted a 100-centimetre refractor, but naturally he could not afford anything of the sort, and moreover there were few people in the world capable of making such a lens. The leading expert was an American, Alvan Clark, but even for him the task would take years, and there was also the bill for the glass blanks, which alone would cost 20,000 dollars.

In 1892 Hale met Charles T. Yerkes, who was immensely rich, and who owned a large part of the city of Chicago. Hale was nothing if not persuasive. Yerkes could afford to pay for the telescope – and he did; the final cost reached 34,900 dollars. Clark made the lens, and the Yerkes refractor was duly completed. It was set up at Williams Bay, over a hundred kilometres from Chicago, and contained many instruments as well as the main telescope. Hale became Director, and within a few years the scientific results had more than justified the cost.

Yet it was clear that with this huge refractor, a limit had been reached. A larger lens would distort under its own weight (as was shown later by the fiasco of the Paris Exposition telescope). But to explore the depths of the universe something with a greater light-grasp than the Yerkes refractor was needed, and accordingly Hale searched for money to build a very large reflector. Again he was lucky. A financial trust known as the Carnegie Foundation (after Andrew Carnegie, one of the few men as wealthy as Yerkes) promised to provide a 153-centimetre reflector, and George Willis Ritchey, unrivalled as a mirror-maker at that time, was put in charge of the optical work.

One important problem was that of a site. The Earth's atmosphere thins out with increasing altitude, and since the dirtiness and unsteadiness of the air is probably the astronomer's worst enemy it is common sense to erect a large observatory at high altitude. Two peaks in California seemed to suit all the requirements: Mount Wilson, and Mount Palomar. Palomar was probably the better, since it was further away from city lights, but on the other hand it was harder to get at, and Hale and his colleagues decided upon Mount Wilson. The 153-centimetre reflector came into use in 1908, and more than fulfilled the hopes of its builders.

DOME OF THE 254-CM REFLECTOR AT MOUNT WILSON

THE 254-CM HOOKER REFLECTOR AT MOUNT WILSON

Even before then, a larger telescope had been planned. The suggestion came from another millionaire, John D. Hooker of Los Angeles, who went to see Hale and asked whether it would be possible to build a 254-centimetre reflector. Hale was confident that the answer was 'yes', and Hooker promised to provide the sum of 45,000 dollars – enough to pay for the mirror, though not for the mounting, which also was bound to be very expensive.

Money was only part of the programme; the work remained to be done. Casting the glass disk gave rise to many anxieties even before the actual grinding could be started, and to shape the mirror to the correct optical curve took Ritchey and his team six years. Meanwhile Andrew Carnegie had visited Mount Wilson, and had been so impressed that he agreed to meet the cost of mounting the 254-centimetre mirror.

One evening in November 1917, while the first world war was raging in Europe, Hale and three companions turned the Hooker reflector to the skies for the first time. They focused carefully on the planet Jupiter, and were appalled at what they saw – a shimmering, blurred image, lacking in any sort of detail. Could the mirror be useless after all, in spite of what Hooker, Carnegie and Ritchey had done?

Hale was not sure. The dome had been opened during the afternoon, and the mirror had become warm. Large glass disks take a long time to cool down and regain their correct shape, and a distortion of a tiny fraction of a millimetre would be too much. Hours later Hale went back to the telescope, and turned it towards the star Vega. This time all was well, and he knew that the telescope was as perfect as he and Ritchey had hoped.

It would be difficult to overestimate the value of the Hooker telescope's contribution to astronomy. It opened up entirely new fields of research; it was so much more powerful than any other instrument of the time that it was in a class of its own. With it, for example, Edwin Hubble was able to prove that the 'spiral nebulæ' really are external galaxies, far beyond our Milky Way system.

Still Hale was not satisfied. 'More light!' was his constant call. The Mount Wilson 254-centimetre reflector had been made, and proved to be successful. Why not a 6-metre mirror, or at least a 5-metre?

THE 508-CM HALE REFLECTOR AT PALOMAR. *This has some special features possible only because of its great size. The observing cage lies at the prime focus, and in the photograph an observer is shown – so that the astronomer is situated inside the telescope. The reflecting surface of the main 508-cm mirror is also shown. The same arrangement is adopted with the 600-cm reflector in the U.S.S.R.*

THE 152-CM TELESCOPE *in the Boyden Observatory, Bloemfontein, South Africa.*

Once again a friendly millionaire or wealthy institute would have to be approached. This time it was the Rockefeller Foundation. Hale marshalled all his arguments, and in the autumn of 1928 the trustees of the Foundation voted six million dollars for 'the erection of a great telescope of the reflector type with a mirror 200 inches in diameter'.

Mount Wilson was no longer suitable as a site; it was too close to the growing city of Los Angeles. Palomar was better, though less accessible. The Observatory Council first met soon after the grant had been approved, with Hale as chairman; it did not take long for the choice to be made, because Palomar had too many advantages to be ignored. It is a huge granite block, rising sharply from the valley and forming a plateau 24 kilometres long by 8 kilometres wide. From its top the view extends down to Mexico, with San Diego in between.

While work on the actual mirror was being started, Hale and several companions went to Palomar to investigate. This was in March 1934. They drove up the track known as Nigger Trail, and finally settled on a site near the middle of the plateau, 1680 metres above sea-level.

It seemed that everything was going well, but many problems lay ahead. Hale had wanted to have the 508-centimetre mirror made of quartz, but after lengthy experiments it was sadly found that no quartz disk of such a size could be cast, and Pyrex had to be used instead. Eventually it was cast. It had to be cooled very gradually, and at one stage floodwater from a nearby river invaded the optical works, so that the supply of electricity had to be cut off for several hours; it was lucky that the cooling block of glass was not damaged. By 1936 it was ready for the main work on it to begin, and it was taken to the Optical Shop in Pasadena, some way from Los Angeles.

Hale died in 1938, and then came the war. All work on the mirror stopped, and did not begin again until 1945. To turn the Pyrex block into a supremely-accurate mirror was difficult by any standards, but by 1947 it had been completed. In November of that year the finished mirror was packed up and taken to the new Observatory. It was a nerve-racking journey, undertaken at a snail's pace; the last sixty kilometres, from the town of Escondido to the Observatory itself, took six hours, and was completed only just before the weather broke.

The Observatory was opened on June 3, 1948 by Dr. Lee Du Bridge, President of the Carnegie Institute. His words are worth remembering: 'This great telescope before us today marks the culmination of over two hundred years of astronomical research. And for generations to come it will be a key instrument in Man's search for knowledge.'

The Hale reflector is truly impressive. Go into the dome, and the first thing you will see is what looks like a giant horseshoe. This is the mounting of the telescope; it weighs 500 tons, and yet it can be moved with absolute smoothness and precision. Inside it is mounted the telescope itself, which has no solid tube; the construction is of the skeleton type, with the mirror at the bottom. The optical system differs from the old-type Newtonian. The light from the object under observation strikes the main mirror and is reflected upwards, to be brought to focus at the observer's cage which is slung inside the tube itself. There is no need for a secondary mirror, and therefore no extra loss of light. The mirror is coated with a very thin layer of aluminium. In fact, the whole telescope exists merely to move around this tiny amount of metal!

THE 102-CM REFLECTOR *at Sutherland, South Africa, photographed by Patrick Moore.*

THE 66-CM REFRACTOR AT THE U.S. NAVAL OBSERVATORY, WASHINGTON. *This telescope is one of the great refractors erected during the latter part of the nineteenth century. At this period large refractors were generally preferred to large reflectors; lens-making had reached a high degree of perfection, whereas the mirror-making methods in use today had not been developed.*

The 66-cm has been used for a variety of purposes, both research and educational. For instance, extensive series of double star measures have been made with it. Its most famous association is with the two dwarf satellites of Mars, Phobos and Deimos, since this was the telescope with which Asaph Hall discovered them in 1877.

THE 122-CM SCHMIDT AT PALOMAR. *A Schmidt telescope, perhaps more appropriately termed a Schmidt camera, can photograph relatively large areas of the sky with excellent definition. Instruments of this sort cannot be used for visual observations.*

PATRICK MOORE'S 12·7-CM REFRACTOR. *An amateur-owned telescope, at Selsey, photographed by Patrick Moore.*

When the Hale reflector was first made, the observer had to spend long nights in the cage, taking photographs and checking continuously to make sure that all was well. Yet it did not take long for the telescope to prove itself; it could 'see' objects which were fainter and further away than any previously known. Using it, Walter Baade showed in 1952 that the universe is twice as large as had been believed. Magnificent photographs of distant galaxies were taken with it, though the main research depended upon photographs of the spectra of very faint objects. Even though it is now more than a quarter of a century old, it remains the second largest telescope in the world.

Times have changed. Photography is being rapidly superseded by electronic devices, and the Hale reflector has been adapted to these new techniques. The observer of today no longer sits in the cage; everything is handled from a control room at the side of the dome, insulated from the telescope itself, and the results are shown on a television screen. Everything has become fully automatic. Moreover, the power of the telescope has been enormously increased by the new electronics.

One problem of large telescopes is that they have small fields of view. To photograph the whole of the sky with the Palomar reflector would taken an impossibly long time. The solution to this problem was found in 1930 by an Estonian optical worker named Bernhard Schmidt, who had originally been concentrating upon explosives, but had blown off one of his arms and had turned to astronomy as being rather safer. Schmidt used a spherical mirror to collect the light, and combined it with a specially-shaped glass plate at the top of the tube to correct the resulting errors. In a Schmidt instrument, a very wide area can be covered with a single exposure. The original Schmidts were purely photographic, but visual modifications of it have since been developed. The main Schmidt instrument at Palomar has an aperture of 123 centimetres.

The one real problem facing Palomar is that of light pollution. It is not only the continued spread of Los Angeles; San Diego is growing too, and the skies are no longer so clear and transparent as they used to be in Hale's time. For the moment the work at the Observatory is not being seriously hampered, but the danger-signals are there.

In the southern hemisphere the main centres are Australia, South Africa and Chile. The main South African telescope, an 188-centimetre reflector, was set up at the Radcliffe Observatory outside Pretoria,* but before long conditions became badly affected by the increasing light pollution, and the decision was made to abandon Pretoria altogether. The Observatory was unceremoniously dismantled, and the reflector was taken to a new site at Sutherland in Cape Province, where it has since been used to the best possible advantage. Indeed, all of South Africa's main telescopes are now at Sutherland apart from the 68-centimetre refractor at Johannesburg, and the telescopes at the Boyden Observatory, Bloemfontein, where

*When the Radcliffe Observatory was established, the city authorities had the bright idea of naming local roads after prominent stars; Arcturus Road, Capella Road and so on. With Canopus they made a mistake, and called the road *Canopsus* Road. One of South Africa's leading amateurs – Jack Bennett, the comet-hunter – wrote to the Council and pointed this out. I have seen their truly glorious reply. It read: 'Our maps have been printed, so it is too late to alter them. Cannot you alter the name of the star?'

conditions remain good. The Boyden Observatory also has a powerful Schmidt instrument. (One of the Boyden domes, by the way, is graced by the presence of a cobra, which lives under the floor. I am delighted to say that I have never seen it, but, as I was told, 'We don't bother him, and he doesn't bother us!')

The largest Chilean telescopes, all internationally controlled, are at Cerro Tololo (401 centimetre mirror), La Cilla (381 centimetres) and Las Campanos (254 centimetres). In Australia, a 381-centimetre reflector was brought into use at Canberra, at the Mount Stromlo Observatory, in 1972. It was followed two years later by the Anglo-Australian Telescope, or A.A.T., at Siding Spring in New South Wales, near the little town of Coonabarabran. The 116-ton tube is 15 metres long; the mirror, 389 centimetres in diameter, weighs 16 tons, and is made of Cervit, a synthetic glass-like material chosen because its shape is almost unaffected by changes in temperature. The mirror's front face has been ground and polished with an accuracy of a few millionths of a centimetre; the aluminium coating is $\frac{1}{10,000}$ of a millimetre thick. There are three workable optical systems. One is the prime focus arrangement; another is a Cassegrain (the central hole in the main mirror is just over a metre across), and the third is the Coudé, in which a series of five mirrors (including the main one) brings the light to a focus lower down in the building. The focus does not move with the telescope, and enables the use of equipment which is too large or heavy to be used at the prime of Cassegrain focal points. The dome weighs 560 tons.

The A.A.T. is as sophisticated as any telescope in the world today. With it, astronomers were able to record the faintest object ever observed – the Vela pulsar, about which I will have more to say later.

The Russians are not likely to be left behind in this branch of scientific research. One of their main observatories, Pulkovo, near Leningrad, was completely destroyed by the Germans during the war, but has now been rebuilt. Another major centre is the Crimean Astrophysical Observatory, with its 264-centimetre reflector, completed in 1960. For some time it remained the largest in the U.S.S.R., but subsequently it was decided to make a real colossus – a 6-metre reflector, almost a metre bigger in aperture than the Hale telescope.

The Soviet Union has no site comparable with California, Australia or Chile, and the best choice seemed to be Mount Semirodriki, near Zelenchukskaya in the Caucasus Mountains, at an altitude of 2080 metres above sea level. Work on the 70-ton mirror was completed in 1974, and regular observing work was started in February 1976 after sixteen years' preparation. In theory, the telescope could detect the light from a single candle at a distance of 24,000 kilometres.

The 6-metre was revolutionary in one sense: it was set upon a massive altazimuth mounting. The trouble about an altazimuth is that the telescope has to be driven in two directions simultaneously; up and down (declination) or *alt*itude, and east to west (right ascension) or *azimuth* – hence the name. This involves two separate driving mechanisms, as against only one for the conventional equatorial. Modern computers can cope with this, and the Russian planners believed that the extra complications of driving were worthwhile. They were certainly correct. It is true that the 6-metre telescope has not yet produced much of value, and it has been plagued by what may be called teething troubles, but at least it has immense light-grasp.

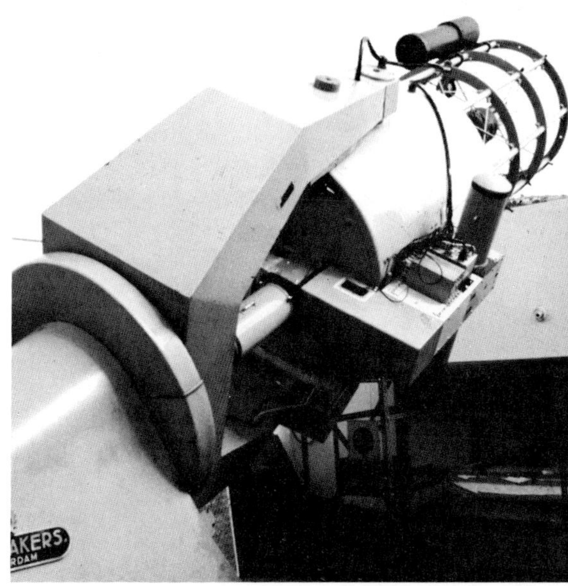

THE 91·4-CM REFLECTOR AT HARTEBEESPOORT, SOUTH AFRICA. *The setting of this telescope is completely automatic. Photograph by Patrick Moore, 1975.*

INSIDE THE AAT CONTROL ROOM. *Photograph by Patrick Moore, 1982.*

DOME OF THE 188-CM REFLECTOR AT THE COMMONWEALTH OBSERVATORY, *at Mount Stromlo in Australia.*

The altazimuth pattern is also to be followed for the main telescope in the new observatory at La Palma, in the Canary Islands. Since this is linked with the story of Greenwich, we must go back for a moment to see what happened at the Royal Observatory after the end of the nineteenth century.

George Biddell Airy had reorganized the Observatory during his long régime. In the time of his successor, Sir William Christie, a 71-centimetre refractor was added to the equipment, and in 1934 a 91-centimetre reflector – known as the Yapp, because it was financed by a wealthy enthusiast named William Yapp. Christie retired in 1910 (he died in 1922) and the next Astronomer Royal was Sir Frank Watson Dyson, who had previously been assistant at Greenwich and then Astronomer Royal for Scotland. Dyson was renowned for his work on stellar movements, and also for his rôle in promoting international co-operation in astronomy. He retired in 1933, and died six years later.

One difficulty became painfully obvious as the twentieth century drew on: Greenwich was no longer a suitable site for an observatory. In the days of Charles II, and even Queen Victoria, it had been a village outside London. Now London had spread, and the smoke, grime and artificial lights hid the stars, so that the telescopes could seldom or never be properly used. There was only one solution. The Royal Observatory must move.

THE DOME OF THE RUSSIAN 600-CM REFLECTOR. *This is now the world's largest telescope; it was completed in 1976, and has been set up in Siberia.*

Near the little Sussex town of Hailsham lies Herstmonceux Castle, where the skies are clearer and there are fewer inconvenient lights. The drastic decision was made to shift the Observatory there lock, stock and barrel. The move took many years to complete, but by 1960 only historical instruments were left at Old Greenwich, which was turned into a museum; it is now called Flamsteed House, and Wren's Octagon Room looks very much as it must have done when it was first built.

The chief organizer of the move was the eleventh Astronomer Royal, Sir Harold Spencer Jones, who had succeeded Dyson. He retired in 1954; then came Sir Richard Woolley, who held office until 1971, after which he went to South Africa to become director of the new observatories there. This led to a break with tradition. Directors of the Royal Greenwich Observatory had always been Astronomers Royal, but Woolley's successor, Dr. Margaret Burbidge, did not hold this title, which went to the Cambridge radio astronomer Sir Martin Ryle, who retired in 1982 and was succeeded by another eminent radio astronomer, Professor F. Graham Smith. Dr. Burbidge soon resigned as Director of the Royal Greenwich Observatory, to be followed in succession by three distinguished astronomers; Dr. Alan Hunter, Professor F. Graham Smith, and (in 1981) Professor Alec Boksenberg.

In 1967 a large reflector was set up at Herstmonceux: the Isaac Newton Reflector. It was always known as 'the 98-inch', though in fact the clear diameter of the mirror was only 96 inches. This is still over 240 centimetres, so that the 'I.N.T.' was one of the largest reflectors in the world for sheer size, but it soon became obvious that things were not entirely satisfactory. Herstmonceux is probably as good a site as can be found anywhere in Britain, but astronomically this is still poor. There is too much cloud and mist, to say nothing of light pollution. The

MOUNT STROMLO OBSERVATORY, AUSTRALIA. *At present the largest reflector in the southern hemisphere is at Mount Stromlo, at Canberra in Australia (the Commonwealth Observatory). The mirror is 188-cm in diameter; the dome is shown in this photograph. The moving parts of the telescope weigh 40·64 tonnes, while the revolving dome has a weight of 101·61 tonnes – and yet can be moved smoothly on its track at the rate of one complete revolution in five minutes. Because of the southern latitude, astronomers at the Commonwealth Observatory can study objects such as the Clouds of Magellan which can never be seen by their colleagues at Palomar or Mount Wilson.*

MIRROR END OF THE 127-CM ZEISS REFLECTOR AT THE CRIMEAN ASTROPHYSICAL OBSERVATORY. *(Right) Photograph by Patrick Moore, 1960.*

DOME OF THE 127-CM REFLECTOR AT THE CRIMEAN ASTROPHYSICAL OBSERVATORY, *taken from the dome of the adjacent 259-cm by Patrick Moore, 1960.*

I.N.T. could seldom be used to its full capacity, and so negotiations were opened with the Spanish Government for a new site in the Canary Isles, at La Palma (not Las Palmas, please note). A new observatory came into being, now called 'El Observatorio del Roque de los Muchachos' or the Observatory of the Rocks of the Two Boys – the Boys (Muchachos) being twin rocks at the summit of the main peak of the island.

The I.N.T. was removed from Herstmonceux, given a new and slightly larger mirror, and re-mounted at La Palma. The second telescope to be installed was a new 100cm reflector. Finally there will be the William Herschel 180-centimetre reflector, mounted on an altazimuth pattern and computer-driven. Within the next decade there will be a new departure: the astronomer using one of the La Palma telescopes will not need to be in the Canaries at all – he will be able to control everything, including the movements of the telescopes, from Britain.

In fact this sort of procedure has already been adopted, and in 1982 an astronomer in the Royal Observatory, Edinburgh, was able to operate the famous UKIRT or United Kingdom Infra-Red Telescope, which is certainly not in Scotland; it is on top of Mauna Kea, the extinct volcano in Hawaii, at an altitude of 4600 metres!

Mauna Kea Observatory was the brainchild of Gerard Peter Kuiper, Dutch by birth but American by adoption. Kuiper was born in Holland, and became assistant astronomer at the Leiden Observatory. In 1933 he emigrated to the United States, and remained there for the rest of his life; he was naturalized in 1937. He worked successively at the observatories of Lick, Harvard, Chicago and Texas before going to Yerkes, where he was Director between 1947 and 1949 and again between 1957 and 1960. Finally he went to Arizona, where he founded the now famous Lunar and Planetary Laboratory. He was particularly noted for his Solar System researches; for instance, he was the first to detect an atmosphere round Titan, the largest satellite of Saturn (1944) and he was one of the main investigators in the programme of exploring the Moon and planets by means of space-craft. It was indeed thoroughly fitting that the first crater to be detected on the surface of Mercury, by the Mariner 10 probe, should have been named in his honour.

355·6-CM REFLECTOR *at the Siding Spring Observatory in Australia.*

But Kuiper was also concerned with stellar astronomy, and he paid special attention to infra-red research. The greatest enemy of the infra-red astronomer is atmospheric water vapour. Kuiper reasoned that the only solution, so far as Earth-based telescopes were concerned, was to climb as high as possible. He made an ascent of Mauna Kea, and liked what he saw. He managed to interest first the University of Hawaii and then NASA, the American space agency; before long Britain joined in, and today there are four large telescopes on the summit of the volcano. The UKIRT has a mirror 380 centimetres in diameter; there are also reflectors operated by NASA (305 centimetres), jointly by Canada, France and Hawaii (366 centimetres) and the University of Hawaii (224 centimetres). The Mauna Kea Observatory was officially opened in 1979. Unfortunately Kuiper did not live to see it; he had died five years earlier.

DOME OF THE 242-CM REFLECTOR AT HERSTMONCEUX, *photographed by Patrick Moore.*

From the summit, ninety per cent of the atmospheric water vapour lies below, so that conditions for infra-red astronomy are excellent. The UKIRT was designed specially for this sort of work. An infra-red telescope, dealing with very long-wavelength radiations, need not be so perfect as an optical telescope, and accordingly the UKIRT mirror was lightweight, weighing only 6½ tonnes instead of the required 16 to 17 tonnes for an ordinary mirror. In fact, it has proved to be so good that it can be used for optical work as well as infra-red, which is sheer bonus. It, too, is fully computerized, and there are special advantages here, because there are many people who cannot tolerate so high an altitude. Even regular observers have to spend several hours at the 'half-way house', Hale Pohaku, just below 2600 metres, before making the final drive to the summit.

Next, I must say something about two telescopes which must be classed as unconventional, or – more properly – revolutionary. These are the Multi-Mirror Telescope or M.M.T. at Mount Hopkins in Arizona, and the solar telescope at Kitt Peak, also in Arizona.

Making a large mirror is both difficult and expensive. Suppose that it were possible to make several smaller mirrors, and combine their light-paths to produce a single image? This is what has been done with the M.M.T. There are six main mirrors, each 183 centimetres in diameter and of 'eggcrate' construction, so that each weighs only one-third as much as a normal mirror of the same size. They work together, and the overall result is equal to a reflector with a single mirror 447 centimetres in diameter. There are only two telescopes in the world larger than that: the Hale reflector, and the Soviet 6-metre.

The M.M.T. does not look much like a telescope. It is short and squat, with a maze of girders and wires, in the midst of which are the six primary mirrors. There is also a separate 76-centimetre mirror which is not connected with the main system, and acts as a guide. All the mirrors are silica-plated, and are accurate even by today's exacting standards.

Keeping even a few mirrors in true alignment is troublesome enough, and it is even worse with the six primaries of the M.M.T. The solution is to use a laser beam. This may be described as a pencil-thin beam of light, known technically as coherent light, which does not spread out, and which can therefore be very accurately positioned. A transmitter is used to send a laser beam through the optical system, producing an artificial star. The required position of this artificial star is known. If the image lies exactly where it ought to be, all is well. If not,

DOMES ON MAUNA KEA, HAWAII. *Photo by Patrick Moore, 1982.*

an automatic adjustment is made until the alignment is perfect. At present there are still problems to be faced (insects flying into the optical system are constant nuisances) but the basic principle is sound enough.

The mounting is altazimuth, and the entire observatory rotates with the telescope. It perches on top of Mount Hopkins, between 2400 and 2800 metres above sea level, and when seen from the distance it looks almost like a child's toy. It is reached by a mountain road calculated to break the nerve of even a skilled driver; the surface is rough, and there is a sheer drop to one side, unshielded by any protective railing!

The M.M.T. has proved to be very satisfactory indeed. It has been built more cheaply than a single 447-centimetre telescope could have been, and it is the first of what we may confidently expect to be a whole new generation of astronomical instruments. Multi-mirror telescopes are being planned now, one of which will be erected on Mauna Kea.

Kitt Peak, near the city of Tucson, is different again. The first dome there was erected in 1957, and today there are seventeen in all, one of which contains a 401-centimetre reflector – the fourth largest in the world (unless we include the M.M.T.). But the most interesting instrument is the tunnel telescope, originally intended for studying the Sun, though in fact it is now used on stellar work as well.

The upper mirror, 203 centimetres in diameter, is called the heliostat. It can be rotated; it catches the Sun's light, and directs the rays down an inclined 183-metre tunnel in a fixed direction. At the bottom of the tunnel there is a curved 152-centimetre mirror, which reflects the rays back to another mirror, flat and about mid-way in the tunnel. This directs the rays down through a hole into the laboratory, where the image is formed. The great advantage of this arrangement is that the sunlight always arrives at the same point, so that the very heavy analytical equipment does not have to be moved at all.

Kitt Peak, now America's national observatory, was yet another new departure. (One of the initial problems concerned the local Papago indians, who occupied the reserve. The sacred mountain Babuquivari,a prominent landmark from the Observatory, is regarded by them as the centre of the universe, with their gods living in nearby caves. After prolonged negotiations a lease was acquired, with the firm promise that the caves would never be disturbed.) The tremendous light-grasps of the telescope means that the light can be widely spread-out or 'dispersed' into a spectrum. Moreover, the telescope can also be used at night-time for stellar work. There are also some very large conventional telescopes, including a 401-centimetre reflector.

I realize that this survey of modern telescopes is hopelessly incomplete, but at least it may serve to underline the tremendous advances which have been made in recent years. Next we may hope for the Space Telescope, a 239-centimetre reflector due to be launched from the Space Shuttle in 1985. It will be a free-orbiting instrument, controlled from the ground, and since it will be free from the atmospheric screening and turbulence it should be far more effective than any telescope previously made.

We have come a long way since the time of Galileo.

THE M.M.T. OR MULTIPLE MIRROR TELESCOPE, *on Mount Hopkins. The top part of the telescope is shown. Photograph by Patrick Moore, 1982.*

DOME OF THE 259-CM REFLECTOR AT THE CRIMEAN ASTROPHYSICAL OBSERVATORY, *(opposite) photographed by Patrick Moore in October 1960; the dome construction was not quite complete, and scaffolding can still be seen.*

26

The Age of Photography

FROM an astronomical point of view, the human eye can be a most infuriating mechanism. It is easily deceived, and it is not reliable. Moreover, it sees everything that it can possibly do very quickly. If the observer keeps on staring, he is inclined to see less and less.

Worse, who can tell exactly what another person sees? The only method is to draw the object being studied, and there are many keen-sighted observers who are poor draughtsmen (Johann Schröter being a good example). There is also the danger of 'seeing' what one expects or hopes to see.The classic case here is that of the canals of Mars, but we also have Saturn's inner Ring D, which was reported by the French astronomer Guèrin in the 1970s and was confirmed by various visual observers – though, as we now know from the Voyager results, it is much too faint to be seen from Earth. Wishful thinking is always a hazard.

The first method of recording a permanent image, unaffected by any personal prejudice, was that of using a camera, and it may be of interest to note that the word 'photography' was coined by no less a person that Sir John Herschel. At first it was a very hit-or-miss process, but its development was comparatively rapid after about1860.

This is no place to tell the story of photography, so I propose to deal only with its purely astronomical connections. It began in the 1830s, with such pioneers as Nièpce and Daguerre, and in 1839 it was the subject of an address by the famous French astronomer François Arago. He was suitably enthusiastic, and went so far as to say that 'we may hope to be able to make photographic maps of the Moon, which means that we will carry out one of the most lengthy, most exacting, most delicate tasks of astronomy in a few minutes'.

Arago's optimism was premature, since the detailed photographic charting of the Moon was not completed until the flights of the Orbiter space-probes more than a hundred and twenty years later; but the power of astronomical photography was undoubted, and it was applied with great energy. To an American, J.W. Draper, must go the honour of being the first man to take a picture of the Moon. This was on March 23, 1840. He used a Daguerreotype process, and the telescope to which he fixed his camera was a refractor with a 12-centimetre objective. The result was an image 2·5 centimetres in diameter, and the exposure time was a full twenty minutes. At least it was a start, and ten years later good Daguerreotypes of the Moon were being taken in the United States by William Cranch Bond a.,d J.A. Whipple. Whipple was interested solely in photography, but Bond was one of the leaders of early American astronomy.

Bond was born at Falmouth, in Maine, in 1789. He became a watchmaker, but his fame as an amateur astronomer spread quickly, and led to his being offered the directorship of the new Harvard College Observatory; as we have noted, he was one of those responsible for collecting sufficient money for the first Harvard telescope. He made several important discoveries, including those of Hyperion, the seventh satellite of Saturn (1848) and also the Crêpe Ring (1850), but he is remembered above all for his pioneer photography. He died in 1859, and was succeeded as Director by his son, George Phillips Bond, who was the first astronomer to photograph a comet. Unfortunately he died only six years after his appointment at Harvard.

The main trouble in those early days was the lack of light, and at first only the Sun and the Moon were bright enough to be recorded photographically. In 1842 Lerebours, in France, made a Daguerreotype of the Sun which showed a certain amount of detail, and three years later H. Fizeau and L. Foucault did rather better. In 1851 Berkowski succeeded in photographing a total solar eclipse, showing the inner corona as well as some prominences – a foretaste of what was to come. By 1854 J.T. Reade in England was able to produce a solar photograph upon which the Sun's image was 23 centimetres in diameter, and showed distinct mottling. Reade used a 61-centimetre reflector, which by the standards of the time was large.

Really good results had to await improved photographic techniques. The Daguerreotypes became obsolete; collodion processes took over, and the change was very marked. So far as the Moon was concerned, the main pioneers were Warren de la Rue, Lewis Rutherfurd and Henry Draper, who recorded a considerable amount of detail even though they could not hope to fulfil Arago's expectations.

Warren de la Rue was a Channel Islander. He was born in Guernsey in 1815, went to England, and amassed a considerable fortune as a paper-maker. He was a man of wide interests, and during the 1850s and 1860s he took some excellent lunar pictures, though it is true that his interest in astronomy waned somewhat in his later years (he died in 1889). Lewis Rutherfurd was born in New York State in 1816, and improved upon de la Rue's lunar results; he also undertook pioneer work in stellar spectroscopy. The third member of the trio was another American: Henry Draper, son of J.W. He hailed from Virginia, where he was born in 1837, and studied medicine before going to Harvard. When he died, in 1882, his widow established a fund to carry out spectroscopic work, and the final result was the great Draper Catalogue of Stars, which was of immense value and is still used today.

Group of galaxies in Leo, *photographed with the 508-cm Hale reflector at Palomar.*

Good photographs of the Sun were being taken before 1860. At the suggestion of Sir John Herschel, a special photographic telescope was set up at the Kew Observatory (built, you may recall, for George III), and daily sunspot pictures were taken. The observatory itself still stands, though it has not been used astronomically for many years now. At present, efforts are being made to preserve it.

The stars were more difficult subjects, because they are so much fainter. Again Bond and Whipple took the lead, and in 1850 they managed to photograph two stars, Vega and Castor. Castor, the senior though fainter of the two Twins, is double, and that first picture showed it as elongated. Seven years later, Bond obtained a picture of the Mizar-Alcor pair in the Great Bear.

Planets presented problems of their own, and time-exposures were essential. This meant using a telescope driven mechanically; hand-guiding is never satisfactory. Equatorially mounted telescopes were already in existence; the first had been Fraunhofer's refractor installed at Dorpat, in Russia*, in 1824. The same need for time exposures applied to dim objects such as nebulæ.

This leads me on to a question which I have been asked many times. Is it possible to hold up an ordinary camera to the eye-end of a telescope, and hope to obtain good pictures of stars, planets or anything else in the sky? Frankly, the answer is 'no'. Real stability is essential. Also, remember that if you magnify an image by (say) 200 times, you are also magnifying its apparent movement by 200 times; and unless your telescope is driven so as to compensate for the Earth's rotation, the object will race out of the field of view. Nowadays it is possible to take good lunar or planetary pictures with exposures of a few seconds, but even then a driven telescope is needed, and this was even more necessary a century ago, when photography was so much less advanced.

The development of astronomical photography was, then, linked inextricably with that of telescopes and telescope-mounts. Well before the end of the last century things had reached a stage in which really good stellar and nebular pictures were obtainable, and in the 1890s the French astronomers Loewy and Puiseux produced a photographic lunar atlas which was a vast improvement on any hand-drawn map. By then, too, it had become possible to take photographs of the spectra of the Sun and the stars. But the most important development of all came as the result of a happy accident.

Sir David Gill had become Cape Astronomer in South Africa, and further improved and modernized the Observatory's equipment. He was enthusiastic about the new methods, and when a bright comet appeared, in 1882, he decided to photograph it. When he examined the picture, he saw that as well as recording the comet, he had shown an amazingly large number of stars – and at once he realized that this was the way to map the sky. There had been many catalogues of stars drawn up by visual methods; as long ago as 1852 Friedrich Wilhelm Argelander had published the positions and magnitudes of 342,198 stars – a truly Herculean task. But no visual measurement could match the accuracy of photography, and Gill proposed an international scheme to produce a photographic atlas of the sky.

This Carte du Ciel proved to take far longer than had been expected, but it was completed eventually; and since then, of course, there have been further atlases – such as the Palomar Sky Survey, which shows so many stars and other objects that to catalogue them all would be a physical impossibility.

The British amateurs Andrew Common and Isaac Roberts were taking splendid stellar pictures before 1900, and by the turn of the century it was fair to say that for most branches of astronomy the photographic plate had replaced the human eye, so that major telescopes were used chiefly as large cameras. Today we are in the midst of another revolution. Photography itself is being superseded by electronic devices, and in a few years' time it seems that electronics will have taken over.

One more point is worth making. Look at a photograph of, say, the Orion Nebula, taken with a telescope such as the Hale reflector. You will see glorious colours. Unfortunately the light-level is too low for these colours to be seen by the observer at the eye-end of a telescope – and so if you fail to see them, do not be disappointed.

---

*Or, more precisely, in Estonia – which after a brief period of independence, is again part of the modern Soviet Union.

27

Exploring the Spectrum

I MUST go back for a moment to the year 1666, when young Isaac Newton used a prism to produce a rainbow *spectrum* of sunlight. This was the first indication that what we call 'white' light is really a mixture of colours, and it proved to be one of the most important discoveries in astronomical history.

In 1802 the English doctor William Hyde Wollaston noticed seven dark lines in the solar spectrum. Wollaston, a native of East Dereham in Norfolk, had taken a medical degree at Cambridge, and had become known as a man with a wide knowledge of all branches of science. Yet on this occasion he missed a great opportunity. He thought that the dark spectral lines merely indicated the boundaries between various colours, and he paid no more attention to them.

Fraunhofer saw the dark lines again in 1814. Unlike Wollaston, he inquired further, and found that the lines never seemed to change; whenever he examined the Sun's spectrum the familiar dark lines appeared, always in exactly the same positions and with the same intensities. Fraunhofer measured the positions of over five hundred of them, and they are still often referred to as the Fraunhofer lines.

THE SOLAR TELESCOPE AT KITT PEAK, ARIZONA.
Photograph by Patrick Moore, 1982.

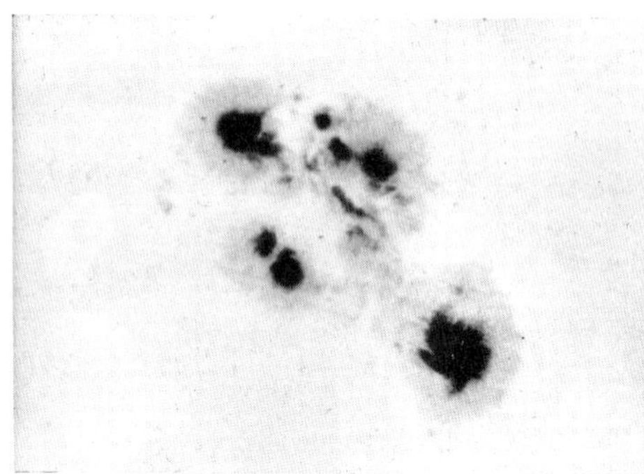

SUNSPOT GROUP, *photographed on March 31, 1960 by W. M. Baxter, using a 10-cm refractor.*

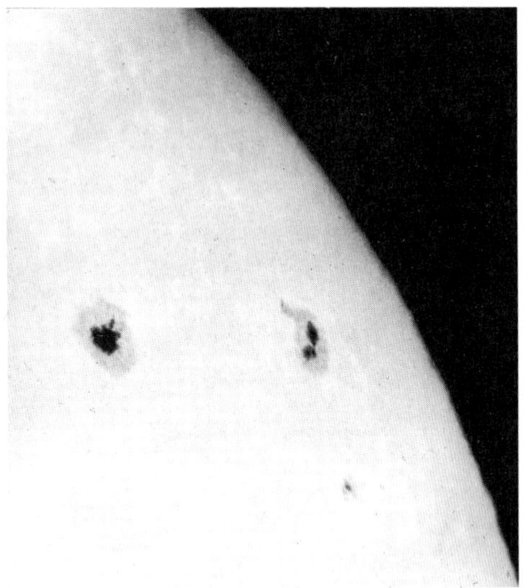

SUNSPOTS, *photographed on October 17, 1959 by W. M. Baxter, using a 10-cm refractor. Note the foreshortening of the group nearer the Sun's limb.*

SOLAR PROMINENCES. *(Opposite above) Prominence 212,000 km high, photographed in calcium light on August 18, 1947. (Opposite, below) Prominence 128,000 km high, photographed in calcium light on August 21, 1909. Photographs by Mount Wilson and Palomar Observatories.*

They showed some particularly interesting characteristics. For instance, the solar spectrum revealed a prominent dark double line in the yellow section. Luminous sodium vapour showed a *bright* double yellow line, and Fraunhofer wondered whether there might be some connection. Given enough time he might well have hit upon the solution, and it was tragic that he died so young.

The mysterious dark lines were explained in 1859 by Gustav Robert Kirchhoff, who had been born at Königsberg in 1824, had been educated at Berlin and Marburg, and had become Professor of Physics at Heidelberg University. For a time he was joined in the research by his colleague Bunsen, of Bunsen burner fame. To show how he reasoned, I must say something about the way in which matter is built up.

All matter is composed of *atoms*, which combine into groups or *molecules*. They are of different types, and are inconceivably small. The simplest atom of all is that of hydrogen, which is the lightest of all substances, and is also more plentiful in the universe than anything else. Its atom may be said to be made up of a central particle or *nucleus*, around which revolves a different sort of particle known as an *electron*. The nucleus itself consists of a *proton*, which carries a unit positive charge of electricity. To balance this, the circling or planetary electron carries a unit negative charge; this cancels out the effect of the proton, so that the complete atom is electrically neutral.

The next lightest substance is helium. This time the nucleus is more complex, and carries two positive charges. There are however two planetary electrons, and once again the complete atom is electrically neutral.

Lithium has three planetary electrons; consequently there are three charges in the nucleus, and so on. Each substance, or *element*, has an extra electron. Oxygen, for instance, has eight. As the number of electrons goes up, the atom becomes more and more complicated, until we come to uranium, with ninety-two electrons.

All the matter known to us is made up of these fundamental elements, which make up a definite sequence. We may be certain that no elements in the series between hydrogen and uranium remain to be discovered; one cannot have half an electron. The series is continued beyond uranium, but most of these highly complex atoms are very unstable.

It may seem surprising that there are so few elements, but of course they may combine in many different ways. Water, for example, is made up of hydrogen and oxygen. Two atoms of hydrogen combine with one of oxygen to make up a water molecule, giving the chemical formula H_2O. There is a rough analogy here with writing; all the thousands upon thousands of words in the English language are made up from only twenty-six fundamental letters of the alphabet, from A to Z.

I must make it quite clear that this picture is hopelessly over-simplified. We cannot think of protons and electrons as tiny solid lumps. But it will suffice for the moment, so let us go back to Gustav Kirchhoff. In 1859 he announced three Laws which still bear his name; these Laws form the basis of all spectroscopy, and give a complete explanation of the Fraunhofer lines.

The first Law is straightforward enough. It states that incandescent solids, liquids, or gases under high pressure produce a rainbow band or *continuous* spectrum. There is a full range of wavelengths, from red

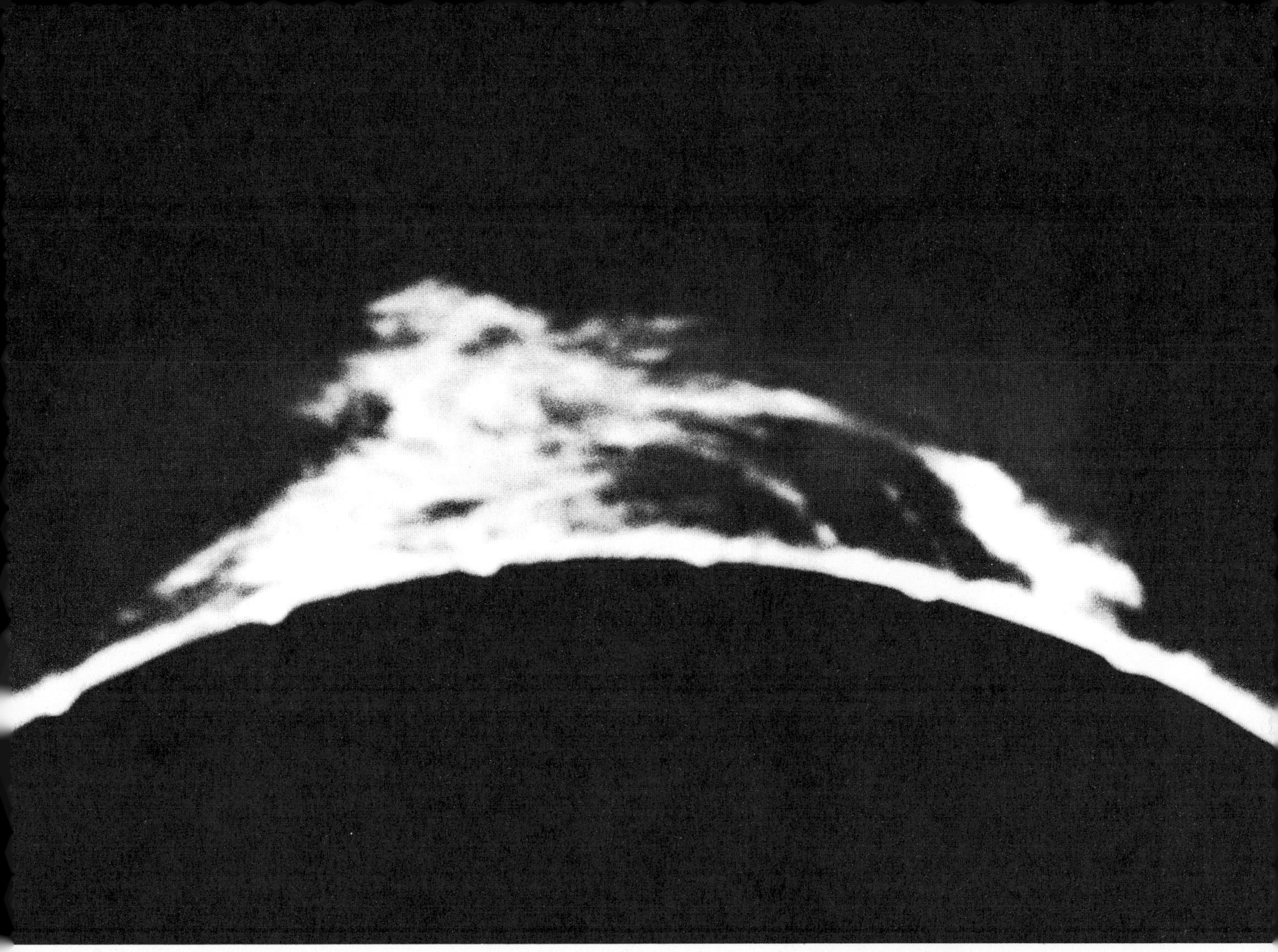

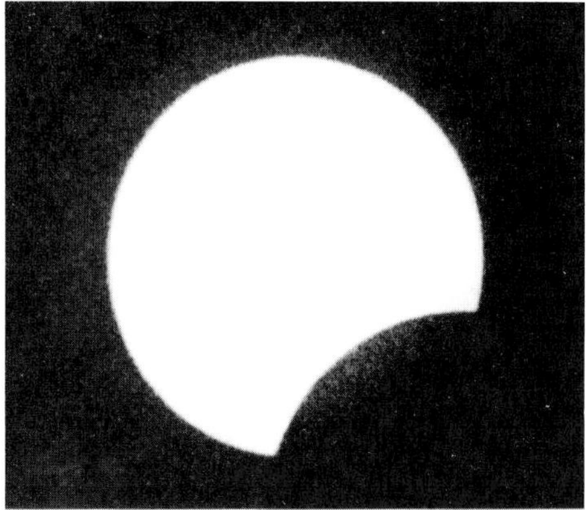

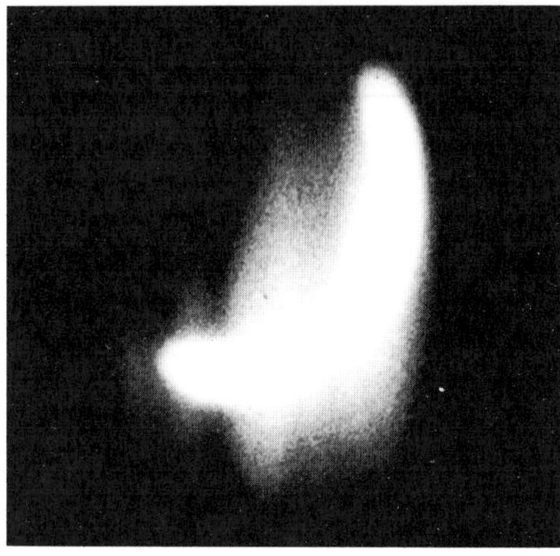

THREE STAGES OF THE SOLAR ECLIPSE OF FEBRUARY 15, 1961. *These photographs were taken from an R.A.F. Shackleton aircraft flying at 3200 m over the Bay of Biscay. The first two pictures were taken before totality, and the third one after totality. Conditions were not good, and some high-altitude cloud shows up on the photographs.*

at the long-wave end of the band down to violet at the short-wave end.

The second Law states that a luminous gas or vapour under low pressure will produce not a rainbow band, but a number of isolated bright lines. This is an *emission* spectrum. The vital fact is that each element (or group of elements) will produce its own distinctive set of lines. The double yellow line of sodium, seen by Fraunhofer, cannot possibly be produced by any element except sodium; lines due to hydrogen cannot be copied by anything except hydrogen, and so on. Each element has its own particular trade-marks, which cannot be duplicated. Often the spectrum is very complicated, and iron alone produces hundreds of lines, but careful measurement will usually enable the research worker to disentangle one line from another.

The crux of the whole dark-line problem is Kirchhoff's third Law. The best way to explain it is to describe a simple experiment. If you burn salt in a flame, you will produce an emission spectrum, including the double yellow line. This line is due to sodium – which is natural enough, because salt is made up of sodium and chlorine. If you next look at the spectrum of an electric light bulb, you will see a rainbow, since the filament of the bulb is an incandescent solid. Now take the bulb and put it behind the flame, so that you are examining the flame against the background of light produced by the bulb. Instead of a rainbow crossed by bright lines, what you will see takes the form of a rainbow crossed by *dark* lines. The atoms in the sodium vapour are removing part of the corresponding portion of the continuous spectrum, producing dark *absorption lines*. As soon as the background bulb is removed, the lines due to the sodium vapour flash out and become brilliant once more. Yet their positions in the spectrum, as well as their intensities, are unchanged.

In the Sun, the luminous surface or photosphere takes the place of the bulb; since it is incandescent high-pressure gas, it produces a rainbow. In front we have the luminous gases above the Sun, known as the chromosphere. By itself, the chromosphere would yield emission lines; but since the rainbow is behind it, these lines are *reversed*, and appear dark. They are nothing more nor less than the Fraunhofer lines. Among them is the celebrated double in the yellow region. Since this must be due to sodium, we can prove that there is sodium in the Sun.

By now, some seventy of the ninety-two naturally-occurring elements have been identified in the solar spectrum. One of them was even found in the Sun before it was known on Earth. In 1869 the English astronomer Norman Lockyer found a Fraunhofer line in the orange-yellow region which he could not identify. He suggested that it might be due to an unknown element, and proposed to name it helium, after the Greek word for 'sun'. A quarter of a century later, W.R. Ramsay detected helium on the Earth. We now know that apart from hydrogen, helium is the most plentiful substance in the universe.

Lockyer was also concerned in another important discovery: the method of observing prominences without waiting for a total eclipse. At totality, remember, the Moon covers the Sun completely, and for a few seconds or a few minutes we can see the chromosphere, the prominences, and the corona. Normally, all these features are overpowered by the brilliance of the photosphere. But the prominences are made up of hydrogen, and at the eclipse of 1868 Lockyer began to wonder whether this could be turned to advantage. He experimented,

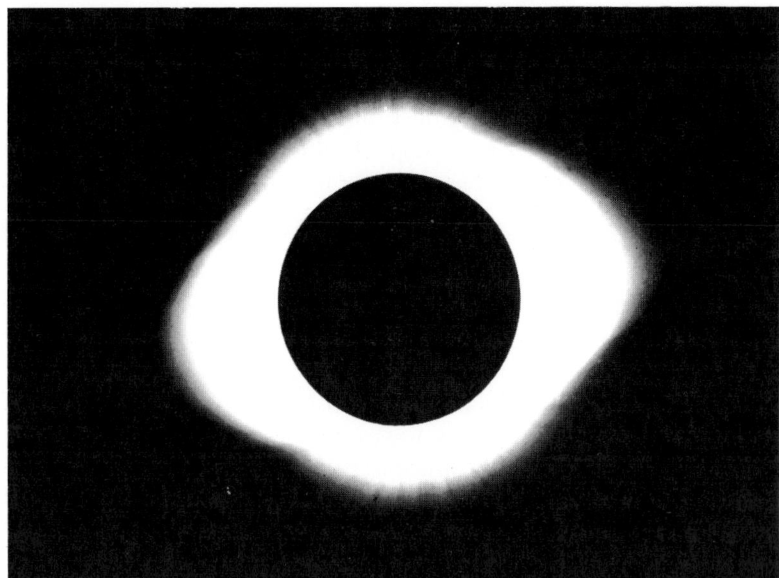

TOTAL ECLIPSE OF THE SUN, JUNE 30, 1954, *photographed from Sweden by Professor Åke Wallenquist. This eclipse was just total from the northernmost Scottish islands, but was seen over most of Britain as a large partial. The line of totality extended across Norway and Sweden.*

and found that by opening the slit of his spectroscope he could isolate hydrogen – and placing the slit tangential to the Sun's limb meant that he could actually see the prominences.

The discovery was made at the same eclipse by a resourceful French astronomer, Pierre Jules Janssen, who later became Director of the Meudon Observatory, outside Paris (the square outside the main entrance to the Observatory is named in his honour, and his statue is there). Janssen was later concerned in a somewhat bizarre episode. In December 1870 Paris was besieged by the Germans; Janssen was anxious to escape in order to observe the total eclipse of December 22, and he did so – in a balloon! He reached Oran safely, only to be thwarted by dense clouds which covered the Sun throughout the eclipse... Janssen's great solar atlas, containing eight thousand high-quality pictures of the Sun's surface, was published in 1904, three years before he died. Meanwhile George Ellery Hale had invented the spectroheliograph, and later came its visual counterpart, the spectrohelioscope. Nowadays, solar observers make great use of what are termed monochromatic filters, which cut out all light except that coming from a selected wavelength, and are particularly well adapted for studying the famous solar flares. The pioneer here was another French astronomer, Bernard Lyot.

Prominences are of two main types. Quiescent prominences may persist for days or even weeks, while eruptive prominences, as their name suggests, are violent and quick-moving; ciné films have been taken of them, showing their growth, development and decay. They may rise to at least 2,000,000 kilometres above the photosphere. By observing at hydrogen wavelengths, prominences may be seen against the bright disk of the Sun as dark filaments or *flocculi* (bright flocculi are due to calcium).

Beyond the chromosphere, where the Fraunhofer lines are produced, comes the corona. The mean temperature is of the order of 2,000,000 degrees Centigrade, but this does not mean that it is 'hot' in the everyday meaning of the word. Scientifically, temperature depends upon the speed at which the various atoms and molecules are

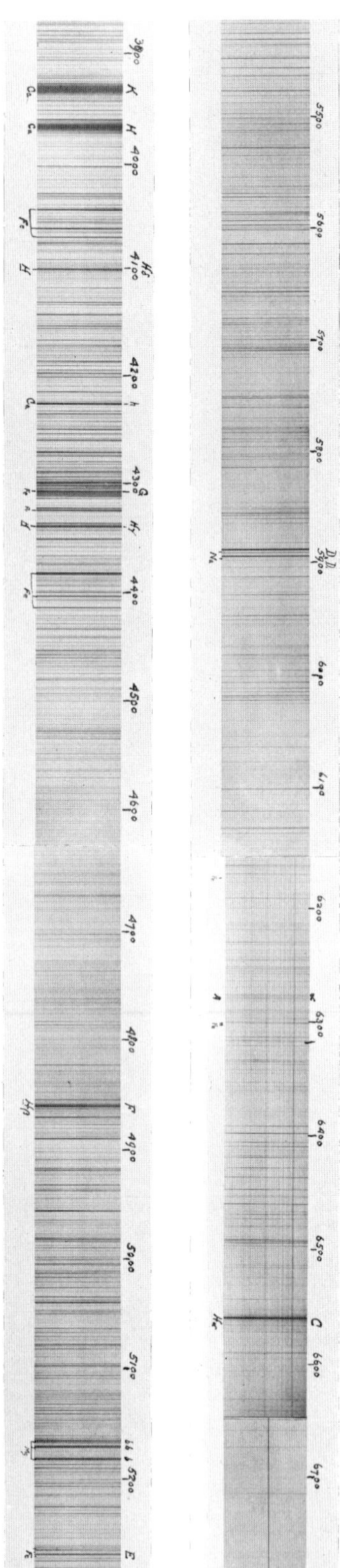

PART OF THE SOLAR SPECTRUM, *taken with the 4 m spectrograph at Palomar. Many lines are shown, and the complexity of the spectrum is very evident. High dispersion or 'spreading-out' is possible, since in the case of the Sun there is plenty of light available.*

moving around; the greater the speeds, the higher the temperatures. In the corona, the velocities are high, and therefore so are the temperatures; but there are so few atoms that there is very little of what we conventionally call 'heat'. The best analogy I can think of is to consider a firework sparkler and a red-hot poker. Each spark of the firework is white-hot, but has so little mass that the firework may safely be held in the hand – though I would not personally care to pick up a poker, whose surface is only red-hot.

In the corona, the temperature rises to between one and two million degrees. The reason for this very high value was a mystery for many years, but the answer has now been found, even though some details remain to be filled in. The Sun's bright surface or photosphere is in a state of constant turmoil, and this is transferred to the chromosphere which lies above it. Pressures waves, which may be termed sound waves, rush through the chromosphere, speeding up the atoms there and so causing a rise in temperature. In its turn, the chromosphere passes this effect on to the corona, which has no definite boundary, but simply thins out until its density is no more than that of the interplanetary medium.

Before the space age, the corona could never be satisfactorily studied except during the fleeting moments of a total eclipse. Bernard Lyot made a noble attempt; he devised a coronagraph, which in effect produced an artificial eclipse inside the telescope itself. By taking it to the clear atmosphere of the high-altitude Pic du Midi Observatory, in the Pyrenees, he was able to study the inner corona. A different sort of tactic was tried forty years later, at the eclipse of June 1973. Dr. John Beckman of Queen Mary College, London, and a team of astronomers flew along the track of totality in a Concorde aircraft, which was fast enough to keep pace with the Moon's shadow as it swept across the Earth, and treated the observers to seventy-two minutes of continuous totality. Of course, the real answer is to observe the corona from space, where there is no obscuring atmosphere. Results from the first true American space-station, Skylab, exceeded all expectations. We know much more about the Sun now than we did a few years ago, though many uncertainties remain.

It would take many pages to give anything like a proper description of the various instruments based upon the principle of the spectroscope. For instance, the splitting-up of light is now generally done not by using a prism, but by means of a very finely-ruled diffraction grating. This was yet another development due to Fraunhofer, but such gratings are not easy to make, since they have to be ruled to many hundreds of lines per centimetre. Meanwhile, it is time for us to turn back to the stars, where the problems are much the same but the methods of study have to be rather different.

The Sun, as we well know, is an ordinary star. If it produces an absorption spectrum, with a continuous background crossed by dark lines, other stars will presumably give similar results. One of the pioneers in this field was Sir William Huggins, who was born in London in 1824, and established a private observatory at Tulse Hill, outside London. Helped by his wife, who was also a skilful astronomer, he examined the spectra of many stars, and in 1863 he identified various elements in Betelgeux and Aldebaran.

Matters are made more difficult by the fact that the stars are relatively faint. There is no question of using tower-type telescopes to

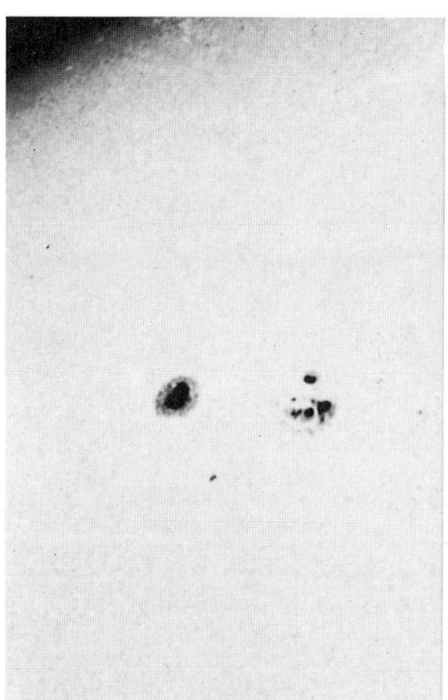
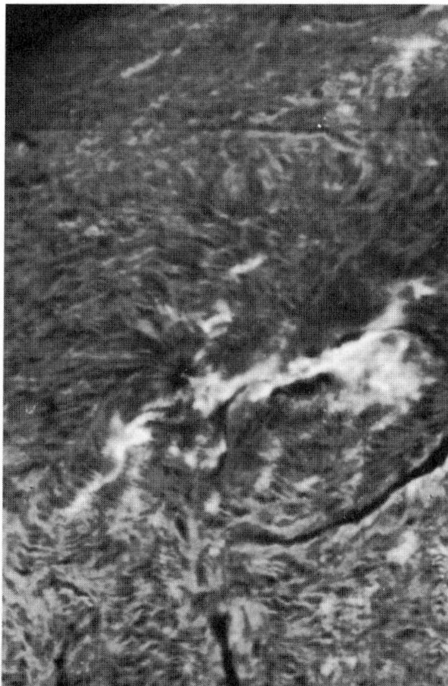
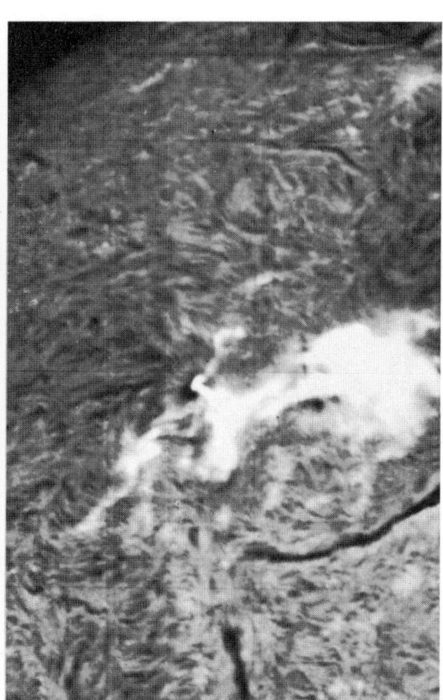

THE SOLAR FLARE OF AUGUST 8, 1937 *as photographed at Mount Wilson.* (Left) *Ordinary photograph of sunspot group.* (Centre) *the same region taken with the red hydrogen line H-alpha, showing the flare.* (Right) *five minutes later, again in H-alpha light, when the flare was at its maximum. Flares are now known to be common phenomena, but very few are visible in ordinary light.*

give spectra of high dispersion; there simply is not enough light. Huggins made another step forward in 1863 when he first photographed the spectrum of a star, but obviously he was facing grave problems. In those days photography was still distinctly primitive.

At about the same time, Angelo Secchi in Italy was carrying out work upon stellar spectra. While Huggins concentrated upon a few stars, and studied them in as much detail as he could, Secchi did his best to examine large numbers of stars and fit them into different spectral types. He knew that not all stars are like the Sun; some are hotter, others cooler. This is shown at once by their obvious differences in colour.

To the casual observer the stars may look white, but closer inspection shows that this is not always so. For instance, compare the two leaders of Orion; Betelgeux is orange-red, while Rigel is white. Among other orange stars there are Antares in the Scorpion, Arcturus in Boötes (the Herdsman) and Gamma Crucis, the third brightest member of the Southern Cross. On the other hand Capella in Auriga (the Charioteer) is yellowish, like the Sun, while Vega in the Lyre is a glorious blue. It is true that few of the colours are very obvious with the naked eye, but binoculars show them well.

It follows that the stars are at different surface temperatures. White heat is hotter than yellow heat, while yellow is hotter than red; therefore Betelgeux and Antares are cooler than the Sun, while Vega is hotter.

Secchi divided the stars into four spectral classes. Type I was made up of white stars, II of yellow or orange, and III and IV of red. In 1890 E.C. Pickering, at Harvard (brother of W.H., the lunar and planetary observer) drew up a more detailed classification, in which he started off by dividing the stars into groups and lettering them A,B,C,D and so on, beginning with white stars and working through yellow, orange and orange-red to red. As so often happens, the letters became out of order, and the final result was alphabetically chaotic, but it is still used.

THE SOLAR CORONA, *photographed by the English Astronomer J. Jackson, from Giggleswick, at the total eclipse of June 29, 1927.*

The 'spectral alphabet' is: O,B,A,F,G,K,M,R,N,S (though of late the last three have more or less fallen into disuse.*

Each type is again divided up into sub-classes, numbered from zero to 9. This gives a smooth sequence. To take part of the order at random, consider Class A, so that we have A1, A2, A3 ... A9. Type A1 is little difference from B9, while A9 is almost the same as FO.

The spectrum of a star is not easy to study, and to work out a really useful system took many years. It was at first thought that the series indicated an evolutionary sequence, so that types W, O, B and A were classed as 'early' and K and M as 'late'. Actually this is not true, but the terms are still commonly used. The Sun is a typical star of Type G2.

The best way to summarize the various spectral types is by means of a table. It is very over-simplified, but it will act as a general guide.

TABLE

Type	Spectrum	Temperature °C	Example(s)	Notes and Colour
W	Emission lines.	Up to 80,000	–	Very rare. Known as Wolf-Rayet stars.
O	Both bright and dark lines.	35,000-40,000	Zeta Puppis	Transition between W and B.
B	Helium lines very prominent.	12,000-30,000	Rigel	Bluish or white.
A	Hydrogen lines very prominent.	8,000-10,000	Sirius, Vega	Bluish or (usually) white.
F	Calcium lines very prominent.	6,000-7,500	Canopus, Polaris	Very slightly yellowish.
G	Numerous metallic lines.	4,200-6,000	Capella, the Sun	Yellow.
K	Very strong metallic lines.	3,000-5,000	Arcturus, Aldebaran	Orange.
M	Very complicated spectra; many bands due to molecules.	3,000-3,400	Betelgeux, Antares	Orange-red.
R	Many molecular lines.	2,600	T Lyræ	Reddish.
N	Strong carbon lines.	2,500	R Leporis	Reddish.
S	Bands of zirconium oxide and titanium oxide prominent.	2,600	Chi Cygni	Red.

*The famous mnemonic is: O Be A Fine Girl Kiss Me Right Now Sweetie (or, alternatively, Smack). Idle minds may amuse themselves by thinking out others?

There are, inevitably, many complications. Bands due to molecules are not found in the hot stars, because the molecules cannot survive there; at such high temperatures they are broken up into their constituent atoms. And though the stars do differ in composition, it would be misleading to say that, for instance, B stars contain more helium than A stars. Everything depends upon the conditions. There are also some stars which do not fit into the general pattern, and have to be classed simply as 'peculiar'. Eta Carinæ, Sir John Herschel's erratic southern variable, is a case in point.

The characteristics of a star's spectrum often give a reliable key to its real luminosity, and hence to its distance. Thus both Vega in Lyra and Deneb in Cygnus have A-type spectra, but Deneb's spectrum tells us that the star is exceptionally luminous; and since it appears fainter than Vega, it must be further away. In fact Vega is only 55 times as powerful as the Sun, but it would take 60,000 Suns to equal one Deneb.

A century and a half ago, a French philosopher named August Comte stated that mankind could never hope to find out anything about the chemistry of the stars. He was soon proved wrong, but stellar spectroscopy led on to all manner of other discoveries as well. In particular, Huggins and (independently) the German astronomer Hermann Vogel, director of a private observatory at Bothkamp, made pioneer measurements of *radial velocities*.

THE TOTAL ECLIPSE OF THE SUN, JUNE 30, 1954, *as seen from Lysekil in Sweden. Conditions were fairly good, though there was a certain amount of very thin cloud. This was one of the most favourable European eclipses of recent years, comparable with the eclipse of 1961. The track of totality extended across Norway and Sweden, into Russia. However, expeditions sent to Scandinavia were in general handicapped by cloud, and from some sites no results were obtainable. It will be many years before another total eclipse will be seen from Scandinavia.*

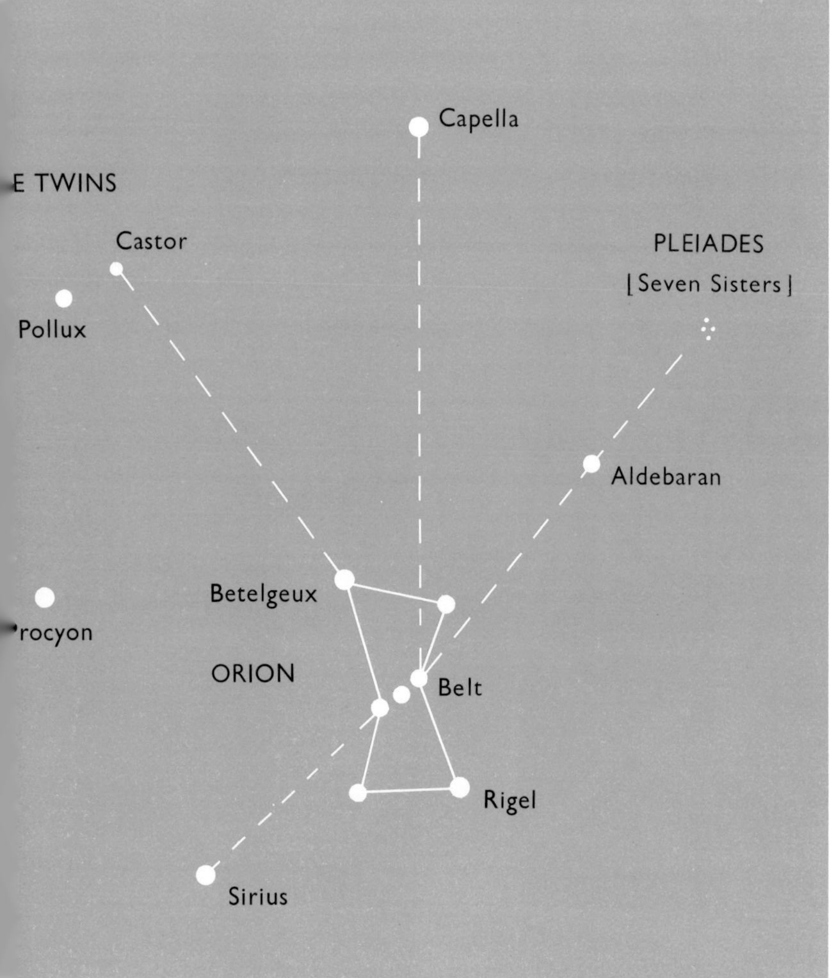

SOLAR PROMINENCE. *The whole edge of the Sun, photographed at Mount Wilson on December 9, 1929, in calcium light (K line).*

ORION AND HIS RETINUE. *(Left) Various types of stars are shown. Betelgeux is a Red Giant of type M; Aldebaran, also red, is of type K; Capella, type G, similar to the Sun; Procyon, slightly yellow, type F; Sirius, white, type A; Rigel, a particularly luminous star of type B. The remaining bright stars in Orion are also of type B. The two Twins are of different spectral type. Pollux is an orange K-star, while Castor is white and of type A. Some of the star-colours are noticeable with the naked eye, but are much better seen with optical aid; a pair of good binoculars will bring them out excellently.*

STELLAR SPECTRA. *(Left) When it is desired to study the spectra of many stars, as is necessary for many astrophysical investigations, it obviously saves a great deal of time to be able to photograph a number of spectra on a single exposure. On this photograph, taken by H. B. Ridley, each star is drawn out into a spectrum. Methods of this sort are extensively used in modern work, and have yielded very satisfactory results, though naturally they have their limitations.*

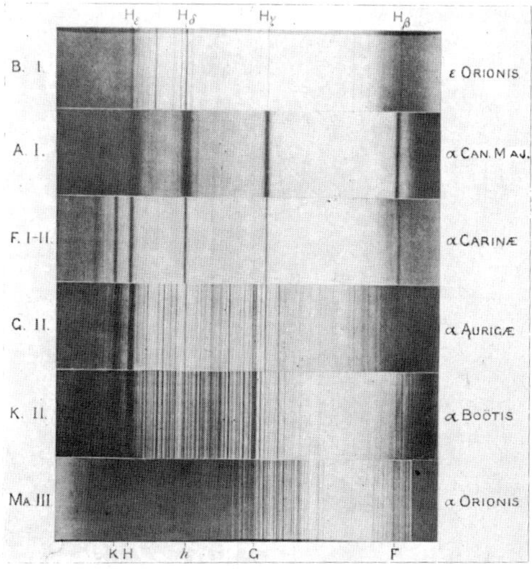

TYPICAL STELLAR SPECTRA, *types B to M. The spectra shown are of Alnilam (Epsilon Orionis, the middle star of Orion's Belt); Sirius (Alpha Canis Majoris), Canopus (Alpha Argûs, alternatively known as Alpha Carinæ); Altair (Alpha Aquilæ); Arcturus (Alpha Boötis) and Betelgeux (Alpha Orionis).*

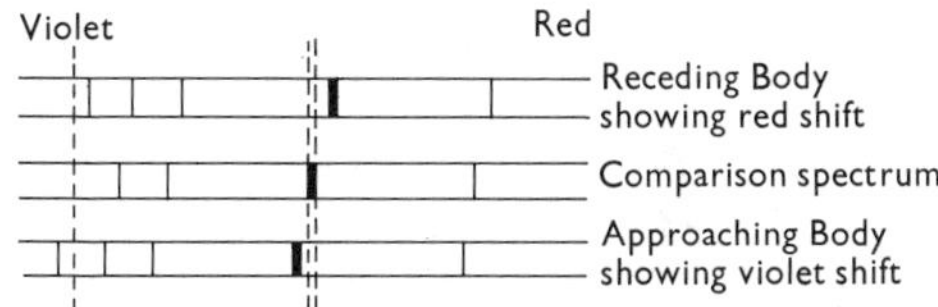

THE DOPPLER EFFECT. *In the upper diagram the light-source is assumed to be receding, and the spectral lines are shifted towards the red. The centre strip is for a stationary light-source, drawn in for comparison; the bottom strip, for an approaching source, shows a shift to the violet. This Doppler Effect is invaluable in astrophysics, since it enables the radial motions of celestial bodies to be measured. It must, of course, be borne in mind that the red and violet shifts are never easy to measure with precision, since there are many complications to be taken into account, but the results obtained are of a high degree of reliability. Doppler effects also enable the rotations of stars to be measured; as the star turns, the light coming from the approaching limb has a violet shift, while the light from the receding limb has a red shift. The result is that the line appears broadened, and the amount of the broadening gives a key as to the rate of rotation.*

It had long been known that the stars show slight but often measurable individual or proper motions, so that over the centuries the constellation patterns will alter. But what about 'towards or away' motions? These would not show up as definite shifts. A man walking straight towards you will not move against his background, but will simply appear larger and larger as he draws near. Similarly, an approaching star will become steadily brighter, but the increase in brilliancy will be too slight to be measured except over periods of thousands or millions of years. Huggins and Vogel made use of a principle which had been discovered in 1842 by the Austrian physicist Christian Doppler, and is still known as the Doppler effect.

If you listen to a train which is approaching you, and sounding its whistle, the note will be high-pitched. More sound-waves per second are entering your ear than would be the case if the train were standing still; the wavelength is apparently shortened, and the note is raised. After the train has passed by, and has begun to recede, fewer sound-waves per second will enter your ear, lengthening the apparent wavelength and causing the pitch of the note to drop. Light may be regarded as a wave-motion, and the effect here is to shift the spectral lines – to the violet or short-wave end if the star is approaching, to the red or long-wave end if the star is receding. Comparing the actual positions of the observed lines with those in a standard laboratory spectrum will show the radial velocity of the star. Thus it has been found that Rigel is moving away from us at 21 kilometres per second, Capella at 28 kilometres per second; on the other hand Arcturus is approaching us at 5 kilometres per second, Polaris at 17 kilometres per second and the brilliant southern Alpha Centauri at 25 kilometres per second. Knowing both the radial velocities and the proper motions, we can work out how the star really is moving through space with respect to the Sun.

It is now well over a hundred years since Kirchhoff laid down his three Laws, but the story began much earlier than that. It began, indeed, on the day when Isaac Newton passed sunlight through a glass prism and produced a coloured rainbow.

THE 46M TOWER TELESCOPE AT MOUNT WILSON, *(opposite) viewed from the north-east. This telescope is designed specially for solar work. Tower telescopes are, in fact, used exclusively for studying the Sun, and are able to produce spectra of very great dispersion. A similar instrument has been set up at Arcetri, in Italy – an observatory which has a great reputation in this field of research – and there are others in America and in Russia.*

28

Stars of Many Kinds

I AM writing a history of astronomy, and this is no place to give a detailed description of what is to be seen in the night sky at different times of the year; there are many books which do that. But before going on to describe the life-stories of the stars, which are much less straightforward than used to be thought a few years ago, I must pause to say something about various interesting stars; and one way to do this is by giving easily-recognizable examples, which can be found from outline maps.

Beware of planets. True, Jupiter and Venus are too brilliant to be mistaken, but Mars and Saturn can look very like stars, and the only safeguard is to check their positions in an almanac such as the annual *Yearbook of Astronomy*. Mercury and Uranus will not be seen unless you are deliberately looking for them, while Neptune and Pluto are well below naked-eye range, so that only Mars and Saturn can cause any real trouble.

Of course, everything depends upon where one lives. The Australian skies are not the same as those of Europe. Therefore I have divided the sky up, and given a few of the most conspicuous groups; once these have been identified, it is surprisingly easy to locate the rest. To make matters as easy as possible I will refer to latitudes such as those of Britain, most of Europe and most of the United States as 'northern'; Australia, South Africa and New Zealand as 'southern'. I realize that this is not precise. For instance, part of the Great Bear drops below the horizon in southern Europe, such as the latitude of Athens, while the brilliant Canopus appears briefly in some of the southern parts of the United States; I have seen it well from Houston, Texas. But the system is good enough to be a general guide, so let us begin our charts.

CHART. 1.

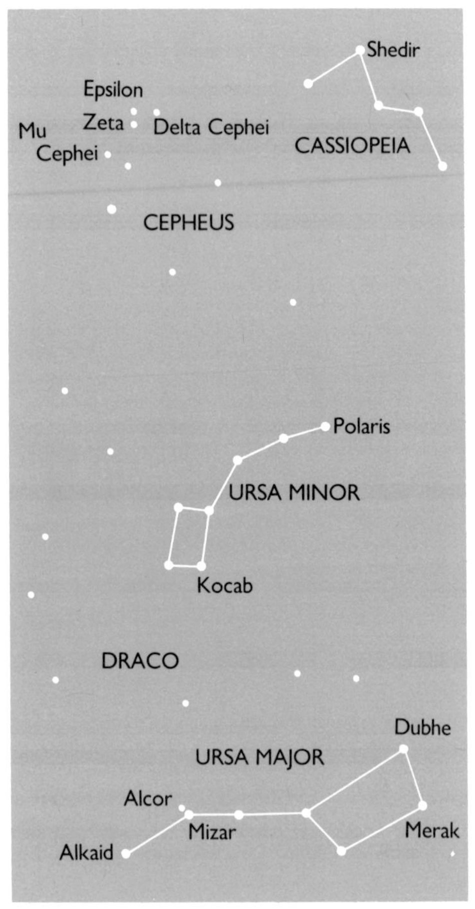

Chart 1. Stars of the Far North.
The main constellations are the two Bears, Ursa Major and Ursa Minor; Cassiopeia, the queen in the Perseus legend; and Cepheus, Cassiopeia's husband. These are always visible from the north, never from the south.

Look first at the two Pointers to the Pole Star, Dubhe and Merak. Dubhe is rather the brighter of the two (magnitude 1.8, as against 2.4) and is decidedly orange, while Merak is white. The colour of the K-type Dubhe is detectable with the naked eye, and obvious with binoculars.

Mizar, the second star in the Great Bear's tail, is a famous binary star. It is known officially as Zeta Ursæ Majoris, but is one of the few stars below the first magnitude to be generally known by its individual name.

Alcor, the naked-eye companion, is easy to see on a clear night without optical aid. A small telescope will show that Mizar itself is double; the senior component (Mizar A) is well over a magnitude brighter than the other (Mizar B). Both are white. The system is a genuine binary, not a mere line of sight effect. The two stars are around 380 astronomical units apart*, and are moving round their common centre of gravity, but they take thousands of years to complete one revolution, so that they look the same over periods of many lifetimes. The distance from Earth is eighty-eight light-years, and the brighter component is almost 100 times as luminous as the Sun.

But Mizar is more complicated than this, as was discovered in 1889, when E.C. Pickering, at Harvard, studied its spectrum. He found that the dark lines of Mizar A became regularly doubled at intervals of about twenty days, staying double for some time before becoming single again. He soon found out the reason. Mizar A is again double, but the two components are so close together that no telescope will separate them.

The doubling is due to that all-important phenomenon, the Doppler Effect. If the two stars of Mizar A are moving round their 'balancing point', there will be times when one component is moving towards us while the other is moving away. Therefore, the spectral lines of the approaching star will be shifted over to the blue or short-wave end of the rainbow band, while those of the receding star will be red-shifted; both sets of lines can be seen, whereas later on they will be superimposed upon each other, and will appear single. Mizar was the first known *spectroscopic binary;* today, many others have been found. (We have to take into account the general motion of the Mizar system, which is moving towards us at about 9 kilometres per second, but this can be allowed for.) Incidentally, Alcor is another spectroscopic binary; the two components have a combined luminosity about fifteen times that of the Sun.

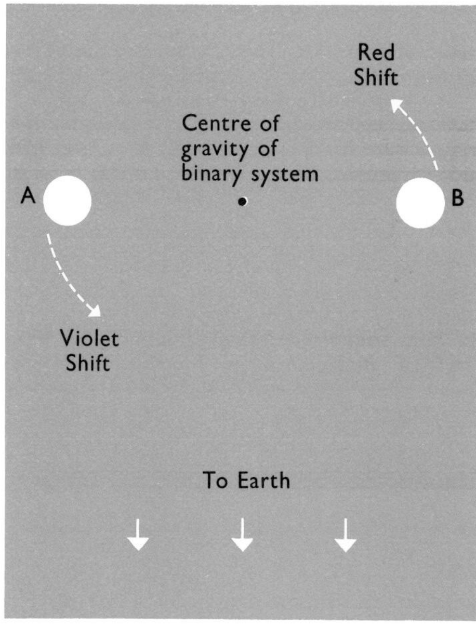

DETECTION OF A SPECTROSCOPIC BINARY. *In the positions shown, component A is approaching and will show a violet shift; B is receding, and will show a red shift.*

Polaris, in the Little Bear (Ursa Minor) may be found from the two Pointers. It lies within one degree of the celestial pole, so that it always remains in virtually the same position in the sky. It is highly luminous, and very slightly variable. It also has a companion, but this companion is much too dim to be seen without a telescope, and may be at least 2,000 astronomical units from Polaris itself. Look next at the other bright star of Ursa Minor, Kocab (magnitude 2.0; the same as Polaris). Kocab, often nicknamed the Guardian of the Pole, is orange, so that its surface is relatively cool.

Cassiopeia, the other really prominent constellation of the far north, is marked by five stars making up a rough W or M; the leaders are just below the second magnitude. The Milky Way is very rich here, though, as we have noted, the lovely shining band is a line of sight effect, and does not mean that the stars in it are really crowded together. Cepheus, which lies more or less between Cassiopeia and the pole, is much less striking, but it does contain two stars of exceptional interest. The first of them has no proper name, and is known simply as Delta Cephei. It never becomes brighter than magnitude 3.5, fainter than any of the seven chief stars of the Great Bear, but it has given its name to a whole class of variable stars: the Cepheids.

The story began in 1784 with some observations made by a most unusual astronomer, John Goodricke. He was then twenty years old, and was deaf and dumb, but there was nothing the matter with his eyesight or his brain, and when watching the stars in Cepheus he found

RICH STAR-FIELD IN THE MILKY WAY REGION

*Remember that one astronomical unit is the distance between the Earth and the Sun; in round numbers, 150,000,000 kilometres.

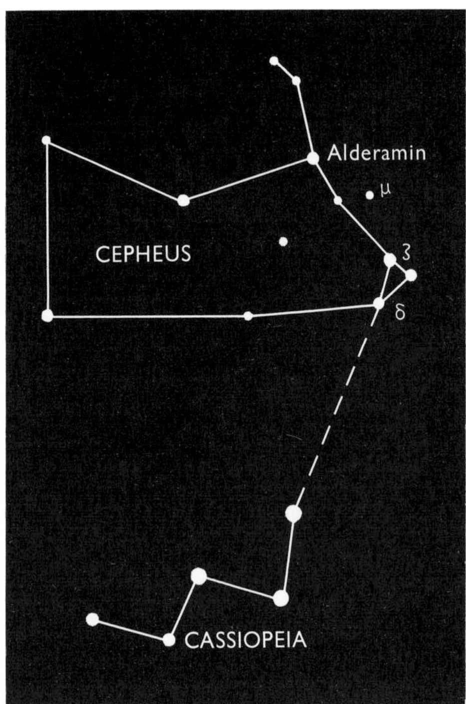

THE CONSTELLATION OF CEPHEUS, *showing the positions of the famous variables Delta and Mu. Cepheus is not a prominent group, but the W of Cassiopeia may be used as a pointer to it; its brightest star is Alderamin. Delta Cephei is not conspicuous, but is always an easy naked-eye object, and may be compared with its two neighbours Epsilon and Zeta, with which it forms a small triangle. Mu is sometimes easy to see without optical aid, but at its faintest it drops to almost the 6th magnitude. The strong red colour which led to Herschel's christening it 'the Garnet Star' is not evident with the naked eye, but any small telescope will show it excellently.*

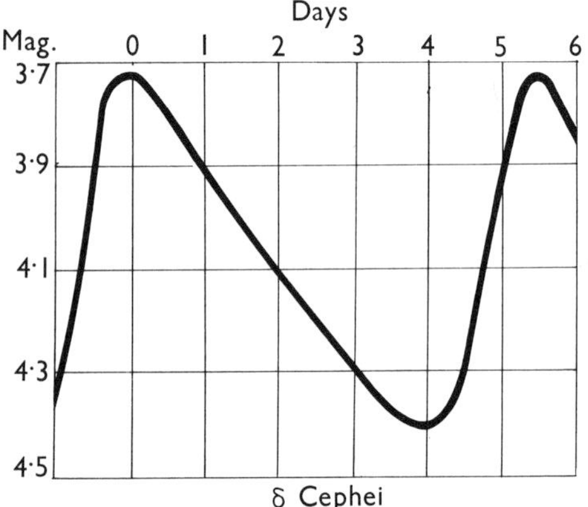

LIGHT-CURVE OF DELTA CEPHEI. *The fluctuations are perfectly regular. This star is the most famous member of its class.*

that Delta was not constant in brightness. It waxed and waned regularly; the period, or interval between one maximum and the next, was 5.4 days, and the fluctuations seemed to be absolutely regular.

Goodricke's method was to compare Delta with two nearby stars, Zeta Cephei (magnitude 3.3) and Epsilon Cephei (4.2). At its peak, Delta was almost as bright as Zeta, while at minimum it was slightly fainter than Epsilon.

In the same year another English astronomer, Pigott, found that Eta Aquilæ, not far from Altair in the Eagle (Chart 4), varied from magnitude 3.7 to 4.5 in a period of just over 7 days. Like Delta Cephei, it was absolutely regular. Others of the same type were found in later years; Zeta Geminorum in the Twins, for instance. Today the known number of Cepheid variables is very great.

Cepheids are unstable stars. They swell and shrink, changing in output as they do so. They are of invaluable help to astronomers, because they act as 'standard candles'; their periods are linked with their real luminosities. The longer the period, the greater the luminosity. Thus Eta Aquilæ, with its period of over 7 days, is more powerful than Delta Cephei, whose period is only 5.4 days.

If we know a star's real luminosity, we can also find out how far away it is – just as we can estimate the distance of a light seen out to sea, provided that we know whether it is really a powerful searchlight or a feeble ship's lamp. And since the Cepheids are very powerful stars, sometimes thousands of times brighter than the Sun, they can be seen over vast distances.

Another variable star not far from Delta in the sky is Mu Cephei, nicknamed 'the Garnet Star' by Sir William Herschel. Unlike Delta, it is not regular; it has no period at all, but it seldom becomes brighter than the fourth magnitude, so that you have to know where to look for it. It is notable because it is fiery red; when observed through a telescope it has been likened to a glowing coal. It is what is termed a red supergiant, of tremendous size, and over 8000 times as luminous as the Sun. Near it is the white star Nu Cephei, which is not variable; its magnitude is only 4.5, but it is a real 'cosmic searchlight', equal to at least 50,000 Suns. If it lay as close to us as Sirius, it would cast shadows. In astronomy, appearances can indeed be deceptive!

Chart 2: Orion and his Retinue.

Orion is crossed by the celestial equator, and is therefore visible from every inhabited continent. In the northern hemisphere it is best seen in winter; in the southern hemisphere it is, of course, a summer constellation.

Orion, the Celestial Hunter, is truly magnificent. There are two particularly brilliant stars, Betelgeux and Rigel, while the line of three stars making up the Hunter's Belt is quite unmistakable. It is not far from the Milky Way, and the whole area is rich.

The brightest star in Orion is Rigel (magnitude 0.1), very nearly the equal of Capella and Vega. It has a B8-type spectrum, and is very powerful; the luminosity is about 60,000 times that of the Sun. It is around 900 light-years away, so that we are now seeing it as it used to be in the time of William the Conqueror. Most of the other leaders of Orion are also hot and white, but Betelgeux is quite different. The spectrum is of Type M; Betelgeux is orange-red, and is variable. Sometimes it nearly equals Rigel, while at others it is not much brighter than another red star, Aldebaran in Taurus (the Bull), which may be found by using the stars of Orion's Belt as a guide.

Unlike Delta Cephei, Betelgeux is not regular in its behaviour. There is a rough period of about five years, but it *is* rough. The star is immensely large, with a diameter of over 400,000,000 kilometres, so that it is big enough to swallow up the whole path of the Earth round the Sun, though its diameter is not quite constant; Betelgeux swells and shrinks. The surface temperature is about 3000 degrees, and the luminosity about 15,000 times that of the Sun.

Betelgeux is a typical red supergiant. It was once believed to be young, but we now know it to be well advanced in its life-story. The mass is probably about five times that of the Sun, though this may be a considerable underestimate.

Betelgeux is several hundreds of light-years away, but it is still one of the very closest of the supergiants, and by using indirect methods it has been possible to measure its apparent diameter, even though no ordinary telescope will show it as anything more than a speck of light. One of the very latest techniques is called 'speckle interferometry'. To go into details is impossible here; but basically it involves taking a large number of short-exposure photographs, with their images enhanced

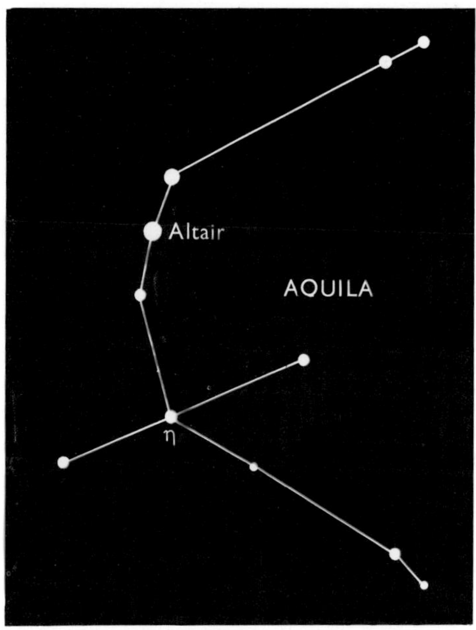

THE CONSTELLATION OF AQUILA, *showing the position of the Cepheid variable Eta Aquilæ.*

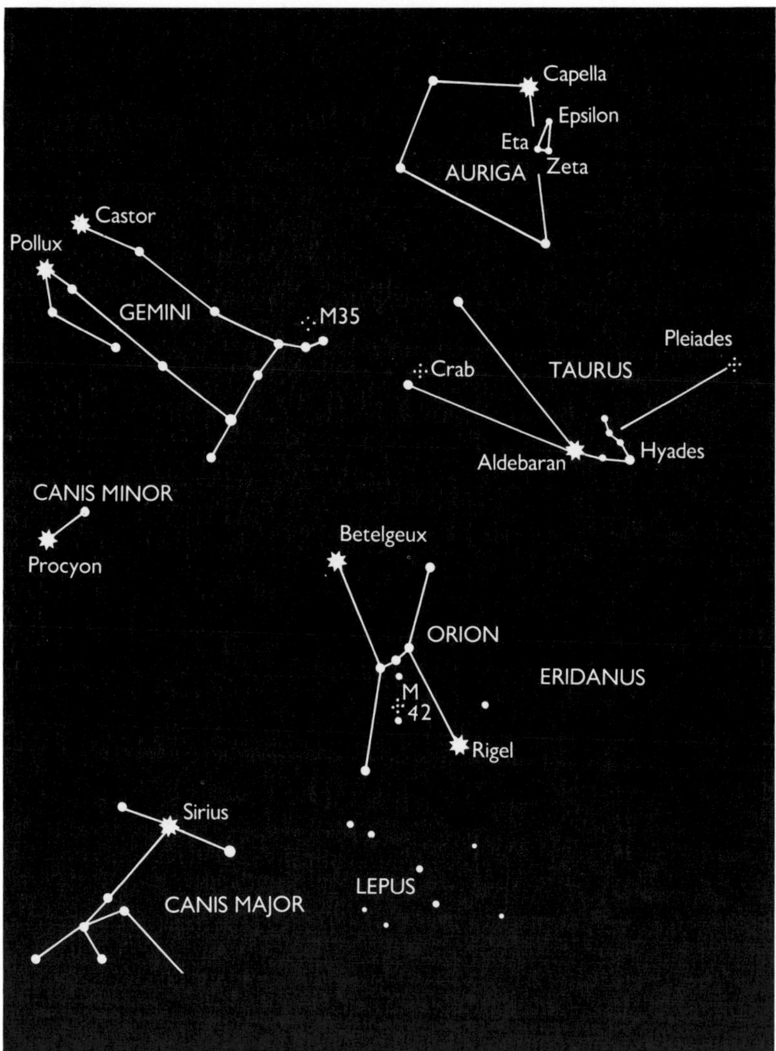

CHART. 2.

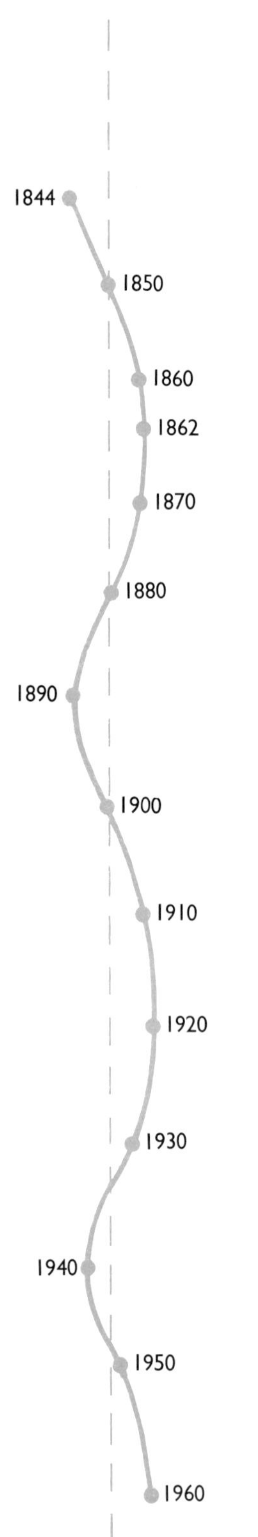

APPARENT MOVEMENT OF SIRIUS IN THE SKY. *The proper motion of the bright component, shown by the thick wavy line, showed that there must be an unseen companion affecting its motion*

electronically, and then using a computer to average them out and obtain the best results. In this way it has been possible to detect patches on the surface of Betelgeux – something which would have seemed out of the question only a short time ago.

Extending away from Orion's Belt is the 'Sword', containing a misty patch which is easily visible with the naked eye. This is the Great Nebula, Messier 42 – a mass of gas (mainly hydrogen) together with 'dust'. Any small telescope will show that in it is a multiple star, Theta Orionis, whose four main components are arranged in a pattern which has led to their being called the Trapezium. All four are very white and hot; they act upon the nebular material, illuminating it and also making it send out a certain amount of light on its own account. This is what is termed an *emission nebula* or H.II region. (The term 'H.II' is used because in the region, atoms of hydrogen are damaged by the effect of the radiation from the hot stars. The planetary electron is broken away from the nucleus, so that the atom is left incomplete, or *ionized.*) There are plenty of emission nebulæ in the sky; they are stellar birthplaces, where fresh stars are being formed out of the thinly-spread material.

Aldebaran, in line with the stars of Orion's Belt, is an orange giant of spectral type K. It looks very like Betelgeux, but is not nearly so remote or powerful. Extending away from it is a group of faint stars arranged in a rough V; these make up the Hyades, and beyond Aldebaran in the same direction is the much more imposing group of the Pleiades or Seven Sisters, the best example of an *open* or *galactic* cluster. Also in Taurus, near the third-magnitude star Zeta, is the gas-patch of the Crab Nebula, all that is left of the supernova of A.D. 1054.

Turn now to Sirius, in Canis Major (the Great Dog), which also is aligned with Orion's Belt, this time in a southerly direction. It is much the brightest star in the sky, with a magnitude of −1·4; it is particularly close to us, at a distance of 8·6 light-years, and it is twenty-six times as luminous as the Sun – powerful enough, but a glowworm when compared with Rigel or Betelgeux. When low down over the horizon it twinkles strongly, flashing various colours. Twinkling has nothing to do with the stars themselves, and is due solely to the Earth's unsteady atmosphere; but Sirius shows the effect particularly well, because of its exceptional brilliance.

Because Sirius is so close, it has a measurable proper or individual motion. In 1500 years it crawls across the background of more remote stars by an amount equal to the apparent diameter of the full moon, which is rapid by stellar standards. In 1834, Friedrich Bessel realized that it was behaving in a peculiar manner. Its motion relative to the background stars was erratic; instead of travelling in a straight line it was 'weaving' its way along, and Bessel decided that there could be only one explanation. Sirius must have a faint companion which was pulling upon it.

Just as Le Verrier and Adams later tracked down Neptune because of its pull upon Uranus, so Bessel worked out where the companion of Sirius ought to be. He was unable to see it; it was hopelessly overpowered by the glare of the bright star, and when Bessel died, in 1844, the Companion was still undiscovered.

Then, in 1862, Clark in America was testing a large new refractor when he saw a dim speck of light close beside Sirius. It proved to be the long-expected Companion, almost exactly where Bessel had said it would be. From the movements of the pair it was estimated that the mass of the Companion was almost equal to that of the Sun, but it had only one ten-thousandth of the luminosity of Sirius, and so it was tacitly assumed to be large, cool and red.

The real shock came in 1915. At Mount Wilson, Walter Sydney Adams examined the spectrum of the Companion, and found to his amazement that the surface was not cool at all. It was white, and its surface temperature was considerably higher than that of the Sun.

Astronomers all over the world were intrigued. The Companion was very faint, so that it sent only relatively little light – and yet it was white-hot! The only way to make the observations fit was to assume that the star was a dwarf, with a diameter of only about 40,000 kilometres. This would make it smaller than a planet such as Uranus or Neptune. But the mass was also known: almost the same as that of the Sun. Therefore, the Companion had to be almost incredibly 'heavy'; the density worked out at something of the order of 60,000 times that of water, so that a cupful of it would weigh many tons.

It is worth looking back at an article written in 1904 by a well-known British astronomer, J. Ellard Gore. He wrote: 'If the faintness were due merely to its small size . . . the density of the Sirian satellite would be over 44,000 times that of water. This is, of course, entirely out of the question.' Yet even a value of 44,000 times that of water proved to be an underestimate. The White Queen in Lewis Carroll's *Through the Looking-Glass* made a habit of believing at least six impossible things before breakfast each day. Astronomers sometimes have to do likewise.

Other 'White Dwarfs' were found, some of them even smaller and denser than the Companion of Sirius; thus a star known as Kuiper's Star in honour of its discoverer, the late Gerard P. Kuiper, is smaller than Mars, but as massive as the Sun. And there are some known White Dwarfs which are smaller than the Moon.

To explain this super-dense matter we must go back to the description of the way in which an atom is built up. There is a central nucleus, around which move planetary electrons (always bearing in mind the fact that one cannot picture these particles as tiny solid lumps). Most of

THE KEYHOLE NEBULA IN ARGO. *Nebulosity in the region of the famous irregular variable Eta Argûs. Unfortunately the area is too far south in the sky to be seen from Europe or the northern United States.*

an atom is empty space (just as most of the Solar System is empty space; the Sun and planets do not take up much room). As we have seen, atoms may be broken-up or ionized by having their planetary electrons knocked away. In a White Dwarf, where the temperature is high and the pressure is tremendous, the ionization is complete. Every planetary electron is removed from every nucleus, and the result is a jumble of atomic nuclei and unattached electrons. Waste space is more or less eliminated, and the broken parts of the atoms may be jammed tightly together, which explains the amazingly high densities.

White Dwarfs are bankrupt stars, which have used up all their nuclear 'fuel' and are near the end of their evolutionary story. I will have more to say about this in Chapter 29. Meanwhile, note that they are very common indeed. Other bright stars, such as Procyon in Canis Minor (the Little Dog) have White Dwarf companions, but many of these senile stars are single. They appear dim because they are genuinely faint – and our own Sun will become a White Dwarf one day.

Next, let me come to the brilliant yellow Capella – actually a spectroscopic binary – which is the leader of Auriga, the Charioteer. As seen from northern latitudes it is almost overhead during winter evenings, but from Australia or South Africa it only appears briefly above the northern horizon near midsummer, and from New Zealand it can barely be seen at all. Close beside it is a triangle of fainter stars (the 'Hædi', or Kids), two of which are remarkable objects. Epsilon Aurigæ, at the apex of the triangle, looks very normal, but again appearances are deceptive. The visible star is a slightly yellowish supergiant, around 60,000 times as powerful as the Sun (and therefore the equal of Rigel) and around 3200 light-years away, so that we see it today as it used to be in the time of the Trojan War. Associated with it is an invisible companion. Every twenty-seven years this secondary object passes in front of the bright star, and dims it down by a magnitude or so, though even at its faintest Epsilon Aurigæ is always easily visible with the naked eye.

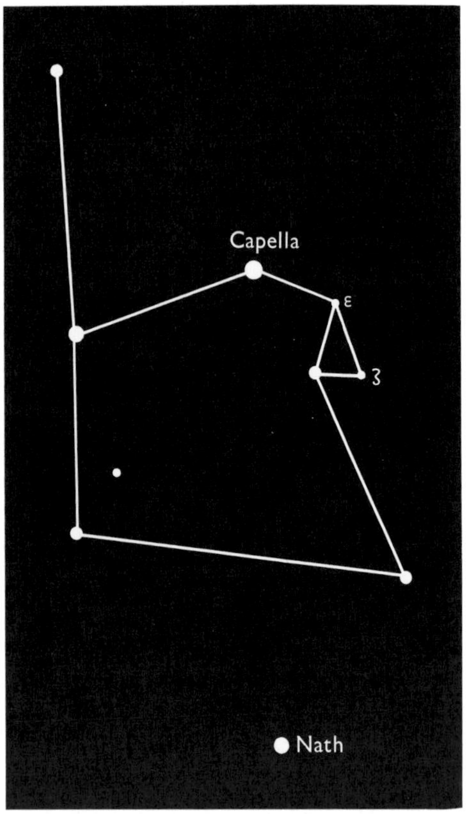

What exactly is the mysterious secondary? It was once assumed to be a very young star, too cool to shine, in which case it would have been the largest star known; it could swallow up all the orbits of the planets round the Sun out to well beyond Saturn. Yet there are problems here; the secondary is presumably semi-transparent, and it could hardly remain stable so close to a massive supergiant. Another interpretation regarded it as a black hole – a region of space round a very old, collapsed star from which not even light could escape (I will say more about this in Chapter 29). Today it seems more likely that the secondary is a small, hot star surrounded by a vast disk-shaped cloud of gas. In any event Epsilon Aurigæ is a remarkable system; the last eclipse began in 1982, and will not end until mid-1984. The third 'Kid', Sadatoni or Zeta Aurigæ, is another huge binary system, though this time the eclipses occur more frequently (once every 972 days) and last for only a few weeks.

Before leaving this chart, look at the two Twins, Castor and Pollux in Gemini. Castor, as we have noted, is a multiple system. There are two main stars, separated by about 14,000 million kilometres and with an orbital period of 350 years. Half a century ago the separation was so great that the pair could be split with any small telescope. Today the separation is less – not because the two Castors have really approached each other, but because we are seeing them from a less favourable

Capella
ε
ζ
● Nath

THE CONSTELLATION OF AURIGA, *showing the positions of the two remarkable eclipsing binaries Epsilon (ε) and Zeta (ζ) Aurigæ.*

angle. Each component is itself a spectroscopic binary, and there is yet another member of the group, Castor C or YY Geminorum, which also is a spectroscopic binary. So Castor is made up of six stars – two bright pairs, and one dim pair. Another interesting object in Gemini is the open cluster Messier 35 (M.35), visible with the naked eye and very conspicuous with binoculars.

Chart 3. Leo, Boötes, Corona and Virgo.
From the northern area this is best seen in spring. From the south, autumn evenings are the best time, though of course the Great Bear can never be seen properly.

Leo (the Lion) can be found by using the Pointers in a 'wrong-way' direction, i.e. away from the Pole Star. The leader is Regulus, of the first magnitude, and the curved line of stars extending from Regulus is known as the Sickle. The rest of Leo is marked by a triangle of stars. The brightest of them, Denebola, was recorded in ancient times as being of the first magnitude; it is now below the second, though whether or not it has genuinely faded is a matter for debate (personally, I doubt it).

Follow round the curve of the Great Bear's tail, and you will come to Arcturus in Boötes (the Herdsman). It is a lovely light orange in colour, with a K-type spectrum, and its magnitude is −0·1, so that it is the fourth brightest star in the sky; only Sirius, Canopus and Alpha

CHART. 3.

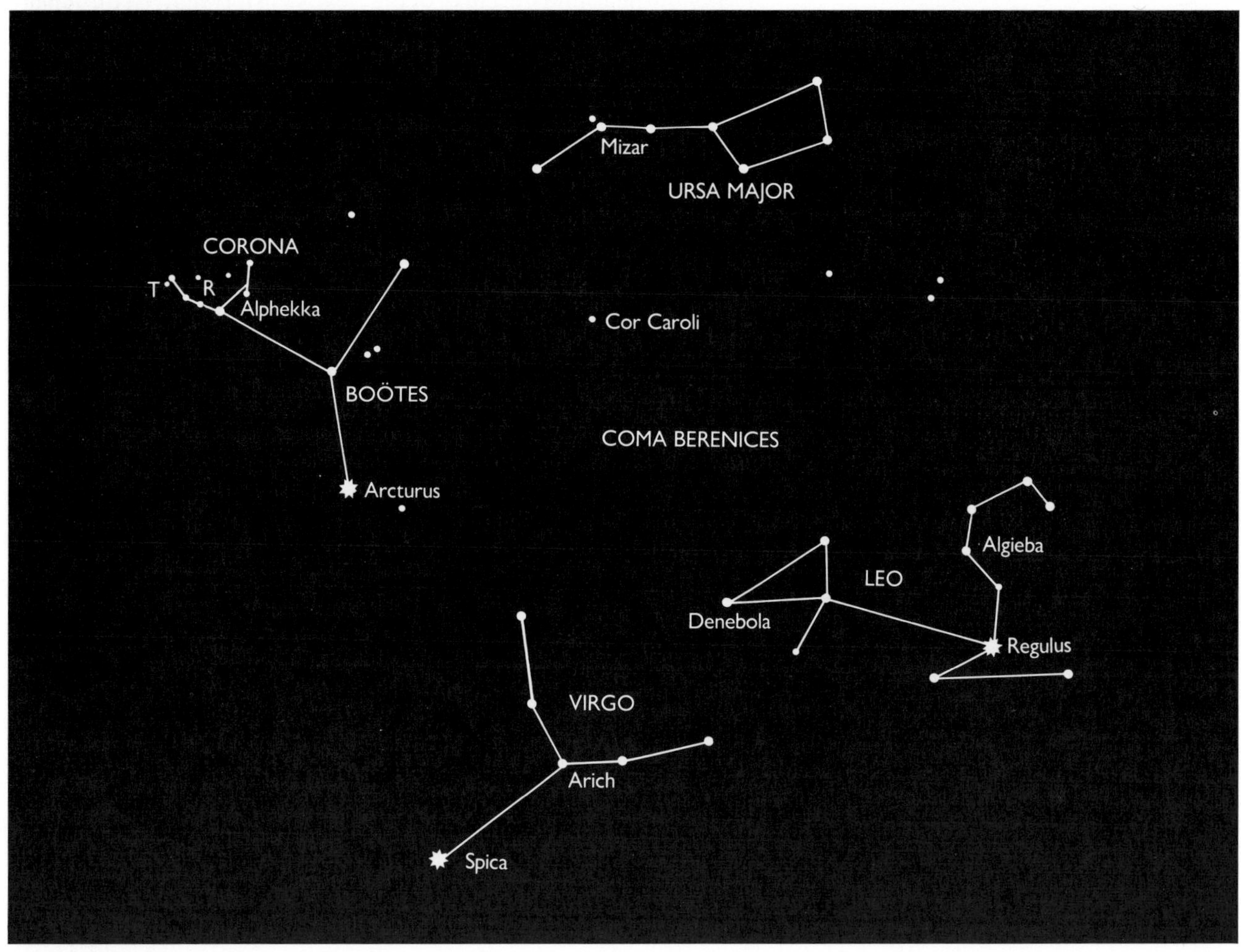

Centauri outshine it. It is thirty-six light-years away, and 115 times as luminous as the Sun. Adjoining Boötes is the small but distinctive semicircle of stars marking Corona Borealis, the Northern Crown. Here we find two remarkable variable stars, R and T Coronæ.

R Coronæ, in the bowl of the crown, is usually on the fringe of naked-eye visibility, and binoculars will show it easily, together with another star in the bowl which is not variable and which therefore makes a good comparison object. At irregular intervals, however, R Coronæ drops in brightness, becoming so dim that a powerful telescope is needed to show it; it remains faint for some time before gradually regaining its normal brilliancy. It is one of a small number of stars of the same type, known as R Coronæ stars. What happens, apparently, is that there is a build-up of carbon particles in the star's atmosphere, blanketing in the light until they are blown away by the star's radiation, so that when R Coronæ drops to minimum it is simply being hidden by a cloud of soot!

T Coronæ, nicknamed the Blaze Star, is different again. Usually it is beyond binocular range, but in 1866 it flared up to the second magnitude for a brief period. The same thing happened in 1946. It is termed a *recurrent nova,* and is a binary, though in a telescope it appears single. Evidently it puts on these displays now and then, but when it will next burst forth remains to be seen. Other recurrent novæ are known, but the Blaze Star is the most spectacular of them.

Following round the curve of the Bear's tail through Arcturus leads on to Spica in Virgo (the Virgin), a B-type first-magnitude star. Virgo is a very large constellation, whose main stars make up a rough Y form. Gamma Virginis or Arich, at the base of the Y bowl, is a binary made up of two equal components – true stellar twins. The revolution period is 180 years. When I first looked at them, in the 1930s, they were so wide apart that any small telescope would divide them, but we are now seeing them less favourably, and by A.D. 2016 they will appear as one star except when seen through a very powerful instrument.

The area enclosed by the Y of Virgo on one side and Denebola in Leo on the other appears devoid of bright stars, but it is very rich in faint galaxies. It is here that we find some of the most interesting features known: the systems of the Virgo Cluster of galaxies, well beyond the Milky Way, and lying at distances of over 60,000,000 light-years.

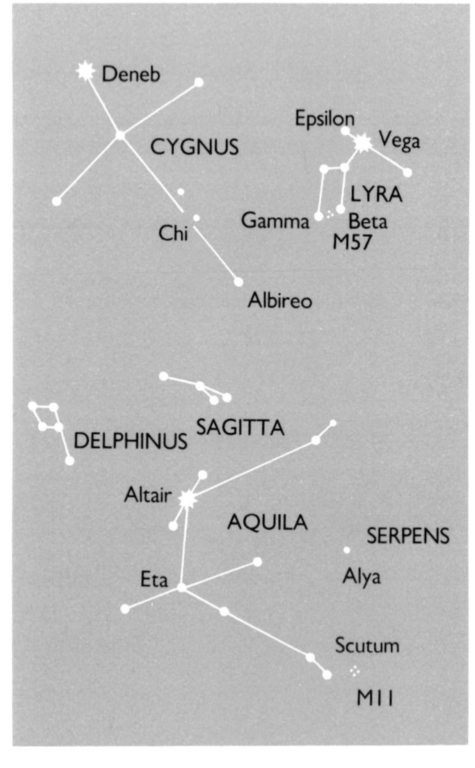

CHART. 4.

Chart 4. The Summer Triangle.

This rather unfortunate and completely unofficial nickname must be laid at my door, because I introduced it in a BBC *Sky at Night* television programme around 1958, and it has come into general use. Of course it applies only to the northern hemisphere, where the three stars making it up dominate the sky during summer evenings; Vega is almost overhead, Deneb and Altair also very high. It is a winter group from the southern hemisphere, and from South Africa or Australia Vega and Deneb are always very low down, though Altair can attain a respectable altitude.

Vega is a brilliant blue A-type star, of magnitude 0·0. It is unmistakable both because of its brightness and its colour. It is the leader of Lyra, the Lyre or Harp, which is a small but interesting constellation. Close beside Vega is a famous multiple star, Epsilon Lyræ. Fairly keen-sighted people can see it as double without optical aid; telescopically each component is again doubled, so that we have a quadruple system. The four components are genuinely associated, but the two main pairs are a fifth of a light-year apart, so that all we can really say is that they share a common motion through space. The distance from us is about 200 light-years.

Beta Lyræ, sometimes known by its proper name of Sheliak, is variable in light, but is one of the 'fake' variables which are nothing more nor less than eclipsing binaries. Beta Lyræ is made up of two stars, so close together that they almost touch, and which certainly cannot be seen separately. As they move round their common centre of gravity they periodically pass in front of each other as seen from Earth, so that the total light we receive from the system varies; but Beta Lyræ is always visible with the naked eye, and its neighbour Gamma Lyræ (magnitude 3·2) makes a useful comparison star. The Beta Lyræ system may well be surrounded by swirling streamers of gas, and if we could see it from nearby it would be magnificent indeed, but it is well over 1000 light-years away from us.

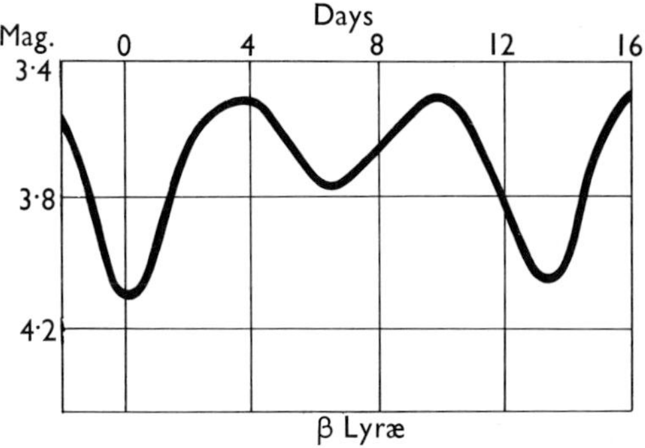

LIGHT-CURVE OF BETA LYRÆ. *The secondary minimum is much more pronounced than in the case of Algol.*

Between Beta and Gamma Lyræ there is another fascinating object, Messier 57, otherwise termed the Ring Nebula. Seen through an adequate telescope it looks like a tiny, glowing cycle-tyre. It is made up of a very small, faint, hot star surrounded by a shell of tenuous gas. Objects of this kind are called *planetary nebulæ*. The name is due to William Herschel, but it is not a good one, because the objects are not planets and are not true nebulæ. More than a hundred of them are known in the Galaxy, but most of them are very faint, and even M.57 is beyond binocular range.

Cygnus, the Swan, is often nicknamed the Northern Cross, for obvious reasons. Deneb, the leader, is another cosmic searchlight, 60,000 times as luminous as the Sun and about 1,600 light-years away, so that we are now seeing it as it used to be when the Romans occupied Britain. The X-form of Cygnus is unmistakable, but the symmetry is rather spoiled by the fact that one member of the pattern, Beta Cygni or Albireo, is fainter than the rest, and further away from the centre. To make up for this, Albireo proves to be unquestionably the loveliest double star in the entire sky. The third-magnitude companion is golden yellow, while the companion is vivid blue. The true distance between the two is about 650,000,000,000 kilometres.

Also in the Swan we find 61 Cygni, the first star to have its distance measured; a famous long-period variable, Chi Cygni; and the remarkable Cygnus X-1, which may include a Black Hole (about which I will have more to say later). But when I went out in the evening of August 29, 1975, soon after dusk, I noticed something else. There, some way from Deneb, was a new star – one that had never been seen before. It was brighter than any of the Cross stars apart from Deneb, and it altered the whole look of that part of the sky. It was a *nova*, or new star.

Strictly speaking, a nova is not a new star at all. What happens is that a formerly very faint object flares up, increasing its brilliance by thousands of times, and remaining bright for a few days, weeks or months before fading back to obscurity. Many have been seen over the ages. Two of our own century – one in Perseus in 1901, the other in Aquila in 1918 – have become really striking; for a while Nova Aquilæ outshone every star in the sky apart from Sirius. Nova Cygni 1975 was not so bright as that, and it faded very quickly. Within a few days it dropped below naked-eye range, and it has now become very dim indeed.

It was once thought that novæ were due to stellar collisions, but this

HR DELPHINI *(upper), photographed by Commander H. R. Hatfield in 1967. The nova was then visible with the naked eye, but is now too faint to be seen without a telescope.* NOVA CYGNI 1975 AT MAXIMUM: *August 30, 1975. Photograph by K. Kennedy, Dundee. The nova (now catalogued as V 1500 Cygni) is arrowed.*

is quite out of the question; the stars are so widely separated in space that collisions can hardly ever occur. In fact all novæ are binary systems, and the causes of the outbursts are linked with the evolutionary careers of the components. At its peak, Nova Cygni 1975 may have shone half a million times more brightly than the Sun.

Adjoining Cygnus is the small but compact constellation of Delphinus, the Dolphin. On July 8, 1967 a well-known English amateur astronomer, George Alcock, was sweeping this area with his powerful, specially-mounted binoculars when he found a star of just below the fifth magnitude which he could not identify. Alcock has spent many years in searching for novæ and comets; he knows the positions and magnitudes of at least 30,000 stars by heart, and he was not likely to make a mistake, so he telephoned various other observers (including myself) and asked for confirmation. Following his instructions, I was able to locate the star at once. It was indeed a nova, now known as HR Delphini. It never became as bright as the third magnitude, but it was exceptionally slow to fade, and it remained a naked-eye object for months. It is still a comparatively easy object; a few nights ago I estimated the magnitude as 11·6. Since we know that its pre-outburst magnitude was 12, it is not likely to fade much further.

Altair in Aquila (the Eagle) is another first-magnitude star, thirteen light-years from us. Aquila is a distinctive constellation; it contains the Cepheid variable Eta Aquilæ, and near the Eagle's tail, in the small constellation of Scutum (the Shield) is a star-cluster, M.11, nicknamed the Wild Duck. A small telescope will show it, and its fan-like shape makes it a beautiful sight.

Chart 5. Scorpio and Sagittarius.

These constellations also are summer objects in the northern hemisphere and winter objects in the southern, but I have given them a separate chart because they lie well south of the celestial equator; they are in fact the two southernmost constellations of the Zodiac.

Scorpio (more accurately, Scorpius) is a truly superb group, and one of the few constellations to give at least a vague idea of the object it is meant to represent; the long curved line of bright stars does conjure up a picture of a scorpion.

En passant, there is a famous mythological legend attached to it. Orion, the Hunter, boasted that he could kill any creature on Earth – but he had forgotten the scorpion, which crawled out of a hole in the ground and stung him fatally in the heel. Subsequently Orion was transferred to the sky; the Scorpion was given a similar honour – but was placed right on the opposite side of the heavens, so that there could be no further unpleasantness!

The leader of Scorpio is Antares, a huge red supergiant of the first magnitude, with a small companion which looks greenish. Antares is recognizable because it is flanked to either side by a fainter star. So, for that matter, is Altair, but the two cannot be confused, because they are a long way apart, and Altair is white rather than fiery red. The name Antares really means 'the rival of Ares' – Ares being the Greek equivalent of the war-god Mars. The Scorpion's sting, always very low as seen from Europe, contains one star, Lambda Scorpii or Shaula, which is only just below the first magnitude. Scorpio lies in the Milky Way, and is exceptionally rich.

Adjoining Scorpio is Sagittarius, the Archer. Here we have no first-magnitude star, but there are more than half a dozen above the second magnitude. Sagittarius does not present a distinctive shape (some people have likened it to a teapot), but is notable because of the splendour of its Milky Way star-clouds which lie between us and the centre of the Galaxy. There are some superb gaseous nebulæ, including M.20 (the Trifid Nebula), M.8 (the Lagoon Nebula) and M.17 (the Omega Nebula). Viewed with small telescopes they are frankly disappointing, but photographs taken with large instruments bring out their vivid colouring.

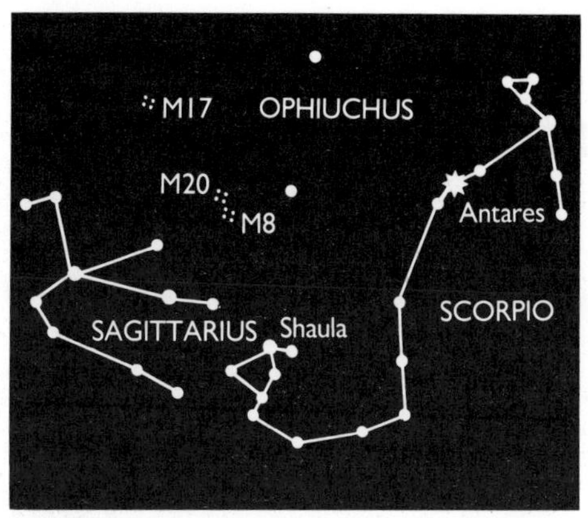

CHART. 5.

Chart 6. Pegasus, Andromeda, Aries, Perseus and Cetus.
Visible from all inhabited countries; best seen during autumn evenings in the northern hemisphere, spring evenings in the southern.

In mythology Pegasus was a flying horse. In the sky his four main stars make up a well-defined square; three of the four are white, but the other (Beta Pegasi or Scheat) is an orange giant, slightly variable in light. The Square can be used to find Fomalhaut in Piscis Australis (the Southern Fish), a white star of magnitude 1·2; it is always very low from Britain, and only southern-hemisphere observers can appreciate how prominent it really is. It is twenty-three light-years away, and

CHART. 6.

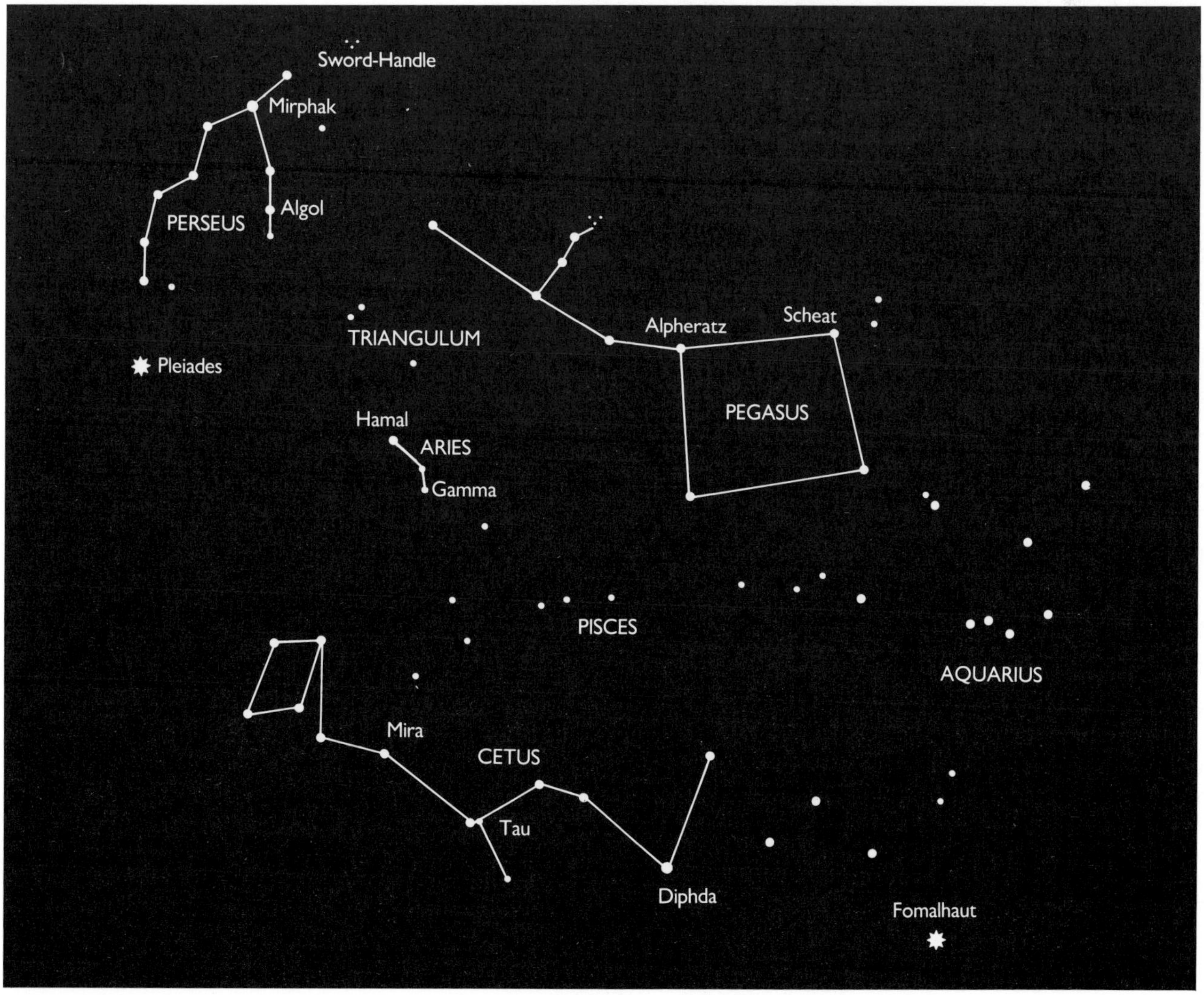

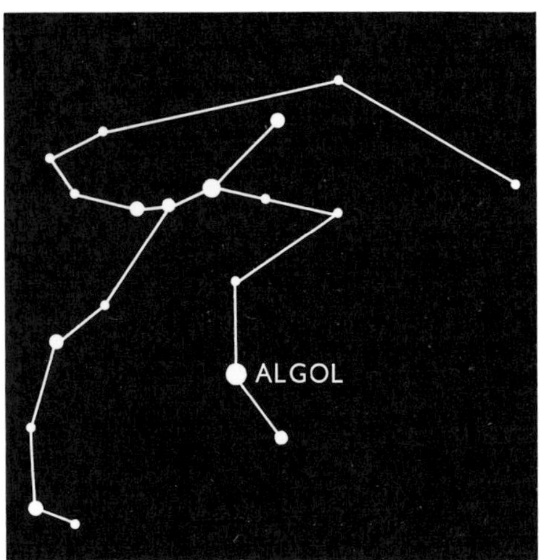

THE CONSTELLATION OF PERSEUS

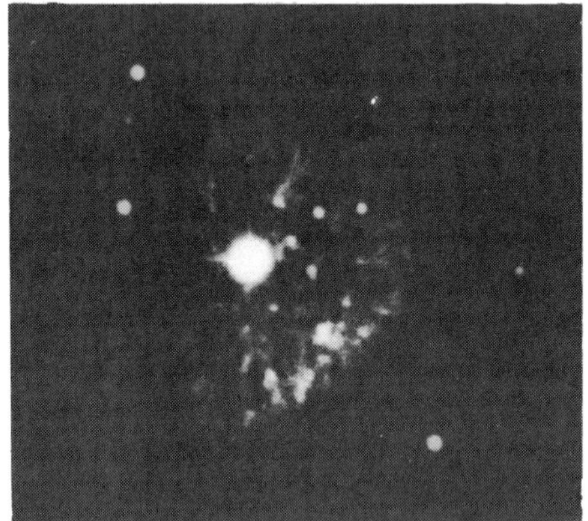

EXPANDING NEBULOSITY ROUND NOVA PERSEI, *photographed with the 508-cm Hale reflector at Palomar.*

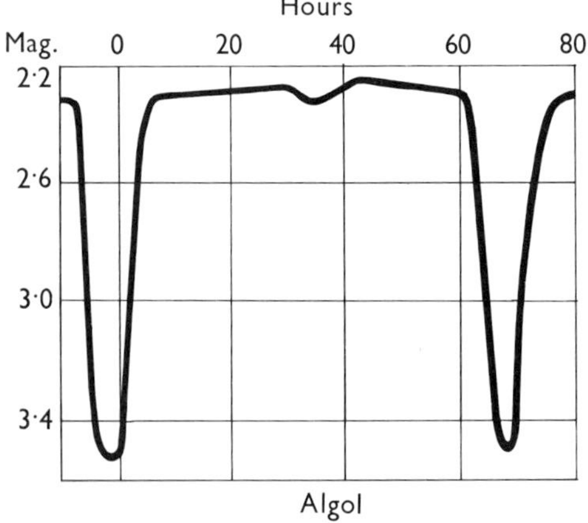

LIGHT-CURVE OF ALGOL. *The small secondary minimum cannot be detected without instruments.*

fifteen times as luminous as the Sun.

One of the stars in the Square, Alpheratz, has been officially transferred to the adjacent constellation of Andromeda, and catalogued as Alpha Andromedæ; I have never understood why, because it so obviously makes up a part of the Pegasus pattern. Andromeda itself consists of a chain of stars, and is notable because it contains the Great Spiral, M.31. It is just visible with the naked eye, and was certainly recognized long ago by Al-Sûfi, but it was then forgotten until more modern times; even Tycho Brahe overlooked it. We now know that it is an independent galaxy, larger than our own. The distance is 2.2 million light-years, and it is therefore the most remote object distinctly visible without optical aid.

Leading off from Andromeda is the constellation of Perseus, which is further north, and never well seen from Australia or South Africa. It is very rich, as the Milky Way crosses it. It has one second-magnitude star, Alpha Persei or Mirphak, and a magnificent pair of clusters, so close together that they lie in the same field of binoculars or a low-power telescope. They were not included in Messier's catalogue (presumably because there was no chance of confusion with a comet), and are known popularly as the Sword-Handle – not to be confused with the Sword of Orion.

However, the most celebrated object in Perseus is the 'Demon Star', Algol or Beta Persei. In 1669 Geminano Montanari, professor of mathematics at the University of Bologna, found that it was variable. Usually it shone steadily as a star of just below the second magnitude, but every 2 days 11 hours it faded down to below magnitude 3, taking five hours to do so; it remained faint for less than half an hour, and then took a further five hours to climb back to maximum.

John Goodricke, painstaking as ever, studied Algol closely, and came to the conclusion that it was not genuinely variable at all. The apparent fluctuations were due to some darker body passing between Algol and ourselves. In this case Algol would be a binary, with one component brighter than the other; when the dimmer star passed in front of its companion it would eclipse the main star, producing the regular 'wink'.

Goodricke, as usual, was right. The bright member of the Algol pair has a B8-type spectrum and a surface temperature of 12,000 degrees, with a diameter of about 4,000,000 kilometres. The fainter component is larger, but cooler and less luminous. When the bright star passes in front of the fainter one there is a small secondary minimum, but naked-eye observers will not detect it.

Though Algol is included in lists of variable stars, it should more properly be called an eclipsing binary. There is no difference between it and an ordinary binary of short period except that the plane of the orbit happens to lie in our line of sight. Were the orbit differently tilted, no eclipse would occur, and Algol's light would remain constant. Many Algol-type stars are known; if both components are bright enough to produce observable spectra, we are more likely to have a system of the Beta Lyræ type.

Adjoining Andromeda is Aries, the Ram, regarded as the first constellation of the Zodiac even though the vernal equinox has now shifted into the much fainter group of Pisces (the Fishes). Alpha Arietis, or Hamal, is of the second magnitude. Gamma Arietis is a wide, easy double – certainly one of the easiest doubles for a beginner,

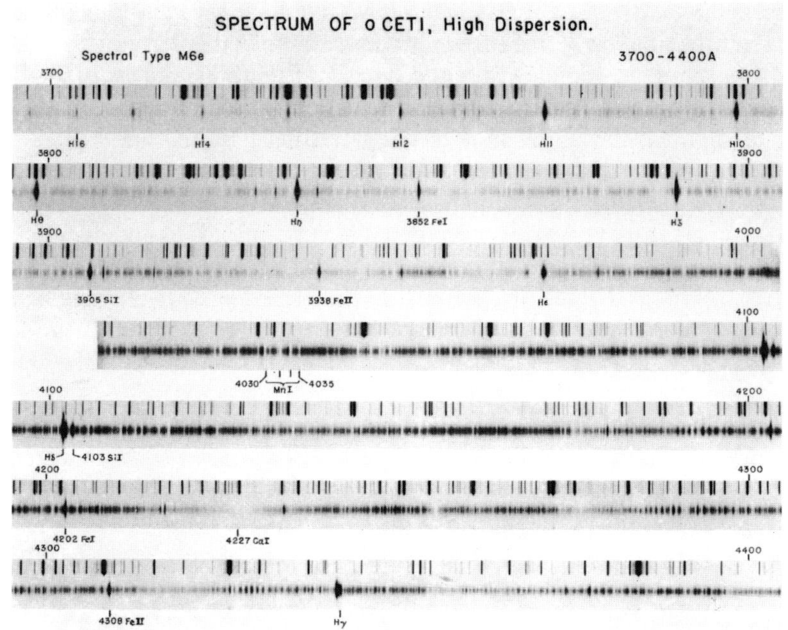

HIGH-DISPERSION SPECTRUM OF MIRA CETI *(left)*.

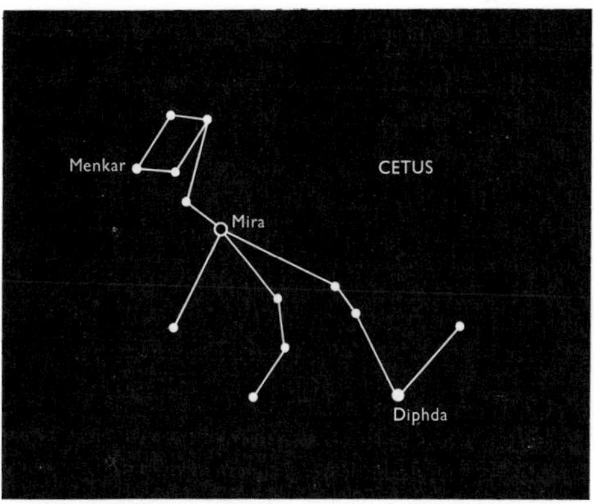

THE CONSTELLATION OF CETUS, *showing the position of the long-period variable Mira.*

since it is convenient to identify and can be separated with a very small telescope.

South of Pegasus and Andromeda is the large, rather dim constellation of Cetus, the Whale, sometimes identified with the sea-monster of the Perseus legend. There is only one second-magnitude star, the orange K-type Beta Ceti or Diphda, but in Cetus we find the best-known variable star in the sky, Mira or 'the Wonderful' Star.

One of the early pioneers of telescopic astronomy was David Fabricius, a Dutch pastor who lived at Osteel in Holland. He was a friend of Tycho and Kepler, and father of Johann Fabricius, who discovered sunspots independently of Galileo and Scheiner, but who died young.

On August 13,1595 David Fabricius was looking at stars in Cetus when he noticed an object of the third magnitude. He paid no particular attention to it, since it looked just like a normal star, but by October 1 it had disappeared. It is rather curious that Fabricius made no attempt to follow the matter up. (He met with an unfortunate end. In 1616 he preached a sermon in which he said that one of his geese had been stolen, and hinted that he knew who was responsible. Evidently he was right, since he was murdered before he could give the name of the thief.)

When Johann Bayer was drawing up his star catalogue, in 1603, he recorded a star in the same place as Fabricius had done. This time it was of the fourth magnitude, and Bayer gave it the Greek letter Omicron. He did not associate it with Fabricius' vanishing star; neither did Wilhelm Schickard, professor of mathematics at Tübingen University, who saw it again in 1631.

Then, in 1638, another Dutchman – Johann Phocylides Holwarda, professor at Franeker University – began a series of observations which showed that Omicron Ceti appeared and vanished regularly. It was a genuine variable star, and further observations by Hevelius from 1648 onwards showed that it had a period of about 331 days. At maximum it may reach the second magnitude, and has been known to

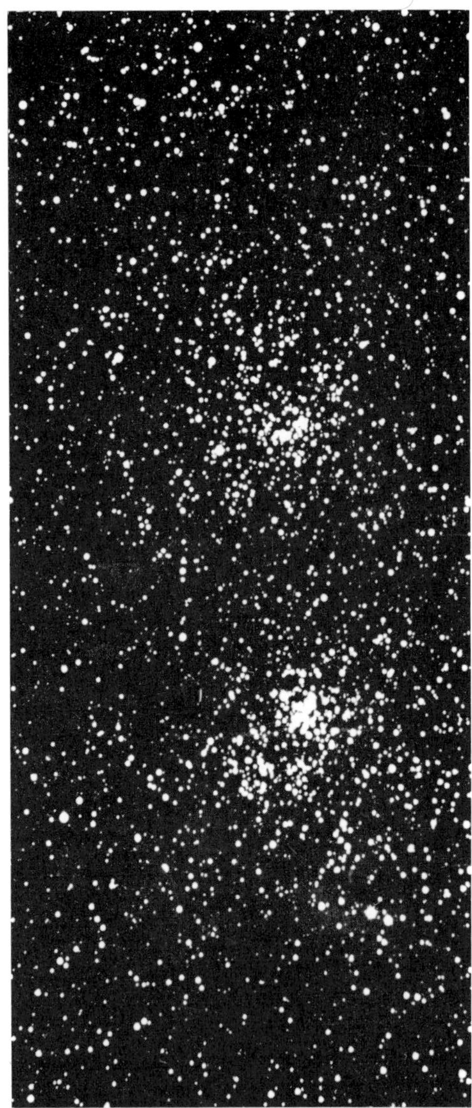

THE DOUBLE STAR-CLUSTER IN PERSEUS. *These objects are not included in Messier's famous catalogue, and are known officially as H. VI.33 and 34, from their numbers in Herschel's catalogue. The photograph given here was taken in 1932 by W. S. Franks, using a 15·24-cm refractor. The clusters, nicknamed the 'Sword-Handle' (not to be confused with the Sword of Orion) are beautiful objects, excellently seen in a small telescope.*

outshine Polaris, but at minimum it drops to magnitude 9½, so that even binoculars will not show it. It has an M-type spectrum, and is strongly orange-red; like Betelgeux, it is a red supergiant. Also, it has a faint companion which may be a White Dwarf.

Omicron Ceti was the first variable star to be identified, and was well named 'Mira'. Neither the period nor the maximum magnitude is constant. In some years Mira has failed to exceed the fifth magnitude, and on average it is visible with the naked eye for only about eighteen weeks out of its forty-seven-week period. The last really bright maximum was that of 1969, when for a time I ranked Mira only slightly inferior to the Pole Star.

Many other long-period variables have been detected, notably Chi Cygni in the Swan, which has a period of 409 days. At its best it is of about the fourth magnitude, and an easy naked-eye object, but at minimum it sinks down to magnitude 14. Like almost all Mira stars, Chi Cygni is red, and of 'late' spectral type. Amateurs make valuable observations of Mira stars; it is fascinating to watch the slow but obvious changes in brilliancy.

Chart 7. Stars of the Far South.

Never visible from Britain; always to be seen from South Africa, Australia or New Zealand.

From the northern-hemisphere observer's point of view, it is a pity that the brilliant stars of the far south do not rise. There are plenty of them. Canopus in Carina, the Keel of the dismembered constellation Argo (the Ship), is the second brightest star in the sky, with a magnitude of −0·7; it has an F-type spectrum, and has been described as yellowish, though I admit that to me it always looks pure white. Also in Carina is the wildly erratic Eta Carinæ, described earlier. Grus, the Crane, has a distinctive line of stars; the two leaders are of the second

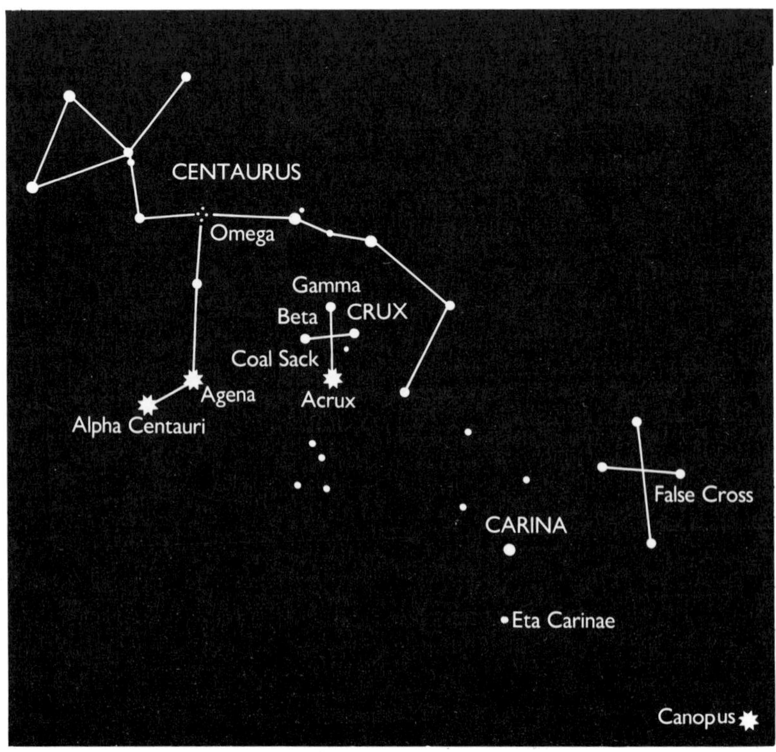

CHART. 7.

magnitude, but while one (Alnair or Alpha Gruis) is white, the other (Beta Gruis) is of a warm orange colour. But of all the southern constellations, by far the best-known is Crux Australis, the Southern Cross.

Frankly, it does not look much like a cross. Its four leading stars make up a pattern which is more like a kite. Crux is actually the smallest constellation in the sky; it was not even accepted as a separate group until 1679, when it was separated from Centaurus by an astronomer named Royer. But it is exceptionally rich, and it contains several objects of special note.

Acrux or Alpha Crucis, the leader, is a superb double; each component is a spectroscopic binary, and there is a third star in the field of a low-power telescope. To the naked eye, Acrux shines as a single star of magnitude 0·8. Next come Beta Crucis, another white, very luminous star (magnitude 1.3) and Gamma Crucis, a red supergiant (1·7). The fourth star of the pattern, Delta Crucis, appears fainter (magnitude 2·8), though it is very remote, and actually more luminous than Gamma. In Crux we find the lovely open cluster round Kappa Crucis, known as the Jewel Box because it contains stars of contrasting colours, and also the dark mass known as the Coal Sack. This is a gaseous nebula, not basically different from a bright nebula such as the Sword of Orion, but it has no suitable stars to make it shine – at least as seeen from our vantage point in the Galaxy – so that it is detectable only because it blots out the light of stars beyond. Many other dark nebulæ are known, several of them in Cygnus, but the Coal Sack is the best example of its type.

The two Pointers to the Cross are Alpha and Beta Centauri. Alpha Centauri has no official proper name, though air navigators sometimes call it 'Rigil Kent'. Its magnitude is −0·3, so that it is surpassed only by Sirius and Canopus, and it has the distinction of being the nearest of the bright stars; its faint companion, the red dwarf Proxima, is the closest star of all – and is likely to retain this title; had we any nearer neighbours, they would almost certainly have been found by now. Alpha itself is a glorious binary, with an orbital period of 80 years. The other Pointer, Beta Centauri or Agena, is quite different. It is nearly 500 light-years away, and more than 10,000 times as powerful as the Sun, so that with the Pointers we have again two stars which appear to be near neighbours, but which are in fact quite unconnected with each other.

Centaurus, the Centaur, is a large, rich constellation, and it contains the most prominent *globular cluster* in the sky: Omega Centauri. Globulars are quite unlike open star-clusters such as the Pleiades or the Jewel Box. They may contain at least a million stars, and are spherical in form; near the centre, the stars are comparatively close together, so that if we lived inside a globular cluster there would be no true darkness. The night sky would be ablaze, with many stars brilliant enough to cast shadows.

Over a hundred globulars are known in our Galaxy. They form a kind of 'outer framework' to the main system, and they provided the first proofs of the size of the Galaxy, because they contain short-period variable stars which 'give away' their luminosities, and hence their distances, by the way in which they behave. To be accurate, the variables in the globular clusters are not classical Cepheids, but the so-called RR Lyræ stars, which have much shorter periods and are all

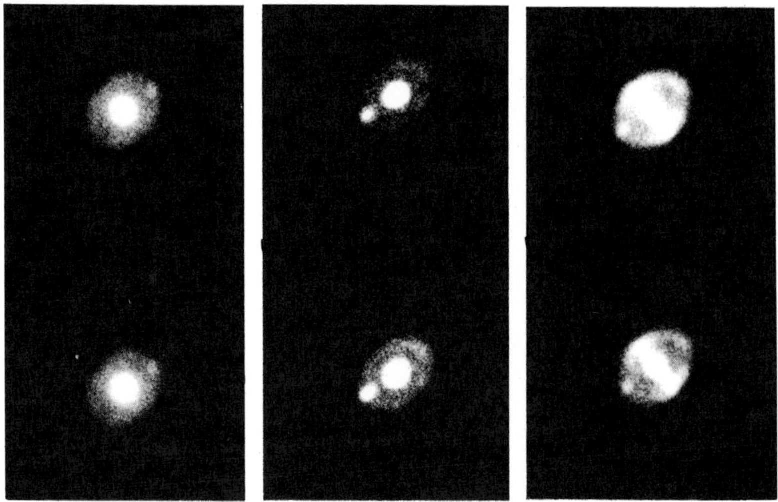

NOVA HERCULIS 1934, *photographed in ultra-violet, green and red light in 1951. Photographs with the 508-cm Hale reflector at Palomar.*

about ninety times as luminous as the Sun; but the principles involved are the same. Only two other globulars are visible with the naked eye. One is 47 Tucanæ, in Tucana (the Toucan), not far from Grus; the other is in the northern constellation of Hercules, and is shown in the Summer Triangle chart.

There is no bright south polar star, but fairly close to the pole are the two Clouds of Magellan – the Nubeculæ or Magellanic Clouds. With the naked eye they look rather like broken-off portions of the Milky Way, but again appearances are misleading. The Clouds are independent systems, over 150,000 light-years away from us. They are the nearest of the bright galaxies, and European astronomers never cease to bemoan the fact that they are so far south of the celestial equator. It is not too much to say that the need for studying the Clouds was an important factor in the decision to set up many of the new large telescopes in the southern hemisphere rather than in the north.

The story of the galaxies goes back a long way. But before reaching out beyond our own system, I must say rather more about the life-histories of the stars.

29

The Life of a Star

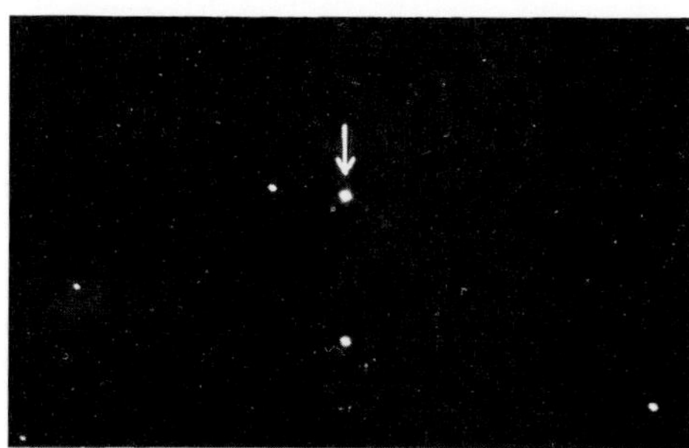

SUPERNOVA IN THE GALAXY I.C. 4168, *photographed with the 259-cm Hooker reflector at Mount Wilson. (Top) August 23, 1937, when the supernova was at maximum. (Centre) November 24, 1938, the supernova is now fainter. (Foot) January 19, 1942; the supernova can no longer be seen. Supernovæ of this sort have often been recorded in external galaxies.*

THE Harvard system of classifying stellar spectra, introduced by E.C. Pickering and his assistant Annie Jump Cannon, proved to be of great importance. Most of the stars fell into types B,A,F,G,K and M; the other types were rare. But what did the sequence mean – and did it have any bearing on a star's evolution? The key was provided by two astronomers independently: Ejnar Hertzsprung of Denmark, and Henry Norris Russell of the United States.

Hertzsprung was born on October 8, 1873 in Frederiksberg, near Copenhagen. He studied engineering and chemistry, but made a hobby of astronomy, and in 1905 and 1907 published two papers which made him turn professional – largely at the instigation of a leading German astronomer, Karl Schwarzschild. Hertzsprung worked at Potsdam, and then went to the Observatory of Leiden in Holland, becoming Director in 1935 and serving as such for ten years before retiring. He made many contributions to astrophysics, and remained active in research until shortly before his death.

Russell was born at Oyster Bay, New York, on October 25, 1877, and was educated at Princeton in New Jersey. After a period working in England, he became Director of the Princeton Observatory in 1908. His main research was concentrated upon the source of a star's energy, but he also made great contributions to theories of the origin of the Solar System and the composition of planetary atmospheres.

The major discovery, announced in Hertzsprung's papers of 1905 and 1907, was that there are two definite classes of stars. Red stars, of type M, are either very large and luminous, or else small and faint. Red stars of about the same luminosity as the Sun, he found, do not exist; there is a very obvious division into 'giants' and 'dwarfs'. The division was less obvious for the orange stars (type K) and less still for the yellow stars (G and F), while for the hot, white stars of types A and B it did not exist at all; but it was unmistakable, and clearly important.

Russell noted the same phenomenon slightly later, and the combined work was used to produce a diagram in which stars are plotted according to their spectral types and their luminosities. Diagrams of this sort are, justifiably, known as Hertzsprung-Russell Diagrams, or H-R Diagrams for short, and their importance in modern astrophysics cannot be over-stressed. Of course, they can be drawn in different ways. Instead of spectral type we can use surface temperature, which comes to much the same thing; after all, the sequence from B to M is one of steadily decreasing temperature. And rather than use luminosity, taking the Sun as unit, we can use what is called *absolute magnitude*. This is the apparent magnitude which a star would have if it could be seen from a standard distance of 32·6 light-years, or 10 parsecs. The Sun's absolute magnitude is +5, so that if observed from this distance it would be a dim naked-eye object. On the other hand the absolute magnitudes of Rigel and Deneb are about −7, so that from the standard distance they would cast shadows.

(A *parsec*, by the way, is the distance that a star would have if its parallax amounted to one second of arc – one *par*allax *sec*ond. Actually, no star apart from the Sun is as close as this; the parallax of Alpha Centauri is 0·76 of a second of arc. Professional astronomers generally talk in terms of parsecs, but for the moment I propose to keep to light-years. If you want to change light-years to parsecs, simply divide by 3·26.)

Plot an HR Diagram, and you will find that most of the stars lie in a band stretching from the upper left (hot, luminous stars) down to the lower right (dim, cool, red stars). This band is called the *Main Sequence*. The Sun is a typical Main Sequence star of Type G2. To the upper right we have the giants and supergiants; thus Capella has the same spectral type as the Sun, but is 150 times more luminous.

Look at the red stars. To the upper right we have Betelgeux, Antares and other supergiants; to the lower left we have stellar glowworms such

THE HERTZSPRUNG-RUSSELL DIAGRAM. *This shows the relationship between a star's spectrum and its luminosity. Supergiants are on the uppermost part of the graph, the giant branch to the right, and White Dwarfs to the lower left. Most of the stars belong to the Main Sequence, which is shown by the white band. For many years it was supposed that this graph showed a strict evolutionary sequence, so that a star descended the Main Sequence from type B to M, but it is now known that matters are much more complex than this. The principle of graphs of this kind was due to the work of H. N. Russell, of the United States, and E. Hertzsprung of Denmark.*

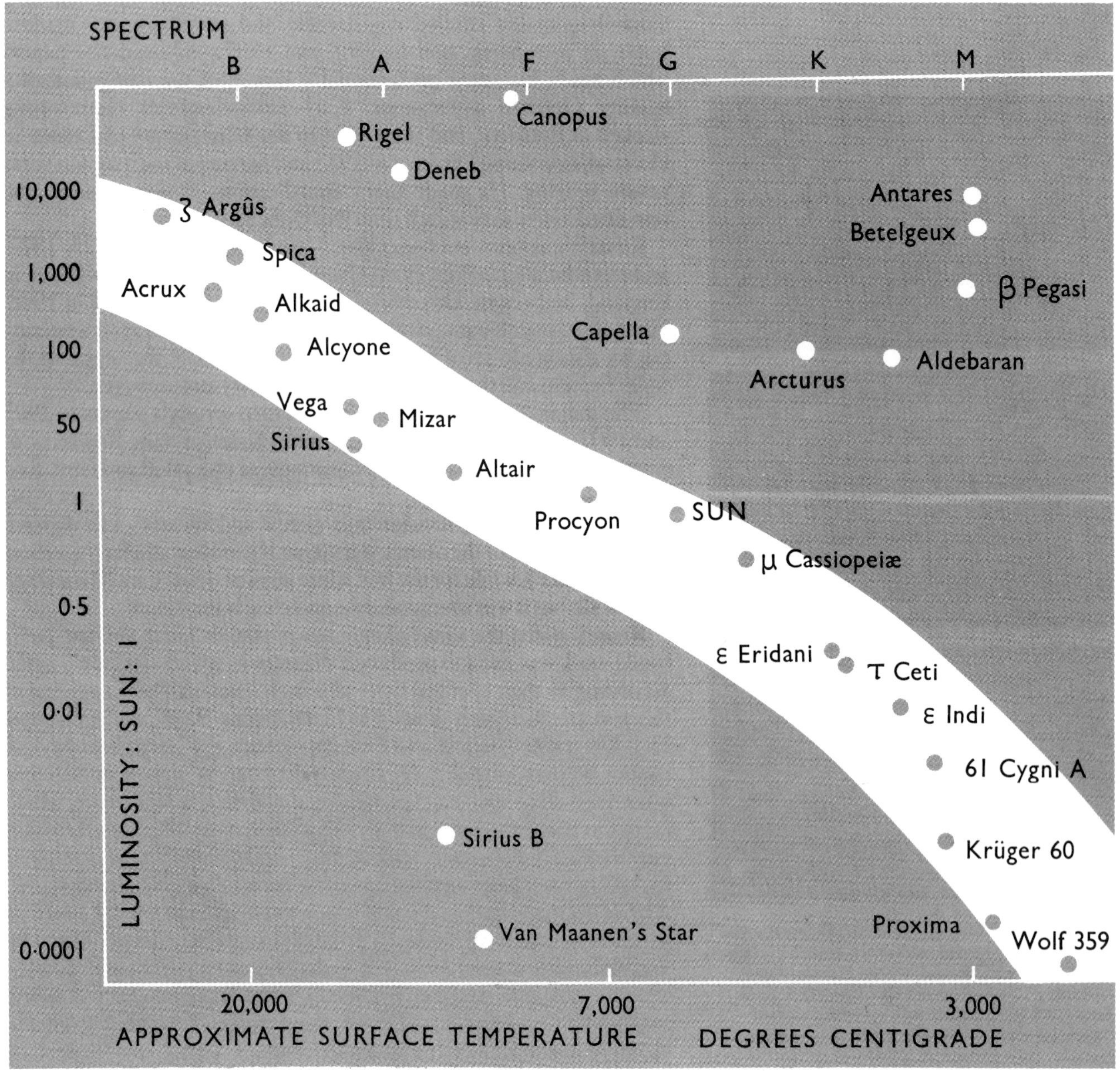

as Barnard's Star, which has only sixteen-per-cent of the Sun's power and is a mere 250,000 kilometres in diameter, less than twice that of Jupiter. But red stars equal to the Sun in luminosity are absent. With stars of Type M, there are no half-measures.

Naturally enough, it was assumed that HR Diagrams might well give some vital clues as to the way in which a star evolves. As a start, let it be said at once that no star is 'burning' in the accepted sense of the term. They are too hot to burn, but in any case there is an easy way of showing that they must draw their energy from some other source. The first man to point this out was not an astronomer, but a physicist: William Thomson, afterwards Lord Kelvin, who lived from 1824 to 1907, and was largely responsible for laying the foundations of modern atomic theory. Thomson showed that if the Sun were burning in the same way as a coal fire, it would last for only a few millions of years at most before it turned itself into ashes. But geologists can tell the age of the Earth with fair certainty; it proves to be about 4,700 million years, and the Sun must be older than that, so that the 'burning' theory is completely out of court. (If we work out a scale according to which the age of the Earth is represented by eighty hours, the whole story of human civilization will have to be crammed into a single second.)

A better idea was put forward by Sir Norman Lockyer in 1890. Lockyer, the pioneer spectroscopist who was so deeply concerned with studies of the Sun, worked out a plan of stellar evolution which sounded delightfully straightforward. We know now that it was incorrect, but it provided a useful working basis.

Lockyer began by supposing that a star condenses out of material scattered in space as dust and gas; in fact, a nebula, which would be a stellar nursery. This part of the theory, at least, is still accepted today.

Lockyer went on to reason that as the material draws together, under the influence of gravity, it will form a spherical mass, and the inner temperature will rise. All the material will tend to draw towards the centre of the sphere, and so the pressure also will increase. At first the star will be a red giant of vast diameter and relatively low surface temperature; as it shrinks and heats up it will become a yellow giant,

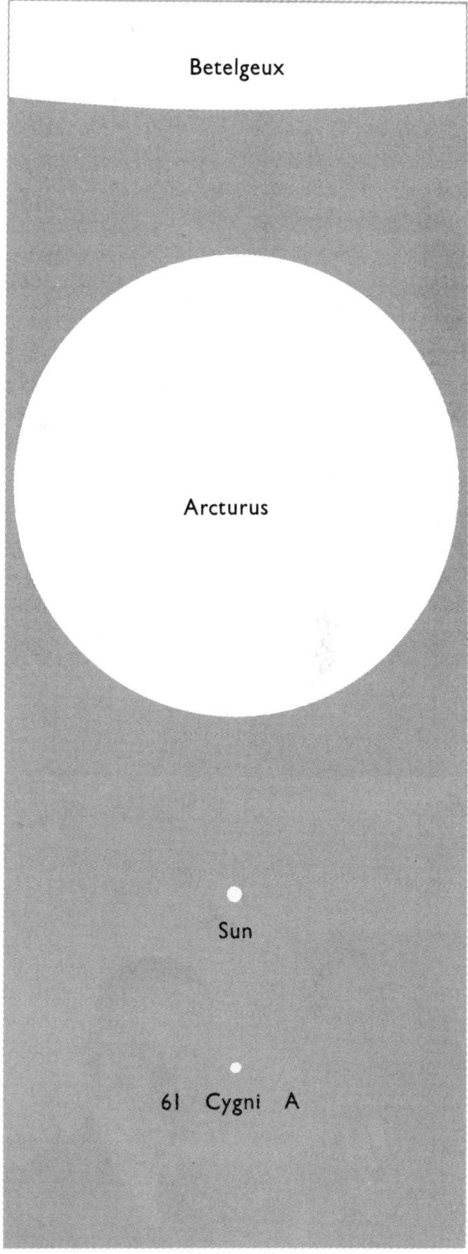

GIANT AND DWARF STARS, *drawn to scale. Betelgeux has a diameter of 400,000,000 km; Arcturus 42,000,000 km; the Sun 1,392,000 m; 61 Cygni A 960,000 km. The supergiant Betelgeux and the red dwarf 61 Cygni A each have M-type spectra. Arcturus is of type K, and the Sun type G.*

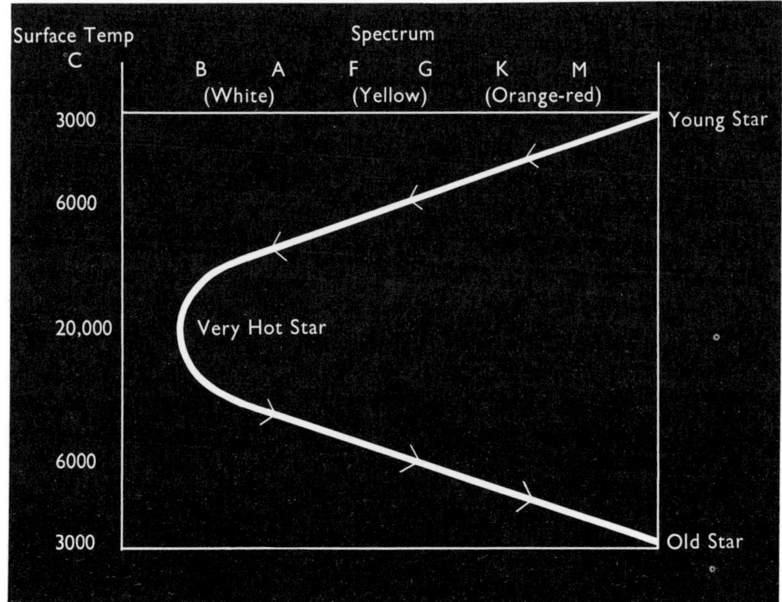

STELLAR EVOLUTION ACCORDING TO LOCKYER. *(Left) In its final development, Lockyer's theory gave what appeared to be a very plausible explanation of stellar evolution. A star would begin its career as a Red Giant such as Betelgeux, pass down the giant branch until becoming a very luminous star of type B (Rigel), and then pass down the Main Sequence, becoming a Red Dwarf (Proxima) and, finally, a cold dark globe. It is now known that the whole theory is incorrect, and matters are in fact much less straightforward. Nevertheless, Lockyer's theory provided a useful working basis.*

and then a smaller but hotter bluish-white star of type B. This will be the peak of its career. The gravitational shrinking will go on, but the temperature will fall, so that the star becomes in succession a yellow dwarf (such as the Sun), an orange dwarf and then a red dwarf of type M. Eventually all its heat will leave it, and it will become a dead globe.

When Lockyer first proposed this scheme, the giant and dwarf divisions had not been recognized. The later work of Hertzsprung and Russell seemed to provide splendid confirmation, but still there was something wrong. A star which radiated only because it was contracting under the influence of gravity could not last for more than 50 million years, which was not enough.

Russell tried to put matters right by introducing the idea of atomic energy. It was known that an atom contains protons, each of which carries a unit charge of positive electricity, and electrons, each of which carries a unit negative charge. If a proton and an electron met, reasoned Russell, they would 'cancel each other out', and both would vanish, with the emission of energy. Could it be that a star shone because it was annihilating its material?

Russell's theory was published in 1913, and it seemed at first to be satisfactory. A star would begin as an M-type red giant, as Lockyer had supposed, and would turn into a hot bluish-white star, after which it would pass down the Main Sequence, ending up as a red dwarf again of type M. Unfortunately there were difficulties of another sort. So much energy would be available that a star could last for at least ten million million years, which was as obviously too long as Lockyer's fifty million years had been too short. To make matters worse, further studies showed that a proton and an electron would not annihilate each other in the way that Russell had assumed.

Astronomers such as Sir Arthur Eddington, who had been born at Kendal in Westmorland in 1882 and had become Professor of Astronomy at Cambridge, made various modifications to the theory; but the problems remained, and almost up to the outbreak of the second world war the chief mystery – the source of stellar energy – was still unsolved.

The main difficulty is that we cannot watch a star as it ages, any more than a visitor from another world could watch a boy turn into a man if his spell on Earth were limited to a day or two. What we can do, however, is to look at stars at different stages of evolution, and work out which are young and which are old. Unfortunately this is not so easy as it might seem, and early theorists, not unnaturally, picked wrong. Neither is it true that the Main Sequence represents an evolutionary track. We must start again.

The starting-point of the old theory may be kept: the condensation of a star out of nebular material. We can even see this process going on. During the 1930s the Dutch-born astronomer Bart J. Bok, who has lived for many years in the United States, drew attention to some small dark patches seen against bright nebulæ. They were named Bok globules; some of them are at least six light-years in diameter, and they may well indicate regions where material is collecting to form a proto-star, though it is true that there are some differences of opinion here. At any rate, there is no doubt that gravitational effects are responsible for the development of a proto-star. As the density increases, the temperature rises, and at last we come to a stage when the shrinking mass may truly be termed a 'star'.

PROFESSOR SIR ARTHUR EDDINGTON. *One of the greatest of British astrophysicists, Sir Arthur Eddington was responsible for many important advances. He was moreover well known as a lecturer and broadcaster, and wrote a number of popular books on astronomy and allied subjects. Modern theories of stellar evolution owe much to his pioneer work in the earlier part of the present century.*

Everything depends upon the original mass of the proto-star, so let us take the different classes in turn, starting with stars of less than one-tenth the mass of the Sun. Here, nothing much happens. The central temperature never becomes high enough to spark off nuclear reactions, and so the star remains dim and red, subsequently fading until no energy is left. One may even say that it is a star which has failed its Common Entrance examination.

Next, consider stars of between one-tenth and 1·4 times the mass of the Sun. The story is very different. The star shrinks, and becomes steadily hotter at its core. It starts to shine – at first unsteadily; it has not settled down, and it varies in an irregular manner. Many of these youthful stars are known; we call them T Tauri stars, after the first-discovered member of the class. But at last, after a somewhat chequered career, our solar-type star will heat up so much that its core temperature reaches about 10,000,000 degrees. And at this point, nuclear reactions begin. The key to the process was discovered in 1938 by two astronomers independently, and almost at the same time: Hans Bethe in America, and Carl von Weizsäcker in Germany. Apparently Bethe had been attending a conference in Washington, and was returning to Cornell University by train when he started wondering whether he could calculate any nuclear reactions which would explain why the Sun shines. Before dinner was served, he had broken the back of the whole problem.

The lightest of all the elements is hydrogen, whose atom is made up of one proton and one orbiting electron. In the universe, hydrogen atoms outnumber all the rest of the elements put together, and it is no surprise to find that the Sun contains a great deal of hydrogen (at least seventy per cent, by mass). At the tremendous temperatures and pressures at the Sun's core, the hydrogen nuclei run together, merging to make up nuclei of the second lightest element, helium. It takes four hydrogen nuclei to make up one helium nucleus, but – and this is the vital point – the single helium nucleus 'weighs' slightly less than the original four hydrogen nuclei. Mass has been lost, and released in the form of energy. It is this energy which keeps the Sun shining, while the mass-loss amounts to 4,000,000 tons per second.

This may sound fantastically great, but I assure you that there is no need for alarm. The Sun has enough hydrogen to keep it radiating much as it does now for the next four to five thousand million years, and since it is at least five thousand million years old it is no more than middle-aged.

Yet even this quantity of hydrogen 'fuel' cannot last for ever, and at last it will begin to run low. Hydrogen 'burning' has to stop; there is no hydrogen left in the core. Gravity takes over, and the star's core, now rich in helium, begins to shrink. But there is still plenty of hydrogen left between the core and the surface, and as the temperature goes on rising this hydrogen becomes hot enough to start reacting. We now have an inert helium core, surrounded by a hydrogen-burning shell. The star swells out as the core goes on contracting, but this state of affairs cannot last. Eventually the core temperature reaches 100 million degrees, and this is enough to make the helium react in its turn, producing the heavier nuclei of carbon and oxygen.

This *helium flash* releases a fresh supply of energy, and the shrinking of the core stops, but the star changes its whole nature. We have a helium-burning core surrounded by a hydrogen-burning shell, beyond which the outer layers of the star are pushed further and further away. This means that the surface temperature drops to something like 3000 degrees. The star has become a red giant; and when the Sun reaches this stage, there can be no hope for the Earth. It and the other inner planets will be vapourized, and the Sun's diameter will reach at least 300 million kilometres, as great as the present diameter of the Earth's orbit.

Once the supply of helium has been used up, the situation changes again. The core is now made up of carbon and oxygen; outside this is a helium-burning shell; outside this again a hydrogen-burning shell, surrounded by the immensely distended and rarefied 'atmosphere'. So far as a star like the Sun is concerned, this is the beginning of the end. The mass is not great enough to make the carbon and oxygen react to form heavier nuclei. The star becomes unstable; it pulsates, and eventually the outer layers are thrown clear of the inert core to produce a planetary nebula such as M.57, the Ring Nebula in Lyra. Planetary nebulæ may look beautiful, but they are signs that a star is dying. As the shells expand, they become even more rarefied, and after a mere 50,000 years or so they cease to shine. The planetary nebula has disappeared; all that is left is the burnt-out remnant of the once

SIZES OF DWARF STARS. *The Sun, with its diameter of 1,392,000 km, is regarded as a dwarf star, but some of the White Dwarfs are much inferior to it in size. Here, the Sun is shown together with the planets Uranus, the Earth and Mars, and two White Dwarfs – Sirius B and Kuiper's Star. It is remarkable to find that Kuiper's Star, with a diameter slightly less than that of Mars, has a mass equal to that of the Sun, so that its matter is remarkably dense.*

powerful star. This remnant shrinks still further, and becomes denser and denser. It has turned into a White Dwarf. It has no reserve of energy left; nothing to prevent its material from pressing down harder and harder. The electrons which have been stripped away from their nuclei prevent the collapse from becoming complete, but a White Dwarf has a bleak future. After an immensely long period it will lose the last of its light and heat, and will become a dead Black Dwarf.

Before going any further, this new viewpoint enables us to explain the outbursts of novæ, such as the 1975 star in Cygnus. A nova is made up of two components; an ordinary Main Sequence star, and a more evolved White Dwarf. The White Dwarf has a very powerful gravitational pull, and actually tears material away from its companion. This material builds up in a ring round the White Dwarf– and, of course, it is made up chiefly of hydrogen. As more and more material reaches the ring, the temperature rises; when it reaches the 'critical point' the familiar hydrogen-to-helium reactions are triggered off – and the result is a violent though short-lived outburst, producing a nova. When the main outburst is over, the system returns to its former state, though there is no reason why the same thing should not happen over and over again; as we have seen, there are recurrent novæ such as the Blaze Star

in Corona, and more conventional novæ may also have outbursts more than once, though the intervals between them are presumably much longer. There is a link, too, with certain variable stars known as dwarf novæ, of which the best-known examples are SS Cygni and U Geminorum. With these stars, the outbursts are relatively mild, and take place more or less predictably; with SS Cygni, the brightest member of the class, the mean period is around 47 days.

Now consider a more massive star, much 'heavier' than the Sun. The early stages of evolution are much the same as before, but everything is accelerated; we are talking now of millions or a few hundreds of millions of years, instead of thousands of millions. For example, the age of Vega has been given as between 200 and 240 million years, while the very hot, luminous star Theta Carinæ, not far from Canopus in the sky, seems to be less than ten million years old. The stories diverge after the building-up of the carbon and oxygen core, with shells of active helium and hydrogen around it. The mass pressing down is so great that the core temperature is forced up to incredible values. At 700 million degrees carbon starts to react, and at 1000 million degrees oxygen follows suit. When these fuels also are used up, we have the usual picture of a shrinking core and rising temperature. Silicon is built up from oxygen nuclei; then, at a core temperature of 3000 million degrees, silicon too starts to react, producing nuclei of iron.

Iron is the key to what happens next, because it will not react, no matter how high the temperature becomes. We reach a stage in which our evolving star has an iron core surrounded by a whole series of shells, burning, in order, silicon, oxygen, carbon, helium and hydrogen. At last the inert iron core can no longer take the strain. Electrons are forced into the iron nuclei, and combine with the positively charged protons in the nuclei themselves. The result of combining a proton and an electron is a different kind of particle, with no electrical charge: a neutron. But the neutrons take up much less space than the original protons and electrons, so that there is a violent implosion, the opposite of an explosion. It is both cataclysmic and brief; the amount of energy set free from the core collapse sends out a shock-wave, and the star is literally torn apart in what we call a supernova explosion.

We have met supernovæ before; in our Galaxy there have been the stars of 1006 in Lupus, 1054 in Taurus, 1572 in Cassiopeia (Tycho's Star) and 1604 in Ophiuchus (Kepler's Star). No doubt many more have occurred, unseen by us because of the fog of interstellar dust in the way. We can at least see the remnants of supernovæ which burst forth in prehistoric times. Such is the Veil Nebula in Cygnus, where there are arcs of nebulosity which are believed to have been produced by shock-waves from the supernova explosion. Another is the Gum Nebula in the southern constellation of Vela (the Sails), named in honour of the Australian astronomer Colin Gum, who died tragically in a skiing accident in 1960. And there is also the Crab Nebula.

The Crab is of supreme importance. It was once said that there are two kinds of astronomy: the astronomy of the Crab Nebula, and the astronomy of everything else! It sends out visible light, radio waves, X-rays, gamma-rays, and in fact radiations at virtually all wavelengths. The expansion of the gas-cloud is measurable over periods of years, and there is absolutely no doubt that the Crab is the wreck of the 1054 supernova, though since it is six thousand light-years away the outburst actually took place six thousand years before the Chinese and Japanese astronomers watched it.

For many years the Crab was a puzzle. It seemed to have some inner 'power-house', but where was it? The answer was found in an unexpected way – not by ordinary observations, but by radio astronomy.

The story of radio astronomy must be deferred until Chapter 31. For the moment, it is enough to say that bodies in the sky send out radiations of various kinds, and those of long wavelength are collected by radio telescopes, which are really in the nature of huge aerials. In November 1967 Miss Jocelyn Bell, at Cambridge (now Dr. Jocelyn Bell-Burnell) was working with the team headed by Sir Martin Ryle when she detected a very unusual kind of radio source. It was invisible optically, but it seemed to be 'ticking', very rapidly and very regularly. The Cambridge team was nonplussed. There was even a suggestion that the signals might be artificial transmissions coming from a world many light-years away.

This attractive theory (known to the team as the LGM or Little Green Men theory!) was soon discarded. It was then suggested that the ticking source might be a rapidly-spinning White Dwarf star, but this would not do either, because a White Dwarf is too large to rotate several times a second. Eventually it was found that Miss Bell's ticking source was nothing more nor less than a neutron star, the wreck of a supernova. It was the first *pulsar*; others were soon found, including one in the Crab Nebula. Then, in 1969, a team at the Steward Observatory in Arizona, using a 91-centimetre reflector, identified the Crab pulsar optically as a very faint, flashing object.

This was the answer; the pulsar was the long-sought power-house. Later, in 1978, astronomers at the Siding Spring Observatory in Australia identified a second pulsar, that in the Gum Nebula; its optical magnitude is below +24, so that it is the faintest object ever observed. Undoubtedly other pulsars also send out a certain amount of light, but they are so faint that they are very hard to locate, and so far only three have been actually seen: those in the Crab Nebula, the Gum Nebula, and the new unusual pulsar in Vulpecula, which is spinning 642 times every second and seems to be in a different category. It was discovered in 1982, and is known by its catalogue number of 1937+21.

A pulsar or neutron star is an incredible object. The 'ticks' range in period from the Crab pulsar (0·3 second) up to about 4 seconds; among 'ordinary' pulsars the Crab is the fastest, and is presumably the youngest, because the pulsars seem to be gradually slowing down as they lose their energy. For example, the pulsar known as CP 1919 is lengthening its period by a thousand-millionth of a second each month.

Apparently the 'ticks' are caused by the rotation of the neutron star. There are immensely powerful magnetic fields, and when electrons at the surface of the neutron star meet the fields at the north and south poles they are beamed outwards, emitting the radio waves that we can detect. The magnetic poles do not coincide with the poles of rotation, and the effect is rather like that of a rotating searchlight. If you stand within range of such a searchlight, you will be briefly illuminated every time the beam passes over you; on Earth, we receive a pulse every time the 'beam' of the spinning neutron star sweeps across us.

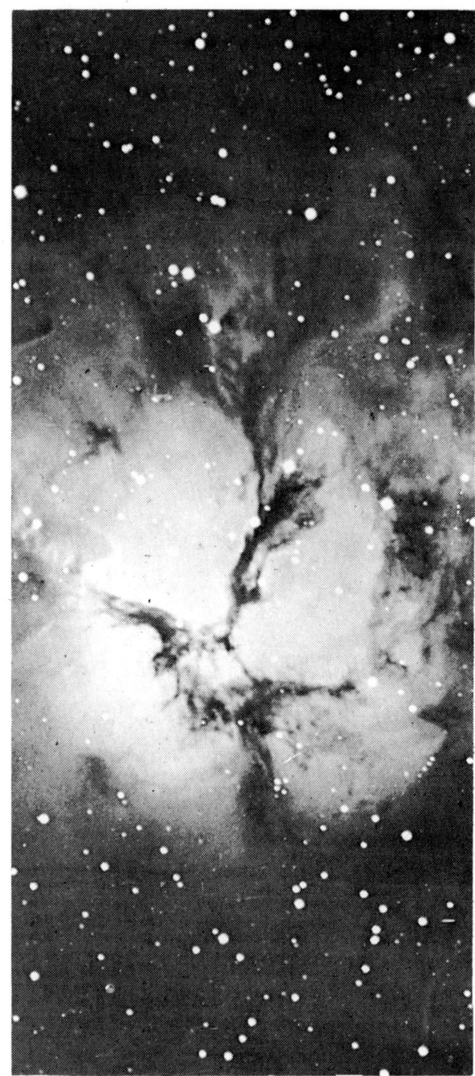

THE TRIFID NEBULA IN SAGITTARIUS, *photographed on June 30, 1921, with the 254-cm Hooker reflector at Mount Wilson; exposure 2½ hours. A nebula of this sort is a mass of dust and gas, and is entirely different from a galaxy such as the Andromeda Spiral, since it is a member of our own Galaxy. The old name for external galaxies, 'spiral nebulæ', is now obsolete. It was in any case somewhat misleading, since by no means all the galaxies are spiral in form.*

What is a neutron star like? Theory indicates that the typical diameter is no more than 20 kilometres, though the mass is equal to that of the Sun. The outer layer may be solid, made of iron; inside this is neutron-rich material, and inside this again a layer made up of neutrons together with a few surviving protons and electrons. The solid core is composed of 'hyperons', which may be more fundamental than neutrons, but about whose nature we can do no more than guess. It has been said that a neutron star is rather like the legendary curate's egg; a thin shell on the outside and several peculiar fluids inside. The density may reach 100 million million times that of water, which is quite beyond our understanding.

All the supernovæ seen in our Galaxy over the past thousand years have been found to be radio sources, and most of the prehistoric remnants as well, though not all seem to contain pulsars (for instance, Tycho's Star does not). Neither can we be sure that all pulsars are supernova remnants. In 1980 Paul Murdin and David Clark, at Siding Spring, carried out a careful study of a very strange object; a star called SS 433, associated with a radio source, W.50. The whole system is most peculiar, and is very variable. It seems to be a binary, made up of two stars. One member of the pair is sending out streamers of gas in diametrically opposite directions, and these jets are rotating rather in the manner of a lawn sprinkler, while the star itself is orbiting its companion in a period of twelve days. SS 433 has been termed a *scintar*, and it could well be a supernova remnant, though as yet it is impossible to be sure.

If the inert core of a stellar remnant is more than two-and-a-half times as massive as the Sun, it cannot remain as a neutron star. Gravity takes over, and is irresistible. Unless the star has managed to shed much of its original mass before the final collapse comes, only one fate awaits it. It becomes a Black Hole.

If neutron stars are bizarre, Black Holes are more bizarre still, and it must be stressed that much of what we think we have found out about them may prove to be badly wrong. There are even a few astronomers who are sceptical about the whole idea. But Black Holes have come more and more into favour during recent years; so let us turn back to Cygnus X-1, not far from the long-period variable Chi.

Cygnus X-1 is so called because it is a source of X-rays. These are very short-wavelength radiations; most people know about them because of their use in hospitals. They cannot reach the Earth's surface from beyond, because layers in the upper atmosphere block them out, and no X-ray astronomy could begin before we learned how to send equipment into space. In 1962, a rocket carrying an X-ray 'telescope' was sent up, and managed to locate a strong source in the constellation of Scorpio. Others were found subsequently, one of which was Cygnus X-1.

Again we are dealing with a binary system. The main star is a B-type giant, very white and very hot, with a diameter twenty-three times that of the Sun (about 18,000,000 kilometres) and a mass thirty times that of the Sun. The companion has fourteen times the Sun's mass. The distance of the pair is 6500 light-years, so that even though the primary is so luminous it is only of the 9th magnitude – well below naked-eye visibility, and none too easy in binoculars. The orbital period is 5·6 days, so that the two components are very close together.

If the companion is so massive, we would expect to see it – but we do not. It is utterly invisible. All we can detect, apart from the light of the primary, is the flow of X-rays.

Now let us go back to the evolution of an extremely massive star. It cannot explode as a supernova, subsequently becoming a neutron star; it is too massive for that. When the collapse starts, nothing can stop it. The star becomes smaller and smaller, denser and denser; even the neutrons cannot withstand the pressure, and the escape velocity becomes steadily greater.

Escape velocity, remember, is the velocity needed for a body to break free from a planet or a star; eleven kilometres per second for the Earth, sixty kilometres per second for Jupiter, 617 kilometres per second for the Sun, and so on. With our massive collapsing star, escape velocity finally attains 300,000 kilometres per second. This is the speed of light. Therefore, light cannot escape – and if light cannot do so, then certainly nothing else can, because light is the fastest thing in the universe. The old star has surrounded itself with a 'forbidden zone' into which material can go, but from which nothing can emerge. The forbidden zone is to all intents and purposes cut off from the rest of the universe, and this is what is meant by a Black Hole.

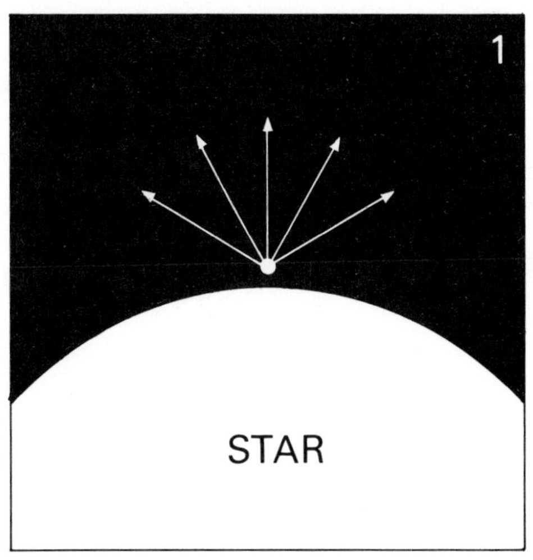

Actually the situation is rather less straightforward than this. The basic idea was suggested as long ago as 1798, by Laplace, but today it is known that 'gravity' is more properly described as being due to the curvature of space itself. I am afraid that this is something which is almost impossible to explain in everyday terms, and I will not attempt it here; it is one of the consequences of Einstein's theory of relativity. But the diagrams may help. In the first, things are normal. In the second, the star has begun to collapse; the curvature of space round it has become marked, and the light-rays are deflected. In the third diagram, the deflection has become so great that no light-rays can leave the star. For the moment, this is much the same thing as saying simply that the escape velocity has become as great as the speed of light.

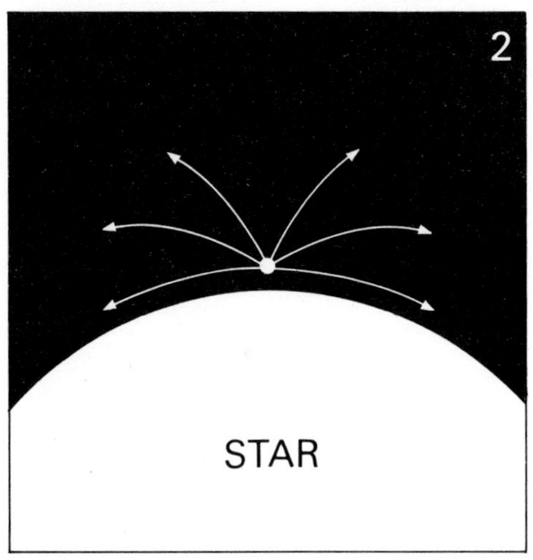

'Well', you may say, 'if nothing can halt the collapse, what eventually happens to the old star?' The plain truth is that nobody really knows. It has been said that the star becomes a *singularity,* which is infinitely dense and infinitely small; or does the star literally crush itself out of existence altogether, leaving nothing but the Black Hole? Again plain English fails us.

The size of the Black Hole will depend upon the mass of the old star. The critical radius of a non-rotating Black Hole is called the *Schwarzschild radius,* after the German astronomer Karl Schwarzschild, who was born in 1873 and became director of the Göttingen Observatory; he died in 1916, as a result of military service during the First World War, but by then he had made tremendous contributions to theoretical astronomy. The boundary of the critical radius is known as the *event horizon.* For a body the mass of the Sun, the Schwarzschild radius would be three kilometres; for the Earth, less than a single centimetre.

There has been a great deal of speculation about conditions inside a Black Hole. For obvious reasons it will never be possible to make direct investigations; we cannot see inside the event horizon. Once a star becomes a Black Hole, it effectively disappears.

This may have happened to the invisible companion of Cygnus X-1. If so, then the immensely powerful pull is dragging material away from the visible star; as this material spirals down toward the event horizon it is violently heated, so sending out the X-rays which we can pick up.

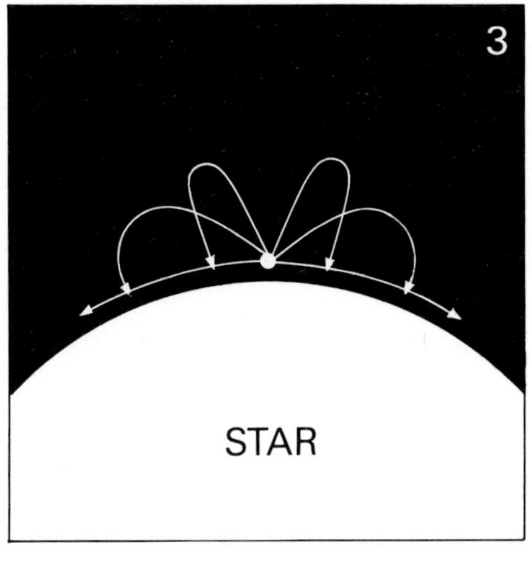

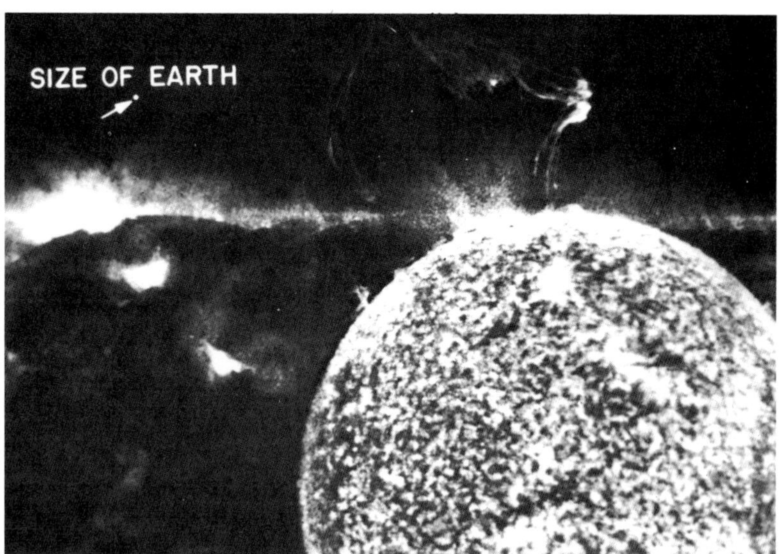

A GREAT SOLAR PROMINENCE. *Skylab picture, NASA.*

Various other Black Hole candidates have been proposed; one is the invisible secondary of the Epsilon Aurigæ system, but on the whole Cygnus X-1 remains the most promising to date.

Exotic theories about Black Holes have been produced in plenty. For instance, it has been claimed that material entering one of these weird objects may emerge either in a different part of our universe or in a completely different universe. I am bound to admit that I view these ideas with profound scepticism, but I may be wrong – and at the moment our research into the nature of Black Holes is at a very early stage. I will have more to say about them when coming on to consider globular clusters, galaxies and quasars.

We have learned a good deal about the life-stories of the stars. If the star is comparable in mass with the Sun, it will become a red giant before turning into a White Dwarf. If more massive, it will die by supernova explosion, leaving a gas-cloud and possibly a pulsar. And in more extreme cases it may become a Black Hole. Fortunately, our Sun is a stable Main Sequence star; nothing much will happen to it for several thousands of millions of years yet. The crisis will come eventually, but at least we will have plenty of warning.

30

Star Clusters and Star Systems

I HAVE several times mentioned the Pleiades or Seven Sisters, No. 45 in Messier's catalogue. Here we have an open or galactic cluster, so conspicuous that it must have been known from prehistoric times. It is much the most striking of the open clusters; all its leading members are hot and white, with early-type spectra, and between its stars there is thinly-spread gas, illuminated to make up what is termed a reflection nebula. The total membership of the cluster exceeds four hundred stars.

The Hyades, round Aldebaran, are much less spectacular, and are rather overpowered by the bright orange-red light of Aldebaran itself. Actually, Aldebaran is not a member of the cluster at all. It is only sixty-eight light-years from us, as against 130 light-years for the Hyades, so that it merely happens to lie in the foreground. Several of the Hyades are visible with the naked eye, making up the familiar V-formation, and one of them (Theta Tauri) is a naked-eye double. As with the Pleiades, the cluster contains around four hundred members, but it appears rather scattered, because it is the closest to us of all the galactic clusters.

Another famous cluster is Præsepe in Cancer (the Crab), sometimes nicknamed the Beehive. It lies more or less between the Twins on one side and the Sickle of Leo on the other; it cannot rival the Pleiades, but it is easy to see without optical aid on a clear night, and binoculars show it well. Another 'binocular cluster' is M.67, also in Cancer; it is notable because it seems to be very old on the cosmical scale, and some of its leaders have already evolved off the Main Sequence. Generally speaking, open clusters are not permanent features; their members will eventually disperse, because of the gravitational effects of non-cluster stars. But M.67 lies over a thousand light-years above the main plane of the Galaxy, so that it is away from the most crowded areas, and has been able to survive as a distinct unit.

In Gemini there is M.35, just visible with the naked eye. I have

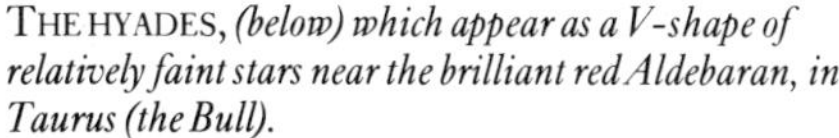

THE HYADES, *(below) which appear as a V-shape of relatively faint stars near the brilliant red Aldebaran, in Taurus (the Bull).*

THE PLEIADES *or 'Seven Sisters'. (Right) All the stars named are visible with the naked eye on a clear night. The brightest of them, Alcyone, is of the 3rd magnitude.*

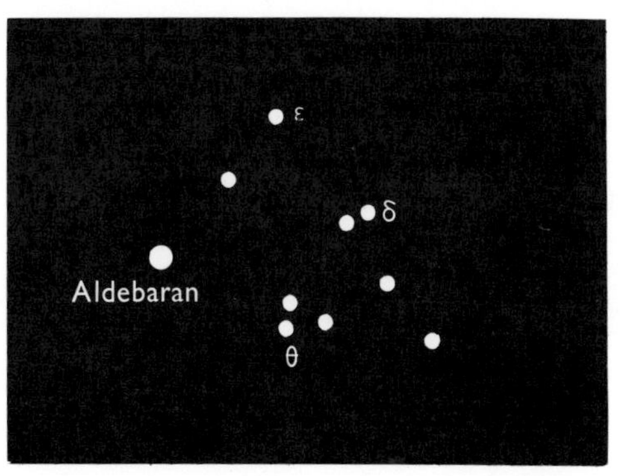

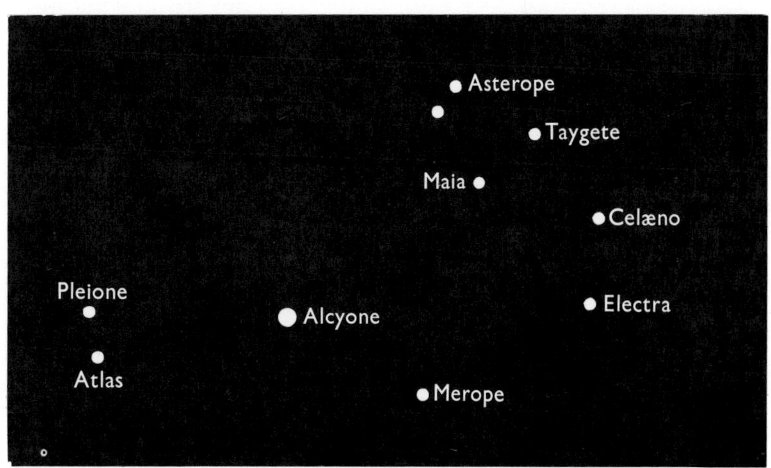

SPIRAL GALAXY N.G.C. 2841, IN URSA MAJOR. *The spiral form is well shown. Like most external galaxies, the object is not well seen with a small or moderate telescope, and photography with very large instruments is needed to bring out the details properly. This picture was taken with the 508-cm Hale reflector at Palomar.*

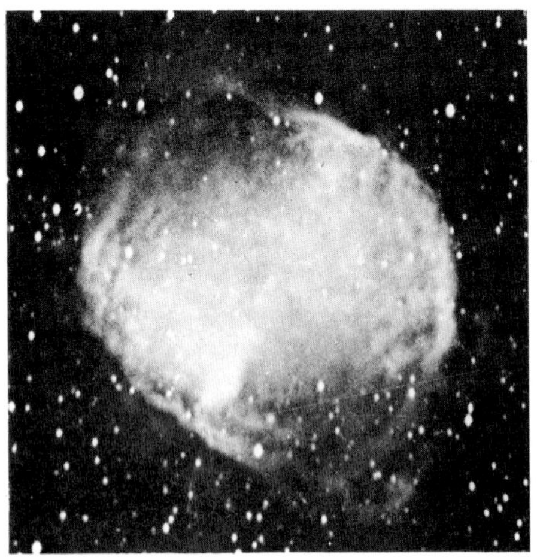

THE DUMB-BELL NEBULA IN VULPECULA, *a famous planetary, photographed at the Lick Observatory.*

already mentioned M.11, the 'Wild Duck' cluster in Scutum, and the Jewel Box, round Kappa Crucis in the Southern Cross. Telescopic clusters are very common indeed; some of them are rich, with hundreds of stars, while others are so sparse as to be scarcely recognizable.

Globular clusters are completely different. Instead of being formless, they are regular, and appear as circular masses; in reality they are spherical and symmetrical. The two brightest, Omega Centauri and 47 Tucanæ, are in the far south, but the globular M.13, in Hercules, is available to northern observers – though it is on the fringe of naked-eye visibility, and I always find it very hard to see without optical aid.

With sufficient magnification, a globular cluster is a superb sight. Stars are seen near the edges, though near the centre of the system they are so packed together that they seem to merge into one mass. This, incidentally, is a case in which we are dealing with genuine crowding, not a line of sight effect, though the individual stars are still too widely spread-out for collisions to occur except on very rare occasions.

Rather more than a hundred globulars are known in our Galaxy. They are faint because they are so remote; even the brightest of them, Omega Centauri, is around 17,000 light-years away. Its declination is −47°, so that it can rise from any latitude on Earth south of +43°, but it is never to be seen from Europe or most of the United States; this is a pity, because it is truly magnificent. The rich central core is 100 light-years across, and the full diameter is about 150 light-years; there may be as many as a million members all told, and close to the centre the average separation between them is only about one-tenth of a light-year. Energetic radiations from some globulars have led to the suggestion that their cores may contain massive Black Holes, and this is certainly a strong possibility.

A detailed study of globular clusters was carried out by Harlow Shapley, at Mount Wilson, between 1914 and 1921. From 1917 he was able to make use of the Hooker reflector, then incomparably the most powerful telescope in the world. He was interested in studying the distances of the globulars, but they are too far away to show

MESSIER 16: NEBULA IN SCUTUM SOBIESKI, *photographed in red light with the 508-cm Hale reflector at Palomar.*

measurable parallaxes, so Shapley turned to less direct methods. He also noted that most of the globulars lie south of the celestial equator, with a marked concentration in the regions of Scorpio and Sagittarius. Such a lop-sided distribution could hardly be due to chance, and Shapley concluded almost at once that the Sun must lie well away from the centre of the Galaxy.

He found the essential clue when he detected RR Lyræ variables inside globular clusters. Remember, RR Lyræ stars are all of about the same luminosity, about 90 times that of the Sun, so that their distances can be found. This in turn gave the distances of the globular clusters in which they lay. (*En passant,* not all RR Lyræ stars are members of clusters – RR Lyræ itself, the prototype star, is not – and the old term of 'cluster-Cepheids' has been dropped.)

Shapley's results were conclusive. The Galaxy turned out to be about 100,000 light-years across,* with a greatest width of 20,000 light-years. The Sun lies well away from the centre, at about 32,000 light-years from the galactic nucleus, but not far from the main plane of the system. The nucleus itself lies beyond the glorious star-clouds in Sagittarius. We cannot see it, because there is too much obscuring material in the way, but radio waves from it can be picked up, and some very interesting results have been obtained in recent years. It has also been shown that the Galaxy is spiral in shape, and that it is rotating; the

*Or approximately so. Some authorities regard this is an appreciable over-estimate.

MESSIER 13, THE GLOBULAR CLUSTER IN HERCULES, *photographed with the 152-cm reflector at Mount Wilson. The first photograph was given an exposure of 15 minutes; the second, 37; the third, 94. The increase in the number of stars shown with increasing exposure is very obvious. In this respect the photographic plate is far superior to the eye; the longer the exposure, the more it records.*

Sun, near the edge of one of the spiral arms, takes about 225,000,000 years to make a full circuit – a period which is sometimes called the 'cosmic year'. One cosmic year ago, the most advanced life-forms on Earth were the amphibians; the Coal Measures were being laid down, and the ferocious dinosaurs had yet to make their entrance. Two cosmic years ago, the only life on Earth consisted of tiny marine creatures. It is interesting to speculate as to what 'men' will be like in one cosmic year from now – if, of course, men still exist.

Nebulæ, on the whole are less conspicuous than clusters, and few of them are visible with the naked eye. Messier catalogued a number of them and so did Herschel, who believed some of them to be 'stellar' and others not. In 1791, Herschel wrote that with the Orion nebula 'our judgement, I venture to say, will be that the nebulosity around the star is not of a starry nature'. In this he was of course, right, but the decisive proof did not come until August 29, 1864, when Sir William Huggins turned his spectroscope toward a nebula in Draco, the Dragon.

A star yields a spectrum with a rainbow background and dark absorption lines. If a nebula were made up of stars, the result would be a jumble of all the star-spectra combined, and the main absorption lines would be recognizable; but Huggins found that the Draco nebula gave an emission spectrum, with isolated bright lines. Actually, the first object he selected was a planetary nebula, now known to be a dying star, but other nebulæ also were found to yield emission lines – M.42, Orion's sword, being one of the first – and so it was clear that these nebulæ really are clouds of gas and dust.

The Orion Nebula is of special interest, as being comparatively close (about 1,400 light-years) and a stellar birthplace, with many stars still in the T Tauri stage as they approach the Main Sequence in the HR Diagram. It is of tremendous size, at least ten light-years in diameter, but it is very rarefied. If you could take a 2·5-centimetre core sample right through it, the total weight of material collected would weigh no more than a new penny. The nebula shines because of the radiation sent out by the hot stars of the Trapezium (Theta Orionis), which lie close to the nearer edge of the nebula. The region is of the H.11 type, with the nebulosity excited to a certain amount of self-luminosity instead of depending entirely upon reflected light, as with the nebulosity in the Pleiades.

We cannot see through to the heart of M.42, but we can tackle the problem by using infra-red radiation, which can pass unchecked through the nebular material. Infra-red astronomy has come very much to the fore recently; the 381-centimetre British telescope at Hawaii was made for this kind of research, and there are many others.

During the 1960s two astronomers at Mount Wilson, Becklin and Neugebauer were carrying out an infra-red survey when they discovered a powerful source deep inside the Orion Nebula. Nothing could be seen visually, or ever will be, because of the intervening material, but the so-called Becklin-Neugebauer or BN object is certainly very powerful, and may be either a highly luminous, well-evolved star or else a young proto-star contracting towards the Main Sequence. Subsequently another powerful infra-red source was found; the KL object, named after its discoverers, Kleinmann and Low.

It was also found that the visible Orion Nebula is only part of a much larger mass of nebulosity which covers Orion, but is invisible optically.

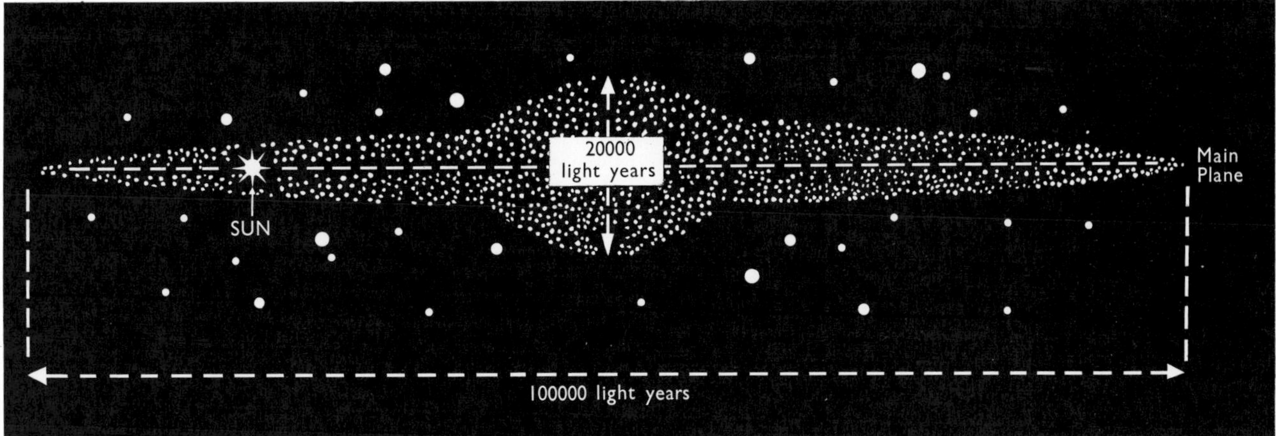

DIAGRAM OF THE GALAXY AS IT WOULD BE SEEN EDGE-ON. *The galactic centre lies about 25,000 light-years away from us. The Sun is therefore well out toward the edge of the system, and is not near the centre, as Herschel believed. When we look along the main plane, many stars are seen in approximately the same direction, and this causes the Milky Way effect.*

It is very cold (around −260 degrees Centigrade) and is made up chiefly of hydrogen molecules. We cannot detect the hydrogen directly, but it acts upon molecules of another substance – carbon monoxide – which we can locate. The carbon monoxide radiates only if it is bombarded by sufficient numbers of other molecules and only hydrogen is plentiful enough for this, so that if we find carbon monoxide it follows that hydrogen must be present too.

There are other lovely galactic nebulæ; the Lagoon, the Trifid, the Omega and so on. Photographs of them taken with large telescopes are spectacular, though I must repeat that direct views of them are apt to be disappointing, since the colours are not vivid enough to show up; the intensity of the light is too low. (If you doubt this, try looking at a red or yellow car by moonlight. You will find that it looks grey.)

Quite apart from the stars, clusters and nebulæ, the Galaxy contains a large quantity of interstellar material. The first indication of this was given by the German astronomer Hartmann in 1904, when he was studying the spectrum of Delta Orionis or Mintaka, one of the stars in the Hunter's Belt. Delta Orionis is a spectroscopic binary, so that its lines show Doppler shifts according to the orbital motion. Hartmann found that some of the absorption lines remained stationary, so that clearly they were due to intervening material rather than the star itself. Then, in 1930 the Swiss astronomer J. Trümpler, working in the United States, found that some of the Milky Way clusters appeared fainter than they ought to have done, so that presumably their light was being dimmed en route.

THE GLOBULAR CLUSTER OMEGA CENTAURI, *photographed at Royal Observatory, Cape, with the 61-cm Victoria telescope on February 19, 1903 (exposure 50 minutes). This is the brightest globular cluster in the sky but unfortunately it lies too far south in the sky to be seen from Europe or the northern United States.*

The gas between the stars is made up chiefly of hydrogen and helium, with smaller amounts of nitrogen, neon, oxygen and others. Since 1968 organic molecules have also been found. These are molecules containing carbon, and include formaldehyde, carbon monoxide, hydrogen cyanide and ethyl alcohol. (One 'cloud' contains enough alcohol to fill the Earth's globe!) This has led Sir Fred Hoyle and others to suggest that life may begin in these interstellar regions rather than on the Earth itself, a point to which I will return later.

Interstellar space also contains dust grains, and it is these which are responsible for the dimming of the light from remote objects. The dust is very thinly-spread, but it may make up as much as one per cent of the total mass of the Galaxy.

Now let me come back to William Herschel's second guess: that some of the so-called nebulæ may be separate systems, far beyond the Milky Way. The discovery of the spirals, by Lord Rosse, made this

GALAXY N.G.C. 4565, *a system seen almost 'edge-on' to us.*

CLUSTER OF GALAXIES IN HYDRA, *photographed with the 200-inch Hale reflector at Palomar. Foreground stars of our own Galaxy are shown, but many of the blurred patches are due to remote external galaxies, each of which is composed of perhaps 100,000 million suns.*

seem rather more likely, but gradually the idea fell out of favour, and by the end of the nineteenth century it had been generally discarded. There seemed no obvious way to solve the problem one way or the other.

Then, in 1920, came a famous debate between two American astronomers, Harlow Shapley and Heber D. Curtis. Shapley had used the short-period variables in globular clusters to measure the size of the Galaxy; he maintained that the spirals were members of our own system. Curtis believed the Galaxy to be much smaller than Shapley had estimated, but that the spirals were separate systems. The result of the debate turned out to be an honourable draw. Shapley's estimate of the size of the Galaxy was correct, but Curtis was right in claiming that the spirals were external.

The next step was to try to locate Cepheid variables inside the spirals, and this was tackled by Edwin Hubble, who was destined to give a final answer to the long-fought battle between those who believed in 'island universes' and those who, like Shapley, did not.

Hubble was born in Missouri on November 20, 1889. He graduated from Chicago University in 1910, where he had distinguished himself not only as a scholar, but as an athlete and as a skilled amateur heavyweight boxer – he once fought an exhibition bout with the famous French boxer Carpentier. Briefly he practised law, but in 1914 turned to astronomy, going to the Yerkes Observatory as a staff member. He was then offered a position at Mount Wilson, but in 1917 the United States joined in the war against Germany, and Hubble enlisted in the Army, rising quickly to the rank of major. In 1919 he was demobilized,

STELLAR POPULATIONS I AND II. *(Left) The Andromeda Galaxy, photographed in blue light, shows giant and supergiant stars at Population I in the spiral arms. The hazy patch at the upper left is composed of unresolved Population II stars. N.G.C. 205, companion of the Andromeda Galaxy, photographed in yellow light, shows stars of Population II; the brightest stars are reddish, and only 1/100 as bright as the blue giants of Population I. The very bright, uniformly distributed stars in both pictures are foreground stars belonging to our own Galaxy. Photographed with the 508-cm Hale reflector at Palomar.*

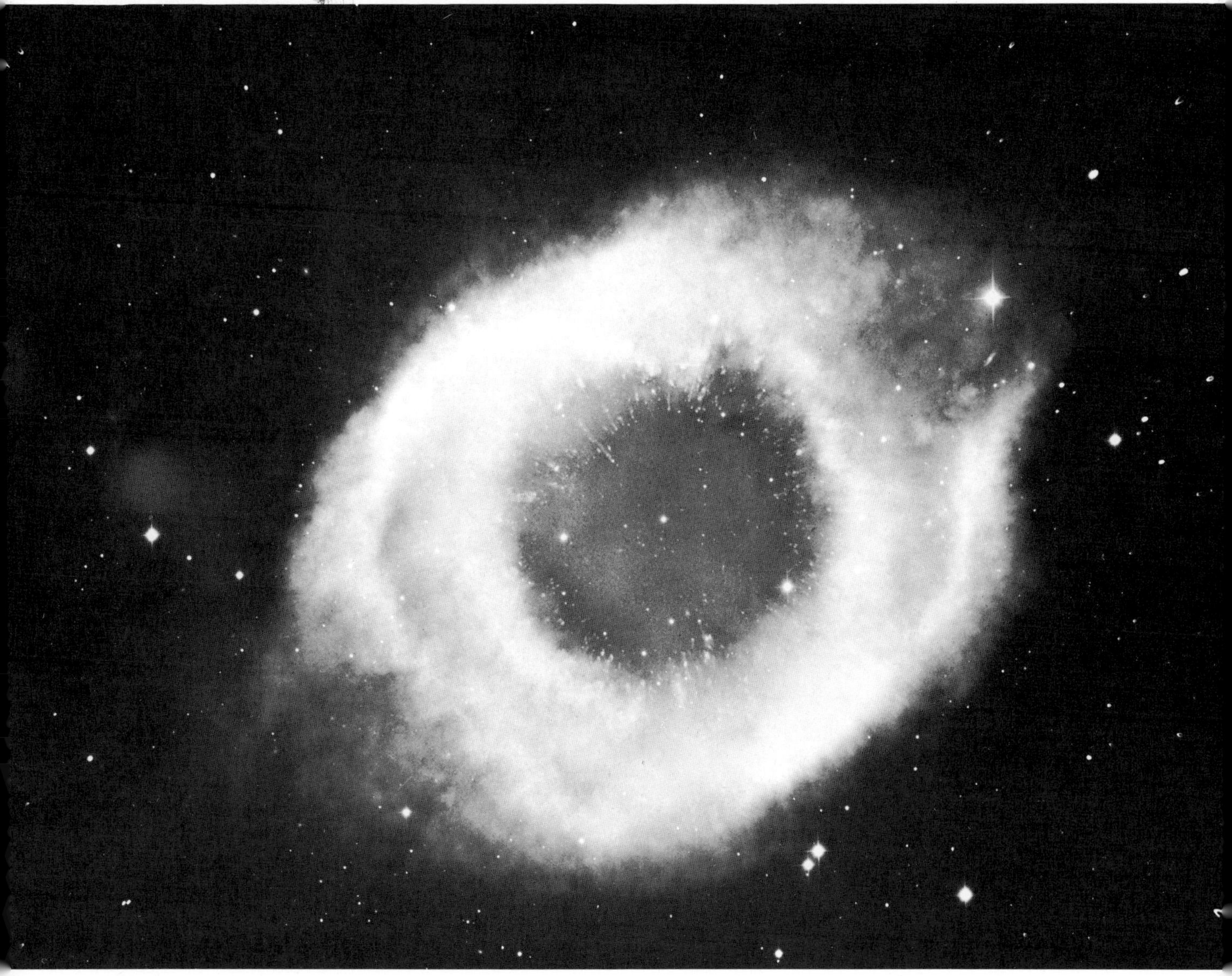

PLANETARY NEBULA N.G.C. 7293 IN AQUARIUS, *photographed in red light with the 508-cm Hale reflector at Palomar.*

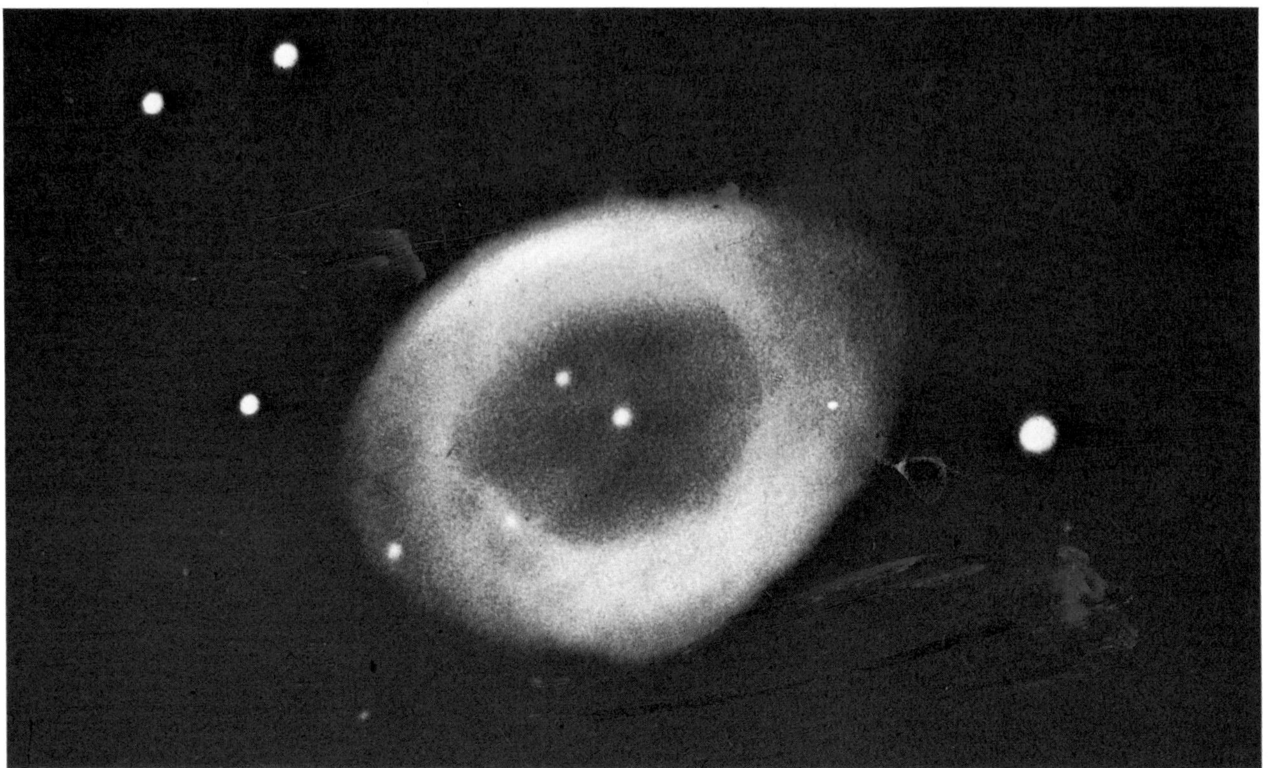

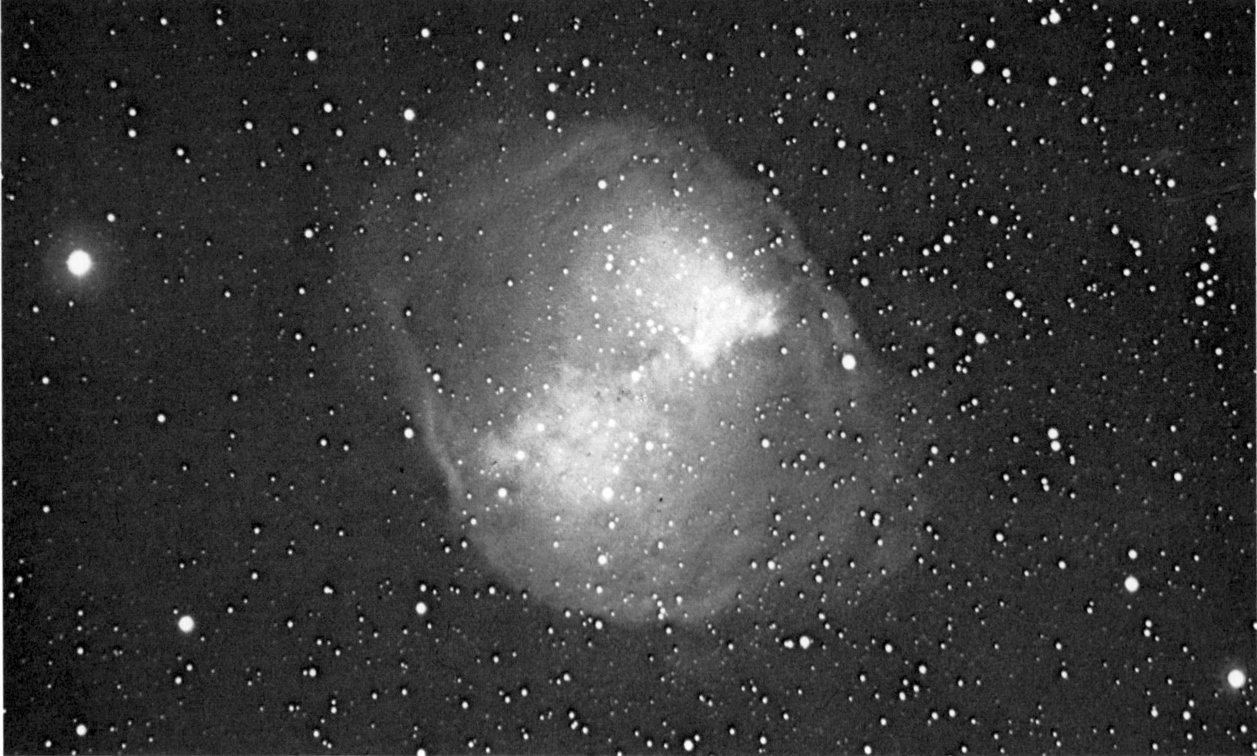

THE RING NEBULA IN LYRA, M57 *(upper). The famous planetary, photographed with the 508-cm Hale reflector.*
THE DUMB-BELL NEBULA IN VULPECULA, M27 *(lower). Another planetary, much less symmetrical than the Ring. 508-cm photograph.*

THE NORTH AMERICA NEBULA IN CYGNUS, N.G.C. 7000 *(opposite, upper). Photograph taken with the 122-cm Schmidt at Palomar.*
THE TRIFID NEBULA IN SAGITTARIUS, M20 *(opposite, lower). 508-cm photograph. This is a gaseous nebula, in our own Galaxy.*

and was able to take up his Mount Wilson appointment. Here he used the Hooker reflector, which was not only the world's largest, but was at that time in a class of its own. Hubble turned his attention to the so-called spiral nebulæ, and in some of them he found Cepheids. There could no longer be any doubt. The variable stars in the Andromeda Spiral were much too remote to be members of our Galaxy. Hubble gave the distance as 750,000 light-years, later increased to 900,000 light-years.

Hubble's work was far from over. Together with his colleague Milton Humason, who had been originally engaged at Mount Wilson as a mule driver but became a superb observer and technician, he measured the velocities of many of the galaxies, and drew up a relationship between distance and recessional velocity – still called Hubble's Law. He again served his country during the Second World War, and came back once more in 1946. He eagerly awaited the completion of the Palomar reflector, but had barely begun his programme with it when he died suddenly on September 28, 1953.

In 1952 there had been another major development. The Hale reflector was now in full operation, and was used to advantage by Walter Baade, German by birth (he was born on March 24, 1893 in Westphalia) but who had emigrated to the United States in 1931. Baade showed that M.31, the Andromeda Spiral, was much further away than had been expected. Instead of being 900,000 light-years

NUCLEUS OF THE ANDROMEDA GALAXY, *showing Population II stars resolved. Photographed with the 508-cm Hale reflector at Palomar.*

NGC 205, *companion of the Andromeda Nebula, photographed in red light with the 508-cm Hale reflector at Palomar.*

SPIRAL GALAXY MESSIER 64 (N.G.C. 4826) IN COMA BERENICES, *(opposite) as photographed with the 152-cm reflector at Mount Wilson.*

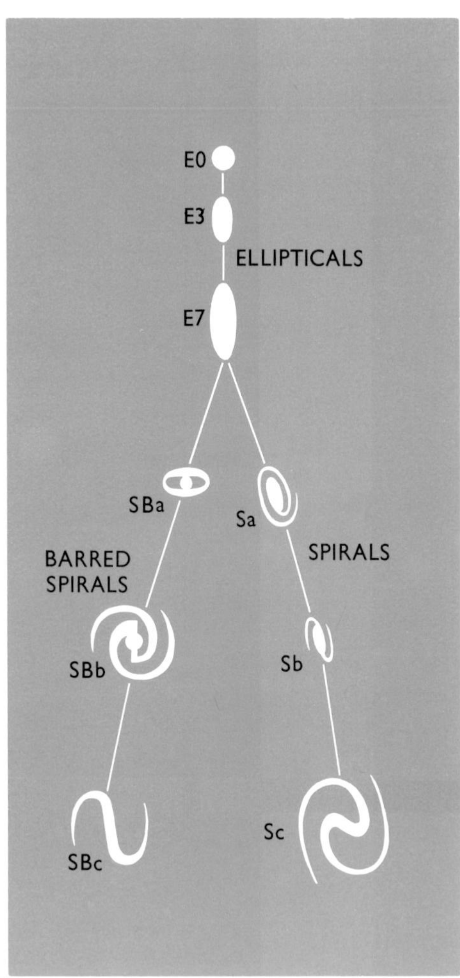

HUBBLE'S CLASSIFICATION OF THE GALAXIES: *ellipticals (E0 to E7), spirals (Sa, Sb, Sc), and barred spirals (SBa, SBb, SBc).*

THE VEIL NEBULA IN CYGNUS. *(Opposite) The colours radiated by these filaments of gas result from their motion through space. Ejected from an exploding star more than 50,000 years ago with an initial velocity of nearly 8,000 km per second, the clouds of gas have now been slowed to a speed of about 120 km per second by constant collisions with atoms in interstellar space. The force of these collisions ionizes the gas and causes it to glow with the characteristic colours shown here. Clouds ejected in the opposite direction at the time of the explosion are now 1248 trillion km from this portion of the nebula. Due to the steady decline in velocity caused by these collisions, the nebula will cease to glow in another 25,000 years. The light captured by the telescope to make this picture left the nebula about 2,500 years ago. Photographed with the 122-cm Schmidt camera at Palomar.*

from us, its distance proved to be 2,200,000 light-years, so that the system is considerably larger than our Galaxy. The same increases were necessary for the distances of other galaxies, so that in fact the universe was twice as large as Hubble had believed.

Years before, Baade himself had pointed out that there are two kinds of *stellar populations.* The brightest stars of Population I are very luminous, with early-type spectra; in these regions there is a great deal of interstellar dust and gas, so that star formation is still going on. With Population II the brightest stars are red supergiants which have already evolved off the Main Sequence, and there is practically no interstellar material. Globular clusters and the centres of galaxies are mainly Population II, and the spiral arms of galaxies are Population I, though inevitably there is a certain amount of overlap.

When mapping the Galaxy, Shapley had used the RR Lyræ stars, not classical Cepheids; and his results were correct, since his estimates of the luminosities of the RR Lyræ stars were valid. What neither he nor anyone else had realized was that there are two types of Cepheid variables, those of Population I and those of Population II, and that there is a considerable difference between them. A Population I Cepheid is more luminous than a Population II Cepheid of the same period.

The Cepheids used by Hubble to measure the distances of the spirals were of Type I, but the distances had been worked out on the assumption that they were Type II. Therefore, since the Cepheids in the spirals were more powerful than had been thought, they were also more remote, and this led to a complete revision of our whole distance-scale. It also explained why no RR Lyræ stars had been seen in M.31. They certainly existed there, but they are less luminous than the classical Cepheids, and had been overlooked.

It also explained the 1885 observation of a new star in the Andromeda Spiral. It became just visible with the naked eye, and was lettered S Andromedæ, but it did not last for long, and soon faded away to invisibility. At the time it was naturally assumed to be in the foreground, but it is now known to have been a supernova. Others have been seen in many external galaxies since then, but S Andromedæ is the only extragalactic supernova to have become bright enough to be seen without optical aid.*

Not all the outer galaxies are spiral, and in fact spirals are in the minority. Some systems, such as the Magellanic Clouds are irregular (though it has been suggested that there are vague indications of spirality in the Large Cloud). Others are spherical, and look at first sight rather like globular clusters; we also find systems which are clearly elliptical; and there are the extraordinary barred spirals, where the spiral arms issue from the ends of a 'bar' passing through the main axis of the system. It is tempting to believe that these different forms – originally classified by Hubble, as shown in the diagram – represent an evolutionary sequence, but this does not seem to be the case. For one

*One of the co-discoverers of S Andromedæ was the Hungarian Baroness de Podmaniczky, who knew little or nothing about astronomy, but who had set up a small telescope on the lawn of her castle to show off stars to her guests. One evening she called attention to a strange star in a nebulous mass, and one of her guests who happened to be astronomically-minded reported it to the nearest observatory. This must be the only case of a supernova being discovered by a Hungarian baroness from a castle lawn!

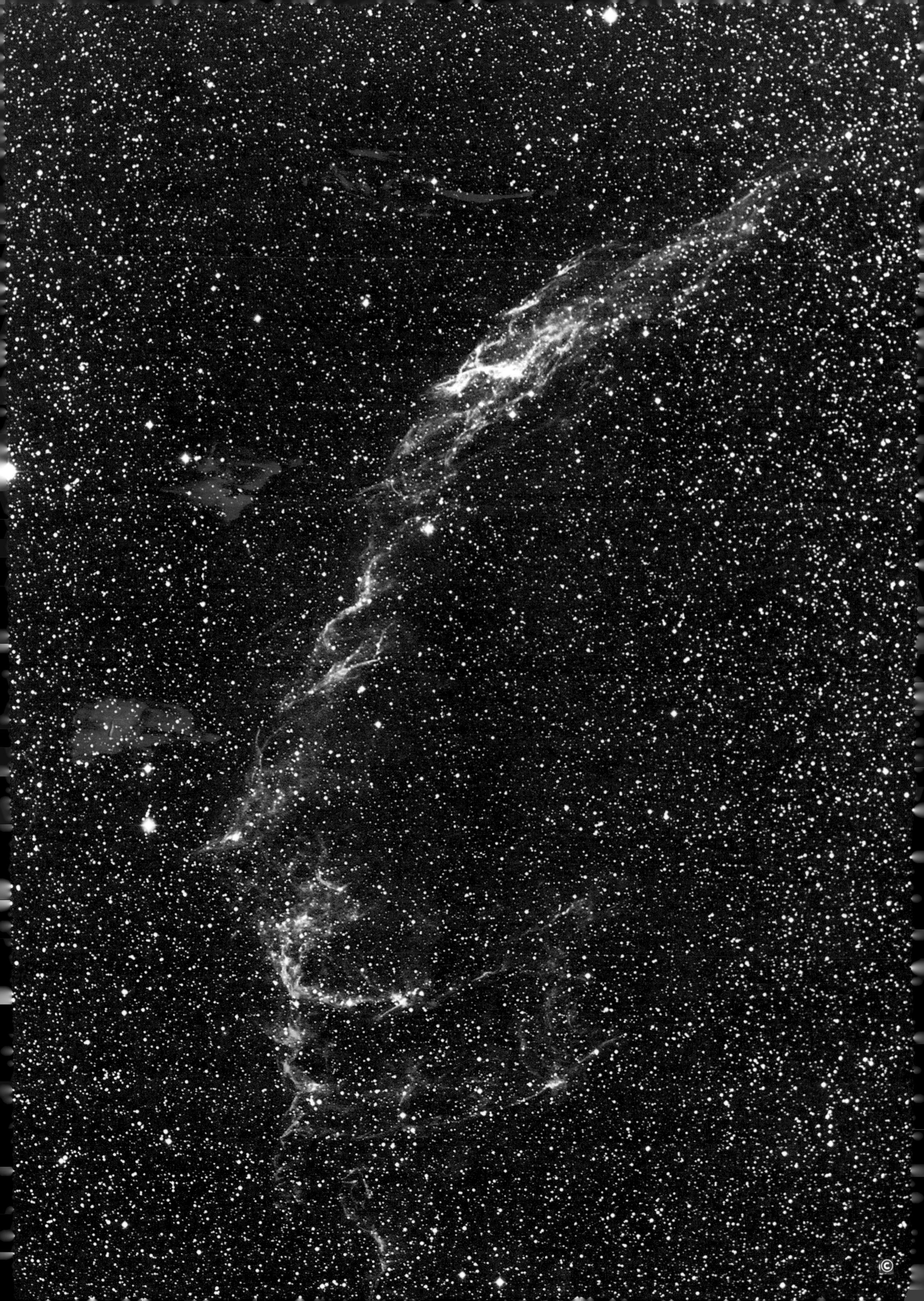

THE NUBECULÆ OR MAGELLANIC CLOUDS. *The Large Cloud is to the right, the Small Cloud upper left. The globular cluster in Tucana is also shown* (lower left).

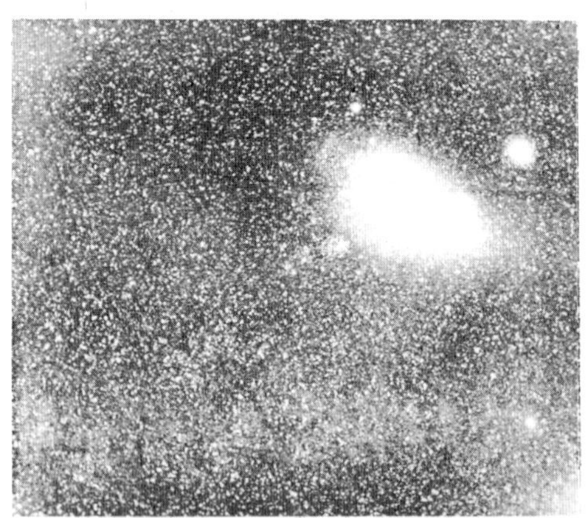

THE NUBECULA MINOR, *or Small Magellanic Cloud.*

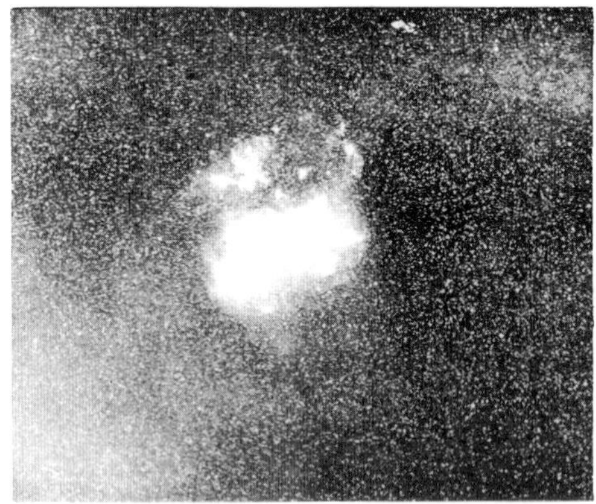

THE NUBECULA MAJOR, *or Large Magellanic Cloud.*

thing, the giant ellipticals are much more massive than any spirals. Thus the elliptical galaxy M.87 in the Virgo cluster is thought to have a mass equal to almost 8,000 million Suns.

What causes the spiral arms of the galaxies? It seems that the clue is to be found in the movement of rotation. Elliptical and globular galaxies have used up all their interstellar material, so that star formation has ceased. With the irregular galaxies there is still a good deal of material spread between the stars, but the systems are not in rapid rotation. With a spiral, there is plenty of star-forming material; as the system rotates, waves of density are set up in the gas, and there is some analogy (not a particularly good one, I admit) with sound waves. As the system spins, the density wave pattern rotates more slowly than the actual stars. The arms begin to 'wind up'; the very massive, luminous stars which have been produced in them soon explode as supernovæ and disappear, or, alternatively, disappear by losing their excess mass more gently. As the brilliant stars vanish, the density wave pattern moves on, and the optimum conditions for star formation move along with it. No spiral arm is a permanent feature; but as old arms disappear as their luminous stars cease to shine, new arms are formed.

Presumably, then, our Sun was formed inside one of these exceptionally dense spiral arms – bearing in mind that by any terrestrial standards the density is still incredibly low. We have now emerged from the arm, though eventually, no doubt, the Sun will again enter one of the regions in which stars are being born.

Our Galaxy is one of a whole group of systems, known as the Local Group. The total number of members is about thirty, including several major systems: our Galaxy, the Magellanic Clouds, the Andromeda Spiral, the Triangulum Spiral M.33, and probably a large elliptical known as Maffei 1 in honour of its discoverer, the Italian astonomer Paolo Maffei. Actually we know little about Maffei 1, because its

THE WHIRLPOOL GALAXY, *Messier 51 in Canes Venatici, photographed by Ritchey with the 152-cm reflector at Mount Wilson.*

THE OWL NEBULA, *Messier 97 in Ursa Major. (Left) Another planetary, photographed with the 152-cm reflector at Mount Wilson.*

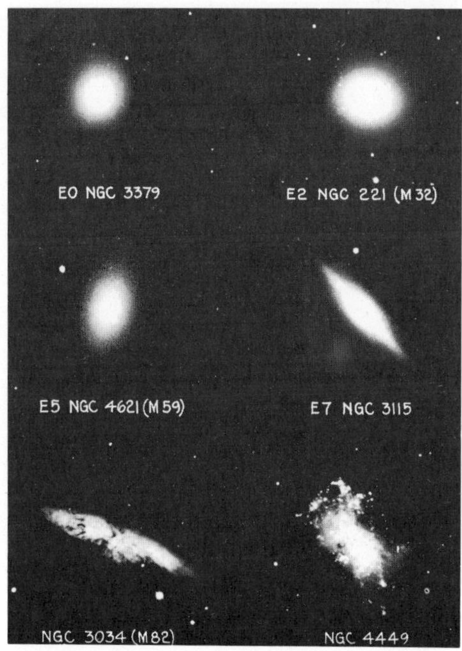

VARIOUS TYPES OF GALAXIES; *elliptical and irregular. Photographed with the 152-cm reflector at Mount Wilson.*

position in Cassiopeia is near the plane of the Milky Way, and Maffei 1 is heavily obscured by interstellar dust – which is why it remained unidentified until recently. (Another galaxy in the same part of the sky, Maffei 2, is even more puzzling. It is a spiral, and seems to be 15,000,000 light-years away. This is too far for it to be a member of the Local Group, but it has the same motion through space as Maffei 1. It is even possible that Maffei 1 is more remote than we think.) The other members of the Local Group are dwarfs.

The Local Group is not unique. In fact, galaxies do tend to occur in clusters, some of them very extensive. The famous Virgo Cluster contains hundreds of members, and is about 65,000,000 light-years away; it is roughly symmetrical, and its leading members are giant elliptical systems. Other clusters of galaxies are less evident, because they are more remote, but there are plenty of them. It has even been suggested that there may be 'superclusters' – clusters of clusters of galaxies, in which case the Local Group would belong to the super-cluster centred on the Virgo system. This may or may not be true; as yet we do not pretend to know.

With the relatively nearby galaxies we can use Cepheids for distance-gauging, but with more distant systems the Cepheids cannot be made out. We can then use the most luminous supergiants, on the very reasonable assumption that the supergiants elsewhere are likely to be of about the same power as the supergiants in our Galaxy. But eventually even this method fails; no individual stars can be seen. We have to resort to less direct methods, and, as always, we come back to the spectroscope.

VARIOUS TYPES OF GALAXIES: *spiral and barred spiral. Photographed with the 152-cm reflector at Mount Wilson.*

GALAXY IN SCULPTOR, *photographed with the 122-cm Schmidt reflector at Mount Wilson.*

SPIRAL GALAXY IN LEO, N.G.C. 2903, *photographed with the 508-cm Hale reflector at Palomar.*

The first serious attempts at studying the spectra of galaxies were made from 1914 onward by V.M. Slipher, at the Lowell Observatory in Arizona. The main absorption lines could be identified, and Slipher found, to his surprise, that apart from members of the Local Group (which, of course, was not then recognized as such), all the galaxies yielded spectra whose lines were shifted towards the red or long-wave end of the band. A Red Shift indicates a velocity of recession, according to the Doppler effect. Therefore, it followed that all the galaxies were receding from us, and when Hubble was able to show

SPIRAL GALAXY IN PEGASUS, *photographed with the 508-cm reflector at Mount Wilson.*

that the galaxies were beyond the Milky Way it followed that the whole universe was expanding.

This seemed unlikely at first sight, but the evidence continued to pile up. Hubble and his colleague Milton Humason established that the Red Shifts were genuine, and that they increased in proportion to the distance of the galaxy concerned. The further away a galaxy was, the faster it was receding. Hubble worked out a relationship between distance and recessional velocity, still called Hubble's Constant; it has been modified since Hubble's death in 1953, and the value now generally accepted is 55 kilometres per second per megaparsec. (One megaparsec is equal to a million parsecs, and a parsec, remember, is 3·26 light-years.)

If Hubble's Constant remains valid, it means that we will eventually come to a distance at which a galaxy is racing away from us at the speed of light, in which case we will never be able to see it. The limiting distance may be of the order of 15,000 million light-years. Unfortunately we cannot yet probe as far as that, but in 1980 new electronic devices used together with powerful telescopes enabled astronomers in the United States to record galaxies which seem to be 10,000 million light-years away, and some of the strange objects known as quasars are thought to be more remote still. One of them, PKS 2000-330, has a recessional velocity of over ninety per cent of the velocity of light.

The fact that the universe is expanding does not mean that we are in the middle of it. We are in no privileged position. Every group of galaxies is moving away from every other group – and if we were observing from inside the Virgo cluster, for instance, the Milky Way galaxy would seem to be racing away from *us*.

It is the Red Shifts which give us a means of measuring the distances of galaxies too remote to show individual stars. Once we know the Red

CLUSTER OF GALAXIES IN COMA BERENICES, *photographed with the 508-cm Hale reflector at Palomar. The cluster is about 40,000,000 light-years away.*

THE 'SATURN GALAXY' N.G.C. 4594, *photographed with the 152-cm reflector at Mount Wilson. The nickname is due to a superficial resemblance to Saturn and its system of rings.*

MESSIER 87, *a globular galaxy in Virgo, photographed with the 508-cm Hale reflector at Palomar.*

Shift, we know the recessional velocity; the distance then follows from Hubble's Constant. But let me add a note of caution. So far as we can tell, the Red Shifts really are Doppler effects, as has been checked from the galaxies in which Cepheids are visible. Alternative explanations, such as light becoming 'tired' and red-shifted because of intergalactic material, have been singularly unconvincing. Yet everything depends upon the shifts of the spectral lines, and there are still a few astronomers who believe that the Doppler effect is not the sole cause. The weight of evidence is strongly against this point of view, but it is not absolutely out of the question, and if we do eventually find that the Red Shifts are due partly to some other mechanism all our ideas will have to be revised. We may not be back in Square One, but we will certainly be back in Square Two.

Optical astronomy has taken us out to thousands of millions of light-years, but today we can probe still further. We can make use of the new techniques which may be called 'invisible astronomy', and to which I must now turn.

31

Invisible Astronomy

To recapitulate: the colour of light depends upon its wavelength, from red through to violet. The unit of length is the Ångström, named in honour of the last-century Swedish physicist Anders Ångström, who followed up the work of Kirchhoff and made many valuable contributions. One Ångström is equal to a hundred-millionth of a centimetre. Red light has a wavelength of about 7600 Å; violet light, about 4000 Å.

It is rather surprising to find how small the range of visible light really is; it makes up only a restricted part of the total range of wavelengths, or *electromagnetic spectrum*. Beyond the ultra-violet, we have X-rays and gamma-rays; beyond infra-red, we have radio waves.

To make matters worse, there are layers in the upper atmosphere which block out most of the electromagnetic spectrum. Consequently most of the radiations coming from space do not reach us at all, but are stopped by the atmosphere. We have the so-called *optical window*, which includes the visible light from red to violet, and also some distance into the near infra-red and ultra-violet; there is also the *radio window*, which allows some of the much longer waves to pass through. Otherwise, we can record nothing – so long as we stay on the surface of the Earth. To study the rest of the radiations, we must send our equipment up above the absorbing layers. This is no problem today, but the situation was very different in the year 1931, when a young research worker named Jansky made a discovery which ranked among the most important of modern times.

Karl Guthe Jansky was born in Oklahoma, in the United States, in 1905. His father was of Czech descent, but had settled permanently in America and had become a professor at the University of Wisconsin. Karl Jansky took his degree in physics at the same university, and then

THE ELECTROMAGNETIC SPECTRUM. *The central strip is the 'optical window', indicating visible light; it is clear that we can 'see' only a very small part of the total electromagnetic spectrum. The Earth's atmosphere will pass only visible light (the optical window) and the radiation in the 'radio window'. This is one reason why research with high-altitude intrumented rockets and space-probes is now so important.*

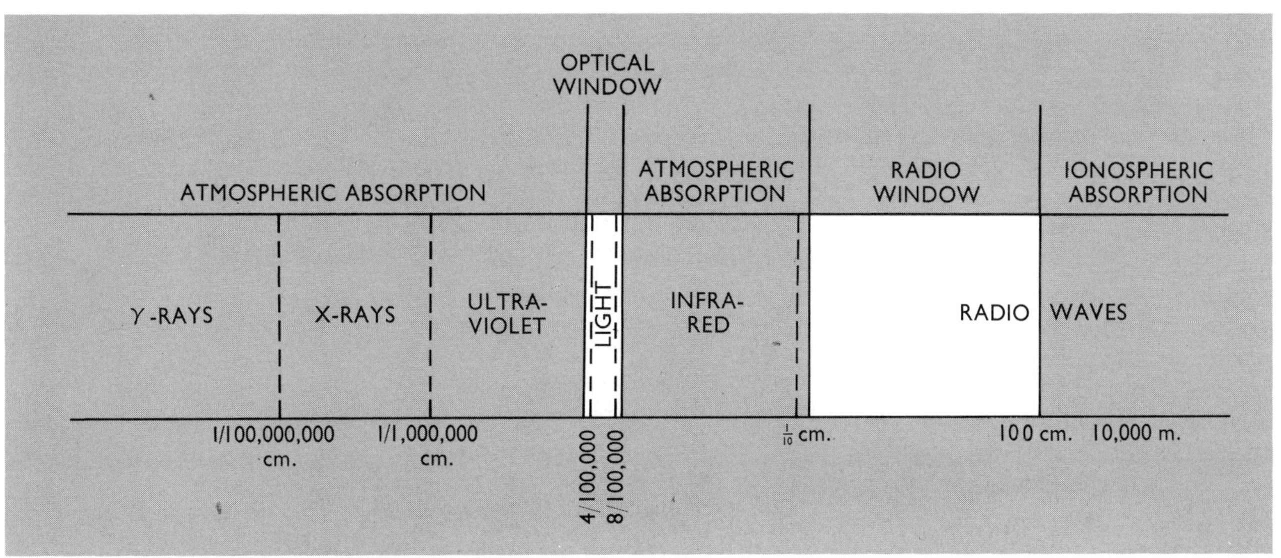

A YAGI ARRAY, *the simplest form of radio telescope.*

KARL JANSKY'S 'MERRY-GO-ROUND', *(right) the first true radio telescope. It was with this equipment that Jansky first detected radio emission from space, and so founded the science of radio astronomy.*

THE 76 M RADIO TELESCOPE OF THE UNIVERSITY OF MANCHESTER, *(opposite) which was opened at Jodrell Bank, in Cheshire, in 1957. It is the largest 'dish'-type radio telescope in the world, and is designed for automatic following of celestial objects across the sky. It is made of steel, and weighs 2,032 tonnes.*

joined the Bell Telephone Laboratories. His main work was to carry out research into problems of short-wave radio communication, and he was particularly concerned with 'static'. Static, known commonly if not scientifically, as 'hissing and crackling', is the radio operator's worst enemy, and the Bell authorities wanted to find out as much about it as they could.

Jansky set up a strange contraption on a farm at Holmden, New Jersey. It was an experimental radio aerial, looking somewhat like the skeleton of an aircraft wing, and it was driven round by means of a motor. The wheels were taken from a dismantled Ford car. Jansky's 'merry-go-round', as it was nicknamed, was designed to investigate static, but it achieved something much greater than anyone could have hoped.

There was plenty of static. There was radio noise due to nearby thunderstorm activity, and more noise due to distant storms over a hundred kilometres away. But there was also a third type of noise: a very weak, steady hiss in the receiver. It was not close at hand, and it seemed to come from a definite point in the sky. The source moved from day to day, and eventually Jansky found the answer. The hiss came from the Milky Way; more precisely, from that part of the Milky Way which lies in Sagittarius.

Let me go back for a moment to the shape of the Galaxy. The Sun is over 30,000 light-years from the centre, and the nucleus lies beyond the Sagittarius star-clouds. This, too, was the position of Jansky's radio source. Could it be that the long-wave radiation picked up by the 'merry-go-round' was coming from the very heart of the Galaxy?

Jansky believed so, and published his results. Technical journals referred to them, and on May 5, 1933, when the news was released, some of the American daily papers carried headlines about it. It is therefore very curious that little attempt was made to follow the matter up. It is even stranger that Jansky himself did almost no further research in radio astronomy, the science which he had created. He published a few more results in 1937, but after that he abandoned the problem altogether. He died in 1949.

Professional astronomers of the time were uninterested, and

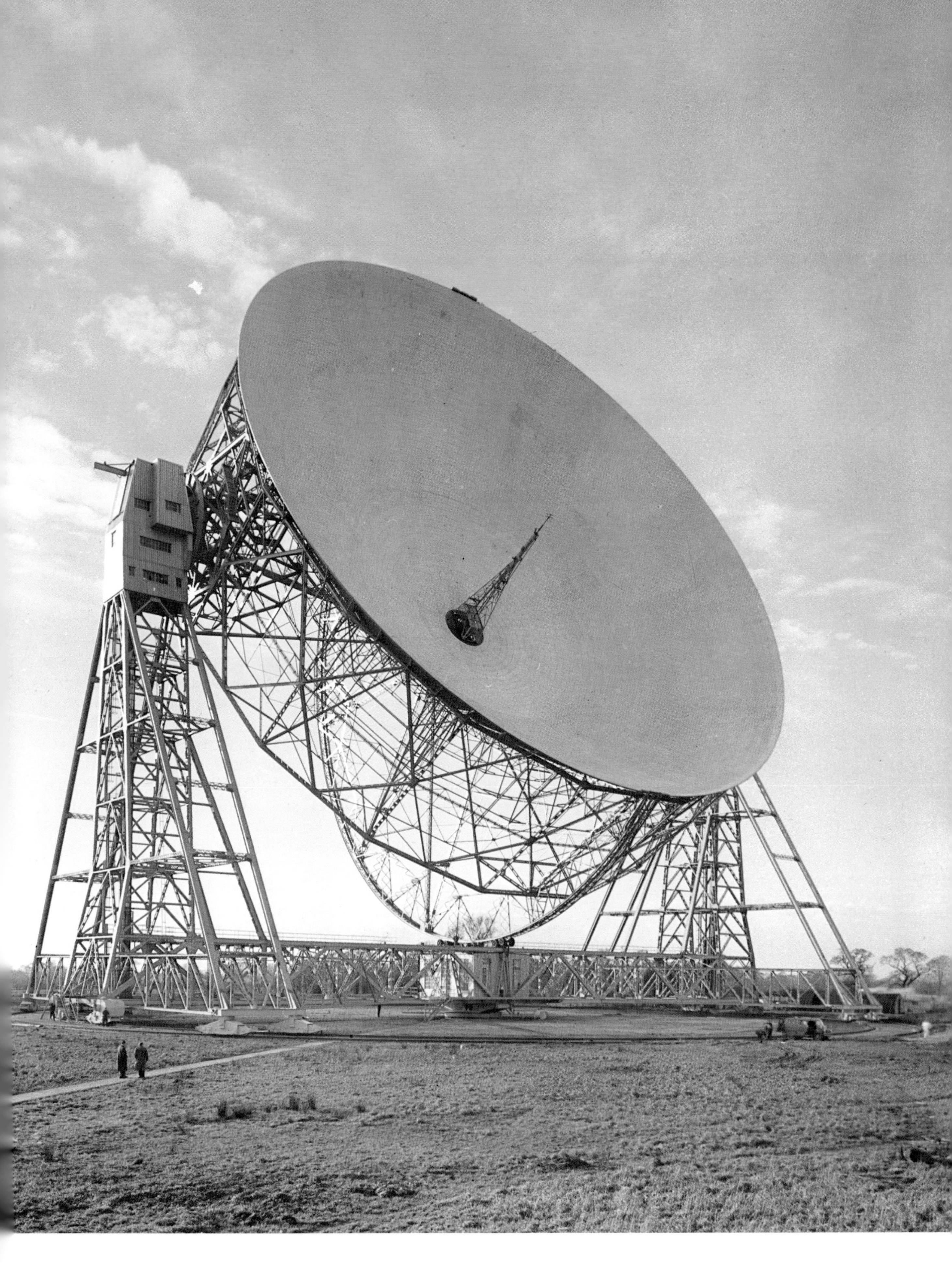

5-M 'DISH'-TYPE RADIO TELESCOPE IN AUSTRALIA. *It operates at a wavelength of 1 to 2 metres.*

Jansky's pioneer work was virtually ignored. Fortunately it was followed up by a brilliant amateur, Grote Reber, who was by profession an electrical engineer. He had been born in Chicago in 1911, and when still a boy he had built powerful transmitters with which he communicated with other radio enthusiasts all over the world.

In 1937 Reber turned his attention to what he called 'cosmic static', and built the first intentional radio telescope. It was a 'dish' or paraboloidal metal mirror, 9·6 metres in diameter operating at a wavelength of about 2 metres. The mirror focused radio waves just as the mirror of an ordinary telescope focuses light waves, but of course no visible image is produced. As I have said, one cannot look through a radio telescope.

Reber confirmed Jansky's discovery of radio waves from the Sagittarius star-clouds. He also found other sources in Cygnus, Cassiopeia and elsewhere. What did surprise him was that these sources did not coincide with the positions of visible stars. For instance, no radio waves could be detected from Sirius, or Vega, or Rigel; yet an area in Cassiopeia, unmarked by any optical object that he could see, emitted strong radio waves.

Reber published some of his findings during the years between 1940 and 1945, when the war was going on. Later he set up some equipment on Haleakala, an extinct volcano in Hawaii, to study radiations of even longer wavelength; subsequently he went to make his home in Tasmania. Modern radio telescopes make his original 'dish' look very small, but there can be no doubt that the first genuine radio astronomer was Grote Reber.

One of Reber's early ideas was to bounce energy pulses off the Moon and record the 'echoes'. He was not successful at the time, but radar developments during the war showed that his basic idea had been completely sound. The best analogy I can give is that of throwing a tennis-ball against a wall and catching it on the rebound. I appreciate that the comparison is very inaccurate, but it does show one thing: if you know the speed of the tennis-ball, and also the time elapsing between the throw and the catch, you can work out the distance of the wall. Actually things are simpler with radar pulses, which move at a constant speed – the same as that of light (300,000 kilometres per second).

Radar, originally known as radiolocation, was developed in Britain during the war years. It allowed the various ground forces to detect enemy aircraft, and to give the range and direction of any attackers; it also led to improvements in navigation, which today have become of paramount importance. To give further details about radar would be beyond my present scope. I will only say that it was discussed by R.A. Watson-Watt as long ago as 1935, so that it is almost as old as radio astronomy itself.

PULKOVO RADIO TELESCOPE. *One of the Russian radio telescopes at Pulkovo Observatory, Leningrad. Its main feature is a 'rail' aerial of 90 separate sheets; it has high resolution, and works at a wavelength of 2 to 5 cm. Photograph by Patrick Moore, 1960.*

Echoes from the Moon were recorded in 1946, both by Z. Bay in Hungary and by an American team. Radar echoes were then received from meteor trails, and showed that there are many daytime meteor showers which, of course, cannot be directly observed. By now, radar has largely superseded old methods of visual meteor-watching, though amateur work is still of value.

One result of the new techniques was a more accurate determination of the length of the astronomical unit, or Earth-Sun distance. Radar pulses were bounced off the planet Venus, and so the interval between the transmission of the pulse and the reception of the echo showed how far the pulse had travelled; this in turn gave the distance of Venus, and the astronomical unit could then be found by applying Kepler's Laws. We can also use radar for mapping purposes; as we will see later, the Pioneer space-probe still orbiting Venus today (1983) carries equipment which can pierce the obscuring clouds, enabling us to draw up an accurate surface map. Radar contact has even been made with the ring-system of distant Saturn.

These are very recent developments, so let me go back to 1942, when radar pioneering was being carried out. British anti-aircraft gunners were already using it to locate German aircraft, but it was found that some mysterious 'jamming' was periodically affecting all the equipment. At first it was thought that the Germans might have found some means of putting the radar out of action, and a research team under J.S. Hey undertook to find out what was happening.

It did not take them long. The jamming did not come from a German transmitter; it came from the Sun.

Years earlier Karl Jansky had searched for radio noise from the Sun, and had failed to find it. This was not Jansky's fault. His work had been done near the time of sunspot minimum, when activity was at its lowest ebb. In the early 1940s, when Hey returned to the problem, there were frequent large sunspots, and flares from these produced bursts of radio noise. Flares, the short-lived violent outbursts associated with active spot-groups, are seldom seen in integrated light; they show up when the Sun can be examined in hydrogen light alone, and are familiar to all solar observers, but they had not previously been associated with radio emission.

When Hey first made his discovery it came under the heading of a military secret, and very little more research in pure radio astronomy was undertaken until the war was over. As soon as possible after peace came, scientists turned back to this new branch of science. The importance of the pioneer work of Jansky and Reber was at last being realized.

THE 26 M RADIO TELESCOPE AT GREEN BANK, VIRGINIA, *operated by the Associated Universities, Inc., under contract with the National Science Foundation. National Radio Astronomy Observatory, Green Bank, West Virginia.*

An American, George Ellery Hale, was mainly responsible for the setting-up of America's largest telescopes. In radio astronomy, Hale's rôle has been played by a Briton: Professor Sir Bernard Lovell, one of the radar pioneers. Hale planned instruments of immense size and power; so did Lovell. The aim this time was a giant radio telescope with a 'dish' 76 metres (250 feet) in diameter.

Lovell had to face immense difficulties. There were the usual money troubles; Britain, after all, had borne the brunt of the war, and was financially crippled. Moreover, radio astronomy was still a relatively new science, and the projected 76-metre dish posed all manner of design problems. By his personal enthusiasm and technical skill, and helped by a brilliant team including another of the radar pioneers, R. Hanbury Brown, Lovell overcame the problems one by one. The 76-metre dish was built, and set up at the Jodrell Bank Research Laboratories near Manchester.

INSIDE THE PARKES RADIO TELESCOPE. *Photograph by Pieter Morpurgo, 1982.*

Jodrell Bank has been to radio astronomy what Mount Wilson and Palomar have been to visual astronomy. Even to list the discoveries made there would take pages. The 76-metre paraboloid has been modified, and a rather smaller 'dish' has been added to the equipment; research is in progress continuously. Of course, Jodrell Bank does not have the monopoly – far from it! The largest fully-steerable paraboloid

THE JODRELL BANK RADIO TELESCOPE, *(opposite) viewed from the control desk. As well as recording radio waves from space, the telescope has been used to track the various artificial satellites and space-probes launched since 1957.*

HAYSTACK, *(right) the American radio telescope, photographed by Patrick Moore in 1964. The equipment is inside the 'dome' which acts as a protective shield.*

MARK II PARABOLOID AT JODRELL BANK *(below). This is slightly different in appearance from the more familiar 76 m. Photograph by Parick Moore, 1967.*
GREEN BANK RADIO TELESCOPE. *(Below right), photographed by Patrick Moore in West Virginia.*

so far built is the 100-metre instrument at Bonn in West Germany, while in Puerto Rico there is a remarkable 1000-foot 'dish' built in a natural bowl. Of course it cannot be steered, but by making use of the rotation of the Earth, and shifting the aerials, much of the sky can be covered. In the southern hemisphere, the 64-metre paraboloid at Parkes, in New South Wales, is of special importance.

A dish-type radio telescope focuses radio waves. The focal point is the tip of a long metal rod or *dipole* which sticks out from the centre of the dish; the radio waves coming from space are reflected to this point, and are amplified before being passed on to an automatic pen which records them on a moving-paper graph.

CAMBRIDGE RADIO TELESCOPE, *with Dr. A. Hewish talking to Patrick Moore, 1974. Dr Hewish was the leader of the team which included Miss Jocelyn Bell, who identified the first pulsar.*

Not all radio telescopes are of the dish type. For instance there is the Mills Cross type, developed by the Australian scientist Bernard Y. Mills. Here we have two rows of aerials set up perpendicular to each other. The first receives the radio waves from one area of the sky; the second receives the waves from another area. If both rows are receiving together, there will be double-strength signals corresponding to the centre of the cross. If the rows are connected so as to work in opposition, the strength in the centre of the cross will be zero, since one row will cancel out the other. By means of a switch, the connections of the rows are rapidly changed from one position to another, and the current in the centre of the cross will vary up and down when a radio source lies in the position defined by the centre of the cross. By this method, the position of the radio source itself can be found.

In the early days, radio telescopes of all types suffered from what is termed low resolution. That is to say, the uncertainty of the position of a radio source was inconveniently large, and two sources close together in the sky would merge into one. The principle of the *radio interferometer* altered matters considerably. Here two radio telescopes, separated by a known distance, are used to study the same source in the sky. Since the receivers are some way apart, there is a slight delay between the arrival of a given radio wave at the first and the second telescope. The waves arriving at any moment at the two receivers will in general

MILLS CROSS. *Not all radio telescopes are dishes. The Jodrell Bank 76 m instrument is of course a paraboloid; the Arcetri radio telescope and the instrument at the Crimea are quite different in nature. This photograph shows a Mills Cross, set up in Australia.*

be out of phase, and will interfere with each other if they are combined. As the source moves across the sky, by virtue of the Earth's rotation, the combined signal is analyzed, and the changing interference pattern will give an accurate position for the source. For a given wavelength, the greater the separation of the two telescopes, the higher the resolving power. Therefore we need very long baselines, and by now we can use telescopes situated in opposite hemispheres of the globe – for instance, Californian radio telescopes have been used as interferometers with telescopes in Australia. Before too long, we may even be able to combine an Earth-based telescope with a similar radio telescope on the Moon. Meanwhile, the resolution problem has been solved inasmuch as it is possible to pinpoint radio sources down to a tiny fraction of a second of arc.

Now let us see what sort of objects send out radio waves.

The Sun is one source, which is hardly surprising. Radio waves come not only from the visible surface, particularly from flares, but also from the corona, which – remember – is at a very high temperature. If our eyes were sensitive to these long-wavelength radiations, the Sun would appear much larger in the sky than it actually does.

In 1955 the American radio astronomers B.F. Burke and F.L. Franklin made the accidential discovery of radio emission from Jupiter. (They were not looking for anything of the kind; Jupiter simply happened to be in the right place at the right moment.) The Jovian radio waves were of several kinds, and were later found to be affected by the position in orbit of the innermost Galilean satellite, Io. But most of the radio sources were found to lie well beyond the Solar System, and one of the first to be identified was none other than our old friend the Crab Nebula. As we have seen, the 'power-house' of the Crab was later tracked down to its central pulsar or neutron star, all that remains of the supernova of 1054. Other supernova remnants were then found to be radio sources, including Tycho's Star, Kepler's Star, and the remnant Cassiopeia A, where the supernova outburst probably occurred only a few centuries ago, but was not observed at the time because of the 'fog' of dust in the way. But there was yet another line of research which turned out to be equally significant.

MILLS CROSS: *looking along the north-south arm.*

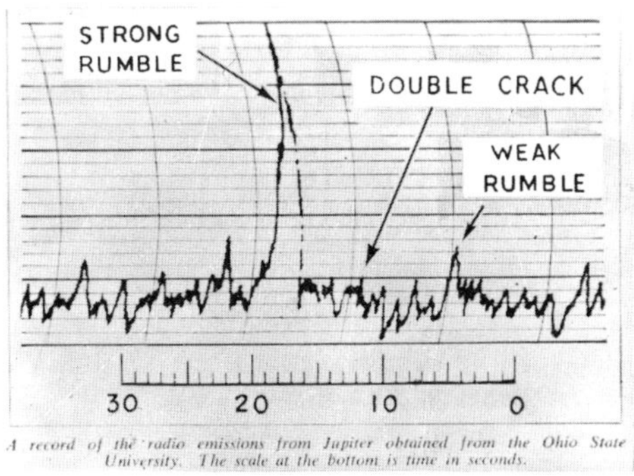

RECORD OF RADIO EMISSIONS FROM JUPITER. *The scale gives time-intervals in seconds.*

Interstellar material sends out radio waves. We know that there is a tremendous amount of rarefied gas spread through the Galaxy; hydrogen makes up most of it, and tends to collect into huge clouds, some thirty light-years across. It is of course extremely cold, and there are less than ten atoms of it per cubic centimetre, corresponding to what we normally call a high laboratory vacuum. Nevertheless, in 1944-5 the Dutch astronomer Hendrik van de Hulst predicted that this hydrogen should be sending out radio emissions at a wavelength of 21·1 centimetres. He knew that it would be weak and hard to detect, but he believed that it must be present.

At that time, van de Hulst was unable to carry out experimental work. The Germans occupied Holland, and all scientific research was at a standstill. It was not until 1951 that H. Ewen and E. Purcell, in the United States, managed to detect the 21-centimetre radiation, but it soon proved to be very informative.

Interstellar gas is most plentiful in the spiral arms of a galaxy (Population I region). If it is moving, it will yield Doppler shifts in the same way as visible light. By plotting the positions of the hydrogen clouds, and observing their Doppler shifts, radio astronomers were

RADIO TELESCOPE AT ARCETRI. *One of the Italian radio telescopes, at Arcetri Observatory near Florence. It is entirely different in type from the 'dish' instrument such as the 76 m paraboloid at Jodrell Bank. Photograph by Colin Ronan, 1961.*

Above right
TOTAL ECLIPSE OF THE SUN, *October 2, 1959, photographed from a McDonnell F.101B Voodoo aircraft at an altitude of 13,716 m. Radio astronomers are particularly interested in solar eclipses.*

able to confirm what had already been suspected: our own Galaxy is a spiral. By now the structure has been traced fairly accurately, and we can even make a start in investigating what lies at the centre of the system, beyond the lovely but obscuring star-clouds in Sagittarius.

Jan Oort, another of Holland's great radio astronomers, found that while in general the arms of our spiral rotate round the nucleus, there is one arm which has a definite velocity of approach; it is moving towards us at fifty kilometres per second, and seems to be around 9000 light-years from the centre of the Galaxy. Oort went on to establish that there are other arms and clouds, closer-in to the centre, which also are expanding outwards at velocities of up to 100 kilometres per second. We also find an immense disk of cool hydrogen which is not expanding; there is a ring of 'clouds' made up of molecules, and inside this ring lies a hydrogen-cloud at a higher temperature, nicknamed the Arc because of its shape. The Arc is moving outward at fifty kilometres per second. And still nearer the centre there is the strange object now known as Sagittarius A.

This is made up of two parts, East and West. Sagittarius A East is an ordinary supernova remnant, but Sagittarius A West is an evenly-lit cloud of ionized hydrogen inside which is yet another disk, perhaps illuminated by a single central source. This disk is about four light-years in diameter, so that it would bridge the gap between ourselves and Alpha Centauri. It is rotating uniformly; it is not expanding, and it is rich in dust and young, energetic stars.

The whole region is completely invisible optically, but as well as radio methods we can also use infra-red techniques – note how all the various branches of astronomy, visible and invisible, are coming together. The region seems to contain about a million stars altogether, only a few light-days away from each other. The total number of stars can be estimated from the amount of radiation emitted, but this leads on to a real problem. The combined mass of these stars is not enough to prevent the disk from dispersing. There should be about five times as

RADIO AERIALS AT CAMBRIDGE, *photographed by Patrick Moore, 1971.*

much. 'Mass' is missing – so where can it be? Apparently there is a central source in the cluster which sends out radio waves, and is several times the diameter of the Solar System, but the main power comes from a very compact region with a diameter no greater than that of the orbit of Jupiter.

If this is correct, we have at last come to the real centre of the Galaxy, and to the source which is illuminating Sagittarius A West. There is only one object which seems to fit all the requirements: a Black Hole. The mass may be about five million times that of the Sun; it could provide enough gravitational pull to make the ionized hydrogen disk stable; and material swirling into it will form a hot, glowing disk, emitting radio waves. It has been said that the Black Hole seems to act rather in the manner of a cosmic plughole.

If our relatively mild Galaxy contains a central Black Hole, what about others? Certainly there are some galaxies which are strangely powerful in the radio range. One of these is M.87, the giant elliptical in Virgo, which shows a prominent 'jet' coming out of it; another is Centaurus A, in the far south. We also have Cygnus A, around 700,000,000 light-years away, where the main emission comes from two regions one on each side of the visible object, at about 100,000 light-years from the centre.

These radio galaxies have proved to be very peculiar indeed. Looking at Centaurus A, for example, one is tempted to believe that it is made up of two separate parts, and this led to the 'colliding galaxies' theory, which was very popular in the early 1960s. It was assumed that in a system such as Centaurus A, two galaxies moving in opposite directions are passing through each other. (This is not theoretically impossible. It is only clusters of galaxies which are receding from each other; individual clusters, such as our Local Group, are not expanding.) The individual stars in a galaxy are so widely spaced that there will be almost no direct collisions, but the interstellar dust and gas will be colliding all the time, and it was thought that this would account for the radio emissions. Systems of the Centaurus A type would represent 'cosmical meetings' on a grand scale, and the radio emission would go on until the collision ended, a process which would take tens of thousands of years at least.

Then, as so often happens, fresh calculations showed that the whole concept would have to be jettisoned. The energy of the radio emission is so great that collisions of interstellar atoms and molecules could not account for it. Galaxies may collide occasionally (there is some evidence that they do), but for the cause of 'radio galaxies' we must look elsewhere.

Nowadays Black Holes seem to give the most logical answer, and it is difficult to think of anything else. There are also galaxies which seem to show visible traces of past explosions, and this brings me on to the strange case of M.82 in the Great Bear.

In 1961 C.R. Lynds, working at the Green Bank radio astronomy observatory in West Virginia, was carrying out a close study of the large spiral M.81. During his research he found that a second, weaker radio source, which he had thought to be associated with this spiral, actually came from its fainter neighbour M.82, which is irregular in form. There seemed to be something unusual about it, and R. Sandage at Palomar, using the Hale reflector, showed that inside M.82 there were complicated hydrogen structures of immense size. It was then found

LONG-ENDURING METEOR TRAILS, *which act as reflectors of radar pulses. By radar studies, our information about meteors has been vastly extended during the last two decades.*

that there are significant movements going on, involving velocities of up to 900 kilometres per second, so that evidently a cataclysmic outburst took place in the nucleus of M.82 in the past. From the present movement of the material, it is possible to fix the time reasonably well; the explosion happened 1½ million years before our modern view of it. Since M.82 is 10,000,000 light-years away, this dates the actual event back to 11½ million years.

Significantly, there are few hot blue stars in M.82, and it is generally thought that the radio waves are due chiefly to what is termed *synchrotron emission*, produced by the acceleration of charged particles in a strong magnetic field. If M.82 is an exploding galaxy (as is widely believed, though of late some astronomers have had second thoughts about it), it must be only one of many.

A QUASAR. *The quasar is shown by the arrows. It looks stellar, but is in fact immensely remote and luminous – assuming that modern theories are approximately correct.*

Radio galaxies are of various kinds. In particular there are the Seyferts, named in honour of the American astronomer Carl Keenan Seyfert, who first drew attention to them in 1942 and who was still working on their problems at the time of his sudden death in 1960, at the age of only forty-nine. Seyfert galaxies have condensed nuclei with very weak spiral arms; almost all of them seem to be active, with radio emissions, and here the idea of a central Black Hole is very plausible. One of the brightest of the Seyferts is M.77, in Cetus, which is visible with a small telescope even though it is over 50,000,000 light-years away. (Incidentally, Lord Rosse, a century ago, commented that M.77 is of an unusual bluish colour – another proof that the optics of the great Birr reflector were very good indeed.)

Radio waves can be picked up from objects so remote that they are quite invisible optically. We can 'see' galaxies out to at least 10,000,000 light-years, but radio telescopes can take us out further still – and this brings me on to those incredible objects known originally as quasi-stellar radio sources (QSOs), and now called quasars.

It is not entirely true to say that quasars are recent finds. Some of them have been known for a century at least, and one of them, referred to as 3C-273 because it is the 273rd object in the third Cambridge catalogue of radio sources, is not particularly faint; its magnitude is about 13. A few nights ago I looked at it through the telescope in my observatory at Selsey. There seemed nothing to single it out from an ordinary star, and had I not been able to identify it from its position I would not have known that there was anything strange about it. No doubt it has been observed frequently during the past hundred years, and mistaken for a normal star.

N.G.C. 4038-9: RADIO GALAXIES, *photographed with the 122-cm Schmidt camera at Palomar.*

The story of quasars really began when the Cambridge researchers tried to identify their radio sources with visible objects, sometimes successfully and sometimes not (in those days, the resolution of radio telescopes was not nearly so good as it is today). In one case, 3C-48, there was what looked like a 16th-magnitude star, with a wisp of nebulosity close by, in the right position. Astronomers were keenly interested, and when they made a careful study of the optical spectrum their interest grew still further. Definite spectral lines were seen, but nobody could identify them. Moreover, the starlike point was decidedly blue.

Early in 1963 Maarten Schmidt, at Palomar, began a study of another similar object, 3C-273. He had the advantage of knowing the position of the radio source very accurately, because C. Hazard, in Australia, had observed its occultation by the Moon. By good fortune,

3C-273 lies in a position which means that the Moon can occasionally pass in front of it, and when this happens the radio waves are naturally cut off; the way in which they fade out gives both the position and the size of the source. Hazard had found that instead of a single radio source there were two, separated by about 20 seconds of arc. Schmidt identified one of the sources with what looked like a star, but when he examined the optical spectrum he had a tremendous shock. The spectrum was most peculiar, and there were emission lines shifted towards the red by a surprisingly large amount. There seemed no doubt about the result; the lines were due to hydrogen, and were well known in laboratory spectra.

Once the clue had been found, 3C-48 was examined again, and this time the lines could be identified; they were red-shifted even more strongly than with 3C-273. Assuming the Red Shifts to be Doppler effects, only one explanation was possible. The objects were not stars at all, but were very remote, incredibly brilliant and comparatively small. They were also racing away at high velocities. These were the first quasars; today some of the new discoveries have been found to be receding at over 90 per cent of the velocity of light. The most remote quasar so far known is PKS 2000-330, discovered by radio astronomers at Parkes in 1982 and confirmed optically at Siding Spring. Its distance is of the order of 13,000 million light-years. Many have been found, by no means all of which are radio emitters. In 1982 P. Harvey and his colleagues at Texas made the first detection of a quasar (3C-345) at infra-red wavelengths.

INSIDE THE HOMESTAKE MINE SOLAR OBSERVATORY *a mile below ground. Drs. Ray Davies and Keith Rowley check the testing apparatus. Photograph by Patrick Moore, 1982.*

What were they? If they were as remote as their Red Shifts indicated, they would have to be far more powerful than any galaxies. Moreover they were small, and this was confirmed when some of them were found to vary in light and radio emission over short periods – proving that they could not be more than a few light-years in diameter at most.

All sorts of theories were put forward. One involved the meeting of ordinary matter with antimatter, with mutual annihilation and tremendous release of energy. Another assumed that the output might be due to chains of supernovæ, but this seemed unlikely from the first; no supernova lasts for more than a few months, and there seemed no conceivable reason why one outburst should trigger off others in

QUASAR HUNTERS! *The two Deep Space Network antennae searching for quasars from Tidbinbilla, Australia. The 34-m antenna is in the foreground, behind it the 64-m antenna. This was the equipment used to pinpoint the position of PKS 2000-330, the most distant quasar so far discovered.*

THE PARKES RADIO TELESCOPE – *now over twenty years old, but still one of the best and most powerful in the world.*

succession. In any case, quasars are far too numerous. There were also the BL Lacertæ objects to be taken into account.

BL Lacertæ itself, a faint telescopic object, had been taken for an ordinary variable star in our Galaxy, but in 1968 Maarten Schmidt identified it with a radio source. He looked at the optical spectrum, and found no lines at all – nothing but a band of light from ultra-violet through to infra-red. Other similar objects were subsequently identified. At last, in 1976, J. Wampler at the Lick Observatory was able to detect some faint spectral lines in a few of the BL Lacertæ objects, and to measure their Red Shifts. It seemed that the 'BL Lacs' were intermediate in luminosity between quasars and ordinary galaxies.

By now it seems very likely that both quasars and BL Lacertæ objects are the nuclei of highly active galaxies, and once again we come back to the idea of a central Black Hole. There is a growing danger of regarding Black Holes as answers to every difficult problem, but it is hard to see what other explanation there can be for the immense power of the quasars – unless, of course, we have made a fundamental error in our distance measurements.

This is not out of the question. In America, Halton C. Arp has drawn attention to lines of objects which contain both galaxies and quasars, which show different Red Shifts. If the members of these lines are really at the same distance from us, then the Red Shifts given for the quasars are misleading, and the quasars themselves could be much closer than is generally believed. This is not a view supported by many astronomers, but the last word has by no means been said, and there are other strange effects too. For example, there is the 'gravitational lens'. It was found that some quasar pairs are really two images of the same remote object, the light of which has been deflected and split into two beams by a massive, invisible object lying between the quasar and ourselves.

Radio astronomy has become an all-important branch of modern research; it is strange to think that only three decades ago it was in its infancy. And at that period, short-wave astronomy had barely begun, owing to the limitations imposed by the Earth's atmosphere; most of the radiations shorter than that of visible light are blocked out. To

THE CRAB NEBULA, MESSIER 1 IN TAURUS. *(Opposite) Here we see the result of a stellar explosion which occurred over 6,000 years ago. Due to its distance from the Earth, the explosion was not observed until 6,000 years after it had taken place. On the morning of July 4, 1054, a star was seen by Chinese observers, shining with a brightness 100 million times greater than previously. After several months it faded below naked-eye visibility. In the place of this star or 'supernova', modern telescopes now reveal this large cloud of gas, still expanding at a rate which increases the diameter of the nebula by nearly 112,000,000 km each day. High-energy electrons, still moving about rapidly as a result of the explosion, cause the centre of the nebula to glow with nearly white light, and at the same time cause the wispy filaments of gas to shine with characteristic colour. In the gas-cloud is a pulsar or neutron star, which is almost certainly the remnant of the supernova itself, and may be termed the 'power-house' of the Crab. Photographed with the 508-cm Hale reflector at Palomar.*

study most of the ultra-violet, and all of the X-rays and gamma-rays, we must go aloft, or at least send our equipment there.

Using rocket-borne detectors, T.R. Burnright, in 1948, managed to detect X-rays coming from the Sun, but it is probably fair to say that X-ray astronomy really began in 1962, when experimenters such as Riccardo Giacconi and Herbert Friedman started to make use of the new, vastly improved rockets. They soon located a strong X-ray source in Scorpio, known today as Scorpio X-1. But rockets have their limitations. They can remain up for only a few minutes, and when they crash-land it is often hard to recover their equipment intact even if it has been separated and brought down by parachute. Some useful knowledge was gained; in 1963 came the first good X-ray 'pictures' of the Sun, and in the following year the team led by G.S. Vaiana showed that X-rays come from the corona as well as the main body of the Sun – which is not unexpected in view of the high temperature of the corona (over a million degrees), which is one of the prime requirements for X-ray emission.

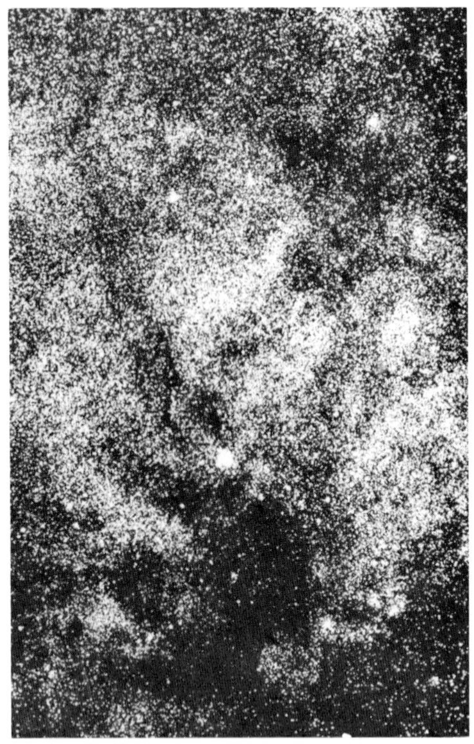

STAR-FIELDS IN THE MILKY WAY, *photographed at Mount Wilson.*

On December 12, 1970 the United States launched a small satellite specially to study X-ray sources. It was sent up from a platform off the coast of Kenya, and was christened Uhuru, which is Swahili for 'freedom'.

Uhuru carried two X-ray telescopes, placed back to back. As the satellite spun round, the telescopes swept over the sky, and when an X-ray source came within range a signal was sent back to the ground. Before its power failed, in 1973, Uhuru had located 339 X-ray sources, and had shown that they are very numerous indeed. Uhuru has been succeeded by several other X-ray satellites; for instance there have been the British Ariel 5, which operated from 1974 until 1980, and the somewhat temperamental Ariel 6, launched on June 2, 1979 from Wallops Island in Virginia. Of special note was the 'Einstein Observatory', an X-ray satellite sent up on November 13, 1978, which proved to be amazingly successful, and went on operating until the early part of 1981. It carried a 58-centimetre X-ray telescope, and was capable of picking up sources ten million times feebler than Scorpio X-1, which shows how far technology has advanced.

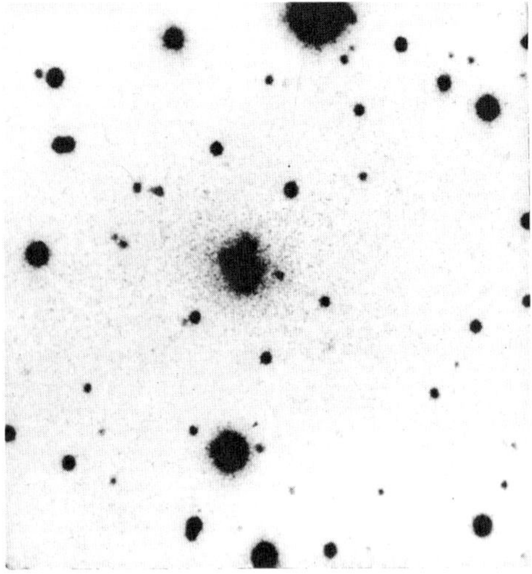

RADIO SOURCE CASSIOPEIA A, *a particularly intense source. (This is a negative photograph.) The emission is due to distributed masses of gas, and there is no bright visual object.*

As early as the 1930s it had been found that X-rays can be polished if they strike a suitable surface at a very shallow angle; this is known as grazing reflection. X-ray telescopes of this kind were designed in the 1950s by Hans Wolter of Kiel University, and were improved by Giacconi and others during the following decade. They are, of course, utterly unlike conventional telescopes. The X-rays pass down a gold-plated tube with reflection surfaces made of low-expansion glass, and accurate in shape to within a thousandth of a wavelength of light; the rays come off at low angles, and the situation has been compared with that of a flat stone skimming over water when performing the manœuvre known popularly as ducks and drakes. Though the analogy is not really a good one, it does give at least a vague idea of what happens. After passing through the tube, the X-rays enter electronic equipment at the far end, where they are recorded and studied. In the Einstein Observatory there were eight reflection surfaces, bringing the grazing photons to a focus 3·4 metres behind the 58-centimetre entering aperture.

The first X-ray source to be optically identified was – what else? – the Crab Nebula. Another to be tracked down in the early days was

Cygnus X-1, which, as we have seen, is probably a binary consisting of a massive normal star and a Black Hole, so that X-radiation is sent out by material which is spiralling down to the event horizon of the Black Hole. Other candidates are equally interesting. V.861 Scorpii is a binary with a visible primary and an invisible companion twelve times as massive as the Sun; like the companion of Cygnus X-1, this is too massive for a White Dwarf or a neutron star, so that a Black Hole seems to be the only logical answer. With Centaurus X-3, the X-rays are almost cut off for several hours every two days, so that presumably the source is being eclipsed by the visible star; a source in Hercules, Hercules X-1, is of the same kind. Dwarf novæ are radio emitters; the first to be identified (in June 1978) was SS Cygni, one of the favourite objects for amateur variable star observers.

Temporary sources can catch the investigators unawares. One was found by Ariel 5 on Christmas Day 1974; it lay near the familiar Centaurus source, and, inevitably, was nicknamed CenXmas. It faded quickly, and eventually disappeared. Another became obtrusive during May 1975 near the Crab Nebula, though apparently unconnected with it; equally inevitably, it became known as Fresh Crab. There are X-ray 'bursters', sudden, violent outbursts whose origin is still uncertain; we receive X-rays from many galaxies as well as quasars. In the United States, van Speybroeck, analyzing the results from the Einstein Observatory, has identified no less than eighty sources in the Andromeda Spiral alone.

It is particularly significant that X-radiation comes not only from stars and galaxies, but also from the thin gas spread between the galaxies of a cluster. Moreover, there have been recent suggestions that the general background X-radiation comes from large numbers of discrete sources, probably quasars. No doubt further advances will be made in the near future.

Gamma-rays are even shorter in wavelength than X-rays, and are more difficult to study. Preliminary experiments were made with balloon-borne equipment in the early 1960s, but it was not until the flight of the pioneer gamma-ray satellite OSO-III, in 1967, that really useful results were obtained. (In case you are wondering what OSO means, it stands for Orbiting Space Observatory.)

To 'catch' gamma-rays, the usual method is to use what is called a spark chamber. This is made up of metal plates which are electrified, so that there is a strong current between them. When a gamma-ray hits the heavy top plate, it is turned into two particles: an electron and its 'opposite', a positron. The particles travel downward through the stack of plates, causing tiny sparks as they jump from one plate to another. The trails of sparks can be photographed, giving the original directions and energies of the gamma-rays. One problem is to disentangle gamma-rays from cosmic rays, and I cannot resist quoting the Gamma-Ray Astronomer's Hymn, which was told me by Dr. Jocelyn Bell-Burnell, discoverer of pulsars:

> Through the night of doubt and sorrow
>   Onward goes the pilgrim band,
> Counting photons very slowly
>   On the fingers of one hand.

(Light, of course, may be regarded as a wave motion, but may also be regarded as a stream of tiny bundles of energy called photons. Broadly

RADIO SOURCE CYGNUS A. *Another intense radio source. This is a negative photograph, obtained at Palomar.*

N.G.C.1275, ONE OF THE MEMBERS OF THE PERSEUS CLUSTER OF GALAXIES, *photographed with the 508-cm Hale reflector at Palomar. This galaxy is a radio source.*

UKIRT, THE UNITED KINGDOM INFRA-RED TELESCOPE AT MAUNA KEA. *Photograph by Patrick Moore, 1982.*

speaking, a gamma-ray photon is a high-energy version of a light photon.)

Needless to say, the Crab Nebula is a gamma-ray source, but even here the latest gamma-ray satellite, CosB, picks up only about five Crab gamma-rays per night. The only gamma-ray source more powerful than the Crab is the Vela pulsar in the Gum Nebula. This is the faintest optical object ever recorded, and yet it is the strongest gamma-ray source in the sky. Astronomy can play some peculiar tricks.

Gamma-ray 'bursts' have been identified; they come from far beyond the Solar System, but as yet we have no real idea of their origin. There are also sources beyond our Galaxy, including Centaurus A, once regarded as a pair of colliding galaxies and celebrated as an X-ray emitter.

There is one branch of gamma-ray astronomy which can be tackled from ground level. When a very high-energy gamma-ray enters the atmosphere, it produces a shower of charged particles which travel downwards, giving out a faint bluish light; as the shower fans out, so does the light, and produces a dim 'pool' a few hundred yards across. This is detectable, and has been studied from Mount Hopkins, in Arizona, with a special reflector. The altitude of the telescope is 2700 metres; the gamma-ray equipment lies below the peak upon which the Multi-Mirror Telescope stands.

Infra-red research is also of tremendous importance, and although very valuable work can be carried out from Earth (particularly at high altitudes, such as Mauna Kea), satellites are playing an ever-increasing part. IRAS, the Infra-Red Astronomical Satellite, launched in 1983 carried detectors which enabled it to survey the whole of the sky, and locate many new discrete sources of infra-red.

All these various branches of invisible astronomy must come together in the end. They cannot supersede visual astronomy; they merely augment it. Eventually, we may hope that they will lead us on to a real understanding of the greatest problem of all: how did the universe begin?

32

Big Bang or Steady State?

THE problems of space and time are closely linked. If we could understand the one we could probably understand the other, but at the moment we are baffled by both, and plain language fails us. We cannot visualize what is meant by a space which has no boundary and it is equally hopeless to form a good mental picture of a space which goes on for every. On one of the few occasions when I met Albert Einstein, I asked him whether he could translate his concept of 'finite but unbounded space' into everyday, non-mathematical terms. He replied that he had never been able to do so with any real success; and where Einstein failed, I am hardly likely to succeed!

With 'time', we again run up against the metaphorical brick wall. If the universe began at a set moment, perhaps 15,000 million years ago, what happened before that? Neither can we fathom a period of time which had no beginning. So I maintain that nobody has ever discussed the *origin* of the universe. We can start with matter which exists, and then build up a complete story, ending with you and me; but this is discussing the *evolution* of the universe, which is by no means the same thing.

A seventeenth-century Church dignitary, Archbishop Ussher of Armagh, solved the problem quite neatly. He maintained that the Earth was created intact, by divine agency, at nine o'clock in the morning of October 23, 4004 B.C.. His method was to add up the ages of the patriarchs and make various other calculations of the same sort, which were no doubt fascinating but which were hardly likely to appeal to scientists. And it is surely right to investigate matters from a purely scientific point of view. It is not religious to bury one's head in the sand, as the Biblical Fundamentalists do.

We have at least something to work on: a time-scale. The Earth is about 4·7 thousand million years old; the Sun, rather over 5 thousand million years; many stars are older than this. If we say that the universe *as we know it* originated about 15,000 million years ago, we are probably not too far wrong, though I agree that we may be several thousands of millions of years out either way.

One theory, known commonly as the Big Bang, was due largely to the pioneer work of a Belgian abbé, Georges Lemaître, though it was modified and made generally known by that great British pioneer of astrophysics, Sir Arthur Eddington. In the late 1920s Lemaître worked out a sequence of events beginning with the abrupt creation of all the matter in the universe at the same moment, in what may be called a primæval atom. This primæval atom exploded, sending its material outwards in all directions. Expansion went on for thousands of millions of years, until the universe had a diameter of about 1,000 million light-years. The expansion then stopped, and meanwhile the original high density had been so greatly reduced that complex atoms had been able to build up from the original simple ones – chiefly hydrogen – which had existed immediately after the Big Bang.

Lemaître also introduced a force, cosmical repulsion, which he assumed to be the opposite of gravitation, so that it grew stronger instead of weaker when the distance between two objects was increased. These two forces more or less balanced in a kind of celestial tug-of-war, until eventually cosmical repulsion gained the upper hand, and the galaxies which were forming out of the primæval material started to move away from each other. As they separated, cosmical repulsion became stronger and gravity weaker, so that the expansion became faster and faster. The end product was the universe we know today.

There have been many modifications of the original theory. In particular, we know much more about the nature of gravity now than Lemaître did, and cosmic repulsion, as a definite repelling force, has been dropped. George Gamow, a Russian-born astronomer living in America, suggested that the present expansion was likely to be due solely to the force of the initial explosion. But on any Big Bang theory, the universe had a definite beginning, is now evolving, and will eventually die. It may be compared with a clock which is running down, and has no chance of being re-wound.

Incidentally, it must be assumed that space, time and matter came into existence simultaneously. There is no point in asking just where the Big Bang occurred. All we can say is that it happened everywhere!

In 1948 Hermann Bondi and Thomas Gold, working at Cambridge, put forward a completely different idea. They rejected the whole concept of creation, and assumed that the universe has always existed; it will exist for ever, and has always looked much the same as it does now. Old galaxies pass over the boundary of our observable universe, but are replaced by new ones, built up from matter which is created spontaneously out of nothingness. Thus in, say, a million million years' time, our region of space will contain about the same number of galaxies as it does now – but they will not be the same galaxies.

Two points must be noted at once. First, the build-up of new galaxies would be a slow process, since the rate of spontaneous creation of matter would also be slow. We have to consider a creation-rate of one new hydrogen atom appearing once every thousand years in a volume of space equal to that of the Earth's globe, and to detect this would be considerably more difficult than locating a new sand-grain in the Sahara Desert, so that there is no chance of checking the theory experimentally. Secondly, this so-called Steady State theory cannot explain just how this matter came into being. It simply appeared – and that was that.

The steady-state theory was popular for some years, and was refined and extended by Sir Fred Hoyle, but unfortunately it has not stood up to further investigations. For instance, the theory assumes that the universe looked the same in the remote past as it does now. We cannot emulate Captain Kirk or Dr. Who in travelling through time, but we can do the next best thing; we can look backwards in time merely by observing systems thousands of millions of light-years away. We see these systems as they used to be thousands of millions of years ago. Also at Cambridge, Sir Martin Ryle and his team showed that the distribution of galaxies 'out there' is not the same as it is nearer home. Therefore the universe is not in a steady state, and the whole theory is wrong.

The final nail in its coffin was hammered in when, in 1965, radio astronomers in the United States detected microwave radiation at a length of 3·2 centimetres, coming in from all directions. This indicates a temperature of about 3 degrees above absolute zero (the coldest state possible). At Princeton University, Robert Dicke and his team had begun a search for this microwave radiation, which had indeed been predicted by Gamow as early as 1948; but they were forestalled by two of their colleagues, A. Penzias and R. Wilson, who had been studying the radio sky for quite a different reason, and had happened upon the 3-degree radiation. Dicke had calculated that the temperature of the universe would have been staggeringly high at the moment of the Big Bang, but would now have fallen to just this overall value. In fact, the 3-degree radiation is the remnant of the Big Bang itself.

Efforts to resuscitate the steady-state theory in various different forms have been notably unsuccessful, and by now it has been abandoned by almost all astronomers. But we still have the problem of deciding whether the universe will go on expanding until all the groups of galaxies have lost touch with each other, and this problem remains open.

Everything depends upon the general density of the universe. If it is below a certain critical level, then the expansion will never stop. But if there is sufficient matter in the universe, the galaxies will not be able to escape; eventually they will slow down, stop, and then start to rush together again, finally meeting up to produce a new Big Bang, after which the expansion will begin anew. This is the Oscillating Universe theory, which I have rather irreverently nicknamed the Concertina Theory. Big Bangs would occur at intervals of perhaps 60,000 to 80,000 million years, and there could have been any number of them in the past.

At present the indications are that the density is lower than the critical value, but we cannot be sure; we come back again to the question of 'missing mass', which could be locked up in various ways – in Black Holes, for example. The last word has by no means been said.

The lesson of all this, surely, is that despite our rapid progress during the last few years, we remain woefully weak in our understanding of fundamentals. We are still rather in the position of a schoolboy who tries to work out complicated sums without having taken the trouble to learn his elementary multiplication tables.

33

Into Space

LAUNCH OF VIKING 1, *August 20, 1975. (Opposite) Viking, a 4-tonne space-craft was sent up from Complex 41 at Cape Canaveral at 5.22 p.m. EDT at the start of its 11-month journey to Mars.*

THROUGHOUT this book I have had to make frequent reference to space research. It is time now to come back to it, and to give an account of what has been happening in our efforts to explore the Solar System.

Actually, the basic idea of space-travel is much older than most people believe. It certainly goes back to around A.D. 150, when a Greek satirist, Lucian of Samosata, wrote a book which he called the *True History* because he admitted quite openly that it was made up of nothing but lies from beginning to end. In it, a party of sailors making their way through the Pillars of Hercules (known to us as the Straits of Gibraltar) were caught up in a waterspout, and hurled upwards so violently that after seven days and seven nights they landed on the Moon – to find that they had been pitchforked into a battle between the forces of the King of the Moon and the King of the Sun, both of whom claimed priority on Venus (the planet, not the goddess). In 1634 a book written by Johannes Kepler, published posthumously, described a lunar journey accomplished by demon power; and the hero of *The Man in the Moone* by Bishop Godwin, published in 1638, was transported moonwards on a raft towed by gansas, or wild swans. On arrival, it was found that the lunar inhabitants were very intelligent and extremely particular; any moonchild who showed signs of latent depravity was hastily exiled to Earth, where there was already so much depravity that a little more would not matter.

Cyrano de Bergerac, famous as an explorer, swordsman and writer, proposed a rather more ingenious method in a novel published in 1657. This time the power-source was simply dew. If you stand outdoors in the early morning and look at a grassy lawn, you will probably see plenty of dew; later, when the Sun rises, the dew disappears, so that the Sun has presumably sucked it up. Cyrano suggested that the best way to go into space was to collect a number of bottles, fill them with dew, fasten them round one's body upside-down, stand out in the morning, and let the Sun do the rest. In another story, however, he introduced the idea of using fire-crackers, which was a distinct improvement!

The first theory with any scientific basis was that of the space-gun, described in a famous novel written by Jules Verne: *From the Earth to the Moon*. It was published in 1864 (its sequel, *Round the Moon*, came out a few years later). The intrepid members of the Baltimore Gun Club, all veterans of the Civil War, build a huge cannon called the Columbiad, and use it to fire a projectile towards the Moon at a speed of eleven kilometres per second. Verne was not a scientist, but he believed in getting his facts as accurate as possible; and he was certainly on the right lines with regard to his suggested velocity, which is, of course, the Earth's escape velocity. He also set his launching site not far from the modern Cape Canaveral, and the telescope which he described on the top of Long's Peak was a preview of the 508-centimetre reflector, though admittedly it was designed more on Lord Rosse's pattern than

LANDING OF THE *COLUMBIAD*, *in Jules Verne's great novel.*

on George Ellery Hale's. Unfortunately the shock of departure at eleven kilometres per second would promptly turn the luckless occupants of the projectile into fine jelly – quite apart from the fact that friction against the atmosphere would burn up the whole vehicle in the manner of a meteor.

The key to space-travel was found by a Russian, Konstantin Eduardovich Tsiolkovskii.* It was he who first seriously proposed using the power of the rocket.

Rockets themselves were not new. They may have been developed originally in China; they were certainly used as war-weapons in a battle fought in 1232 between the Chinese and the Mongols, and later military rockets were made by the Indians, who turned them against the British forces at Guntoor in 1781. War-rockets were developed by Sir William Congreve for use during the Napoleonic Wars; Boulogne was bombarded in 1806, and Copenhagen in the following year, when the luckless Danes had been forced into the war on the wrong side. At least half the city was burned down. Then, in 1812, a small British force fired rockets against American troops at Bladensburg, and the immediate result was spectacular; Washington was taken, and the White House destroyed. And in 1814, rockets were fired against Fort McHenry. The bombardment was hardly a success, but it was at least noisy, and it was then that one of the defenders, Francis Scott Key, wrote the famous line about 'the rockets' red glare' which is part of the American anthem.

That, for the moment, was more or less the end of the military rocket. Some European armies retained their rocket corps, but im-

*The late Willy Ley, historian of space research, spelled this name 'Ziolkovsky', which is phonetic and is certainly more logical; but I have reluctantly followed the current trend. Other variants are 'Tsiolkovský' and 'Tsiolkovski'.

PREPARATIONS FOR FIRING VANGUARD *(left)*.

PIONEER LAUNCHER *(above). A 4-stage Thor-Able rocket blasting off from Cape Canaveral (now Cape Kennedy), Florida, carrying Pioneer V in its nose. The 43 kg sphere, packed with scientific instruments, was launched into a 311-day orbit around the Sun between the orbits of the Earth and Venus. This was the third of the artificial planets; the first two were the American vehicle Pioneer 4, and the Soviet Lunik 1.*

provements in more conventional weapons soon made them obsolete, and so far as I know they were last used by the Russians during a long and tedious campaign in Turkestan in 1881. Subsequently rockets were put to better use, both as fireworks and for sending lifelines between the shore and ships in distress.

So far as manned rocketry is concerned, I cannot resist mentioning a tale about a Chinese gentleman with the rather improbable name of Wan-Hoo. According to reports, Wan-Hoo became interested in rockets as early as the year 1500, and decided to carry out a personal test. He built a framework of bamboo with a chair in the middle, and fastened forty-seven rockets to it. When all was ready, he sat in the chair and ordered his retainers to light all forty-seven rockets at once, the idea being that he would be lifted into the air and would land gently by using a primitive form of parachute. Unfortunately, all that happened was a tremendous bang, accompanied by clouds of black smoke. When the smoke cleared away, Wan-Hoo was no longer to be seen...

I have no idea whether this report is reliable; I have an inner feeling that it is not. I am also dubious about the claim that an early 19th-century Italian inventor named Ruggieri sent up small animals such as mice in cages attached to rockets, and rescued them alive. But before going any further, I must pause to explain just why rockets are so essential in space research.

Ordinary flying machines, such as aeroplanes and jet-aircraft, cannot function unless there is air around them, but rockets depend upon what Newton called the principle of reaction: 'every action has an

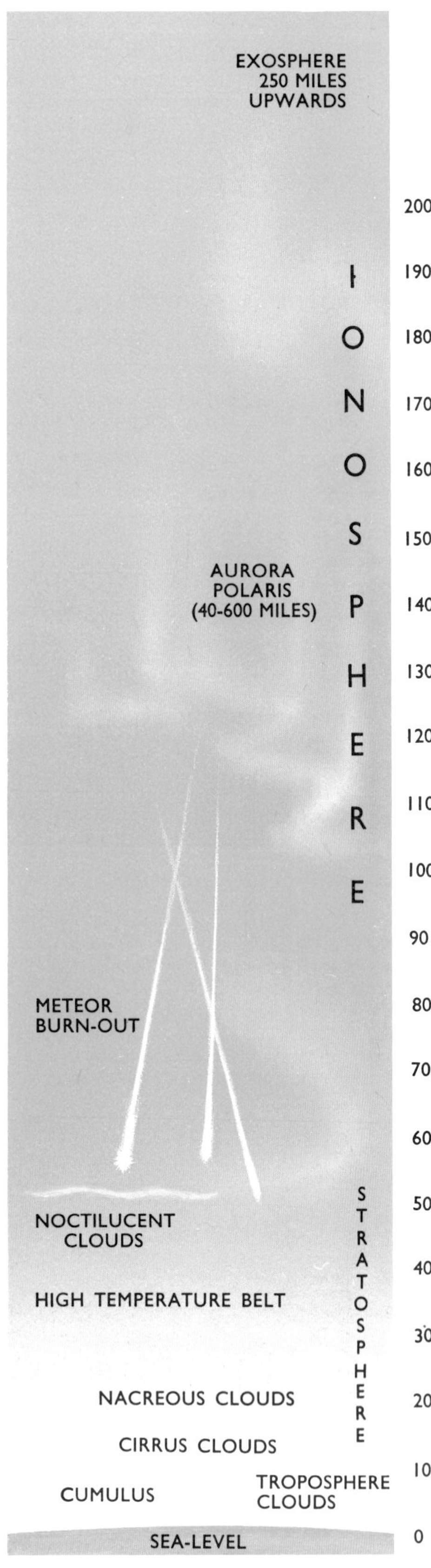

CROSS-SECTION OF THE EARTH'S ATMOSPHERE

equal and opposite reaction'. Consider a rocket of the kind used in firework displays. It is made up of a hollow tube filled with gunpowder, with a stick attached to provide stability. When you 'light the blue touch-paper and retire immediately', the gunpowder starts to burn. As it does so, it gives off hot gas; this gas tries to escape from the rocket – and can do so in only one direction, through the exhaust. In rushing out, it 'kicks' the rocket tube in the opposite direction, and the rocket flies. It does not matter whether or not there is atmosphere around it; in fact air is actually a handicap, because it sets up friction and has to be forced out of the way.

In strict historical sequence, two imaginative researchers preceded Tsiolkovskii. One of them, Nikolai Ivanovich Kibaltchitch, went so far as to draw up a preliminary design of what he termed a rocket aircraft, but as he had been unwise enough to make the bomb which was used to assassinate the Czar of Russia his career came to an untimely end in 1881. Then, ten years later, came a public lecture delivered by a German named Hermann Ganswindt, who seems to have been the archetypal crazy inventor; he produced such devices as an engineless helicopter and a fire-engine worked by treadles – though let me add that he really did invent the free-wheeling mechanism for bicycles. His spaceship was a weird contraption in which the power was to be given by cartridges, fired from the lower compartment to knock against the top of the vehicle and propel it upward in a series of jerks. One is entitled to doubt whether any sane astronaut would entrust his life to it, but Ganswindt did at least realize that a space-man moving in free fall would appear to be weightless, and he suggested spinning his vehicle so as to produce artificial gravity. Unfortunately he had neither the knowledge nor the patience to take matters any further, and the stage was set for Konstantin Tsiolkovskii.

Tsiolkovskii was born at Ijevsk, a village south-west of Moscow, in 1857. He was delicate and deaf; he had little mathematical training, and became a country schoolmaster. He realized that the rocket, and only the rocket, can function in space, where there is no atmosphere, but he was not naïve enough to think that gunpowder rockets would take men to the Moon or anywhere else. Instead, he planned to use liquid propellents. In this sort of arrangement, two liquids – a fuel, and an oxidant – are forced into a combustion chamber; they react together, and produce hot gas, which is sent out through the exhaust in the conventional way. A liquid-propellent rocket is controllable, and it can also begin its journey slowly, accelerating to full velocity only when it is high enough not to be burned away by friction against the dense lower atmosphere.

Tsiolkovskii also knew that no single rocket could carry enough propellent for a lunar trip. Accordingly, he suggested making a compound vehicle, consisting of several rockets mounted one on top of the other. When the large lower stage had exhausted its propellent, it would break away and fall back to earth, leaving the second stage to continue the journey by using its own motors – having been given what may be called a running start. This is perfectly sound, and all modern space-probes have used precisely this system. At the moment we have nothing better.

Thirdly, Tsiolkovskii gave an accurate description of weightlessness, or zero gravity. The best analogy I have ever been able to work out is to picture what happens when a coin is put on to a card, and the card allowed to fall. As the two are dropping, the coin ceases to press on the card; the two are moving in the same direction at the same rate, so that there is no mutual pressure. It is just the same with an astronaut in a space-ship, and so there is no sensation of 'weight'. Note that this is not the same as 'getting out of gravity', as some people still fondly imagine.

Finally, Tsiolkovskii realized that once a vehicle has been taken above the atmosphere and put into a closed orbit round the Earth, it will not fall down, but will behave in exactly the same manner as a natural astronomical body would do. After all, we do not have to fasten motors on to the Earth to keep it in its path round the Sun.

Tsiolkovskii was purely a theorist; he never fired a rocket in his life, and since his papers appeared in obscure Russian journals (the first one in 1895) few people heard about them. However, by the time he died, in 1934, he had become something of a national hero, and the Russians are justified in calling him 'the father of space-travel'.

Meanwhile, events were moving elsewhere. In the United States, Robert Hutchings Goddard of Clark University had been experimenting with rockets (quite independently; he had never heard of Tsiolkovskii at that time). In 1919 he even suggested that it might be possible to send a vehicle to the Moon, and the result was widespread ridicule which made him wary of the Press for the rest of his life. In 1926 he fired the first liquid-propellent rocket in history. It was a modest affair, travelling for a mere 53 metres at a top speed of 96 kilometres per hour, but it was the direct ancestor of the space-ships which have now flown to the depths of the Solar System. It was sad that Goddard did not live to see the start of the Space Age; he died in 1945.

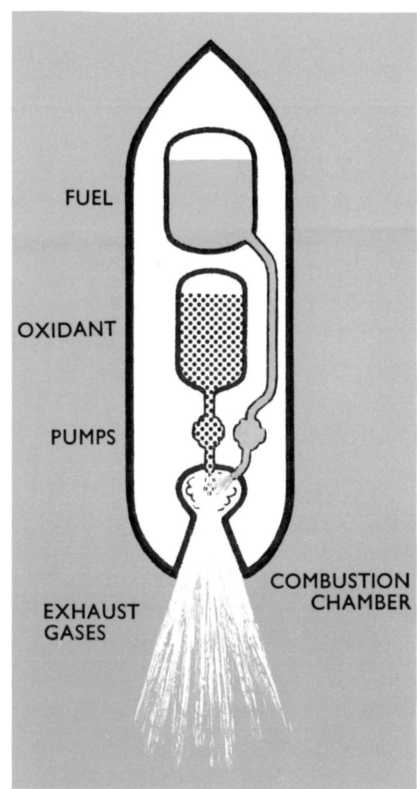

Principle of the rocket

Next came another theoretical pioneer, the Roumanian mathematician Hermann Oberth, born in Transylvania in 1894. He produced a book, the English title of which is *The Rocket into Interplanetary Space*, which was in fact the first really scientific treatise on astronautics. It came out in 1924, and though much of it was too technical for the general reader it became something of a best-seller. The immediate result was that in 1927 the first interplanetary society was formed in Germany. Its title was the Verein für Raumschiffart, or Society for Space Travel (VfR for short), and within a year it had a membership of over five hundred. Oberth joined it, and before long a full-scale experimental base had been established outside Berlin. This was the famous Raketenflugplatz, or Rocket Flying Field.

Working independently, a German rocketeer named Winkler had fired the first European liquid-propellent rocket; since nothing much had been said about Goddard's work, it was for the moment assumed to be the first of all. But the VfR workers were not far behind, and before long they too had launched rockets, some of which worked while others either blew up on the ground or nose-dived immediately after take-off. One of the early VfR leaders, Wernher von Braun, was destined to play a vital rôle later on.

There was one episode which is certainly worth re-telling, because it had its humorous side. This was the Magdeburg Experiment of 1933, which sounds almost unbelievable today.

It began with a member of the City Council of Magdeburg, who believed in an extraordinary theory according to which the surface of the Earth is the inside of a hollow sphere, so that from Europe Australia is almost overhead; the Sun lies in the middle of the sphere, surrounded by a crystal shell upon which the stars are etched, while

Goddard's rocket launcher at Roswell, 1934. *Goddard, the first man to fire a liquid-propelled rocket, was one of the greatest of the pioneers in this field.*

beneath our feet the Earth extends infinitely in all directions. It would seem, then, that a rocket launched straight upwards with sufficient velocity would end up in the Antipodes. The VfR was approached by the City Council, and asked to prepare a rocket which would test the theory. Neither von Braun nor anyone else at the Raketenflugplatz had the slightest faith in hollow earths, but at least the test would mean a financial grant, and two rockets were built and fired. Alas, the first failed to rise beyond its launching rack, while the second took off horizontally and ended its career in a ploughed field...

The Magdeburg Experiment was pure comedy, but there was little else in Germany to laugh at. The Nazi Party, under Adolf Hitler, realized that rockets could be used as weapons of war. The VfR was taken over, and soon ceased to exist; von Braun and many of his colleagues departed for the Baltic islands of Peenemünde, where they set to work on military projects.

WERNHER VON BRAUN

Peenemünde became a hive of activity. From it, in 1944, came the V2 weapons which were sent against England until the Germans surrendered in the following year. The V2s were not designed to go beyond the atmosphere, and indeed could not do so, but they marked a tremendous advance on any rockets previously built. The primitive firework had been developed into a highly sophisticated vehicle containing thousands of separate parts; it was powered by alcohol (the fuel) and oxygen (the oxidant), and it could carry a considerable payload of explosives. It was fortunate that the V2 was not perfected earlier. With the end of the European war, Peenemünde was stormed by the advancing Red Army; von Braun and most of his colleagues gave themselves up to the Allies and went to America, where the emphasis soon shifted from weaponry to space research.

A proving ground was established at White Sands, in New Mexico, and captured V2s were used as test vehicles. The last of them, launched on August 22, 1951, reached the dizzy height of 212 kilometres, but this was not a record; it had been surpassed two years earlier by the first successful step-rocket, a small WAC Corporal mounted on top of a V2. The WAC Corporal had attained no less than 392·5 kilometres, and once again Tsiolkovskii's theories had been vindicated.

White Sands was too small to be convenient,* and the main work was moved to Cape Canaveral in Florida. There, in 1955, the United States Government announced definite plans for sending up a man-made moon, or artificial satellite. Operation Vanguard was born.

As everyone will know, the Americans were forestalled. On October 4, 1957 the Russians launched Sputnik 1, which was not much larger than a football and which carried very little apart from a radio transmitter, but which was put into an orbit taking it from 135 to 590 kilometres above ground level, and which had a revolution period of ninety-six minutes. It was not completely above the resisting atmosphere, and eventually friction dragged it down to destruction, but it lasted until the first week of 1958. By then a larger, heavier Sputnik had been launched, and the Americans were becoming frankly alarmed. It seemed that in any 'space race' they were lagging disastrously behind.

SPUTNIK I. *The first artificial satellite, launched by Soviet scientists on October 4, 1957.*

*Moreover, there was also the problem that some of the early rockets were liable to go wildly off course. One of them crash-landed in Cuba, and killed a cow. The cow was then given an official State funeral as a victim of Imperialist aggression!

The situation soon changed. Explorer 1, the first United States satellite, went up on February 1, 1958, master-minded by Wernher von Braun, who had long since become an American citizen. Though it was small, with a weight of only sixty-six kilogrammes, it proved to be of tremendous importance, since its instruments led to the discovery of the Van Allen radiation zones surrounding the Earth. By the end of 1959 no less than eighteen satellites had been put into orbit, of which only three were Russian.

The histories of astronomy and astronautics are intertwined, but all I can do here is to list a few of the most important developments of the next few years. In 1960 Tiros 1 became the first satellite to be launched for purely meteorological purposes; many others have followed, and have improved weather forecasting considerably. They have also saved lives, by giving advance warning of dangerous tropical storms developing far out to sea. In 1962 came Telstar, the first satellite able to act as a television relay between Europe and America; today it would seem very strange without relay vehicles of this kind. Other vehicles carried equipment to detect and analyze cosmic rays, which are not really rays at all, but high-speed atomic particles coming from all directions, and about which there is still a great deal that we do not know. X-rays and gamma rays were also studied, and tiny meteoritic particles counted. There were also the 'Earth Resources' satellites, which were of tremendous value to geologists and to agriculturalists. The list of uses for artificial satellites seems endless.

TRAIL OF SPUTNIK-2. *Dunfermline Abbey is shown in this photograph by Morris Allan.*

A novel idea was proposed in 1945 by the British scientist and science-fiction writer, Arthur C. Clarke. His article appeared in the periodical *Wireless World*, and caused no excitement at the time, though it is now regarded with some reverence. Clarke pointed out that if a satellite is put into an orbit 35,880 kilometres above the ground, over the equator, it will have a period of twenty-four hours. As the Earth spins, this 'synchronous satellite' will keep pace with it, and will remain in a fixed position in the sky as seen from the surface of the Earth, so that several such vehicles, suitably spaced out, can provide full television and radio coverage throughout the world. Within forty years after this prediction, synchronous satellites were indeed in orbit, being used in precisely the way that Arthur Clarke had suggested.

Some of the satellites have been bright naked-eye objects, and almost everyone must have seen them as they crawl across the sky like slow-moving stars (and, of course, give rise to innumerable flying saucer reports). In the early days amateur observers of 'Operation Moonwatch' carried out invaluable work in keeping track of them.

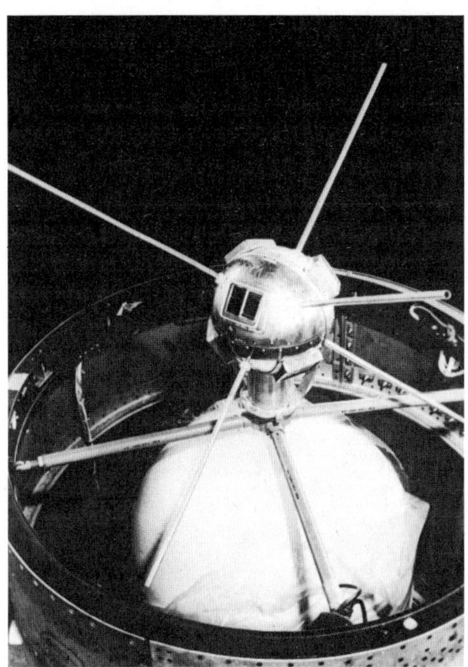

THE FIRST AMERICAN EXPLORER SATELLITE. *Though much smaller than the Russian vehicles, early American satellites carried complex instrumentation, and provided a great deal of valuable scientific information.*

Just as the Russians had been the first to launch an artificial satellite, so they were the first to send a man into space. The pioneer was Yuri Gagarin, of the Red Air Force. On April 12, 1961 he went up in his tiny, cramped vehicle, Vostok 1, and made a full circuit of the Earth before landing safely in the prearranged position; the whole flight took 1 hour 48 minutes. Gagarin was a worthy cosmonaut. There had been many prophets of doom; it had been claimed, for instance, that a man would be killed by cosmic radiation immediately upon emerging from the atmosphere, and that in any case he would be overcome by space-sickness, while the vehicle itself was liable to be battered to pieces by meteorites. Gagarin showed that these pessimists were wrong, and that even zero gravity was not uncomfortable, though it was certainly unfamiliar.

SKYLAB 1 IN ORBIT

MAJOR YURI GAGARIN. *The first space-traveller, Major Yuri Gagarin, made his pioneer flight on April 12, 1961 and orbited the Earth in the vehicle* Vostok, *landing safely in the prearranged position. He suffered no ill-effects from his flight, and the whole experiment was a complete success.*

I once asked Gagarin whether he hoped to go to the Moon. He replied that he did, but, tragically, he was killed in an ordinary aircraft crash a few years later.

The first American in space was Commander (now Admiral) Alan Shepard, who made a brief sub-orbital 'up and down' hop, lasting for fifteen-and-a-half minutes, on May 5, 1961; the first American to complete a full orbit was John Glenn, on February 20 of the following year. As the 1960s drew on, more and more flights were made. There were 'space-walks', initially by Alexei Leonov from Vostok 2 in 1965; there were space-rendezvous, space dockings and even space-stations, of which the most ambitious to date has been the United States' Skylab. It was launched in 1973 and remained in orbit until 1979, when it came down – rather prematurely, it must be admitted, scattering fragments over Western Australia. But Skylab has an honoured place in the history of astronomy; from it, three successive crews carried out long-continued observations of the Sun as well as other objects, notably Kohoutek's Comet of 1973, which was found to be surrounded by an immense halo of tenuous hydrogen. The Russian Salyut stations have been equally successful in their own way.

We now know that a man can remain under conditions of zero gravity for several months at a time without suffering any ill-effects. Whether this is long enough for a journey to Mars or Venus remains to be seen. But the epic flight of Yuri Gagarin showed conclusively that lunar travel, at least, was possible, and after 1961 not even the most hardened pessimist could seriously doubt that the Moon was within reach.

Docking of the Soyuz spacecraft with the orbital Salyut station. *(Above) Russian drawing.*

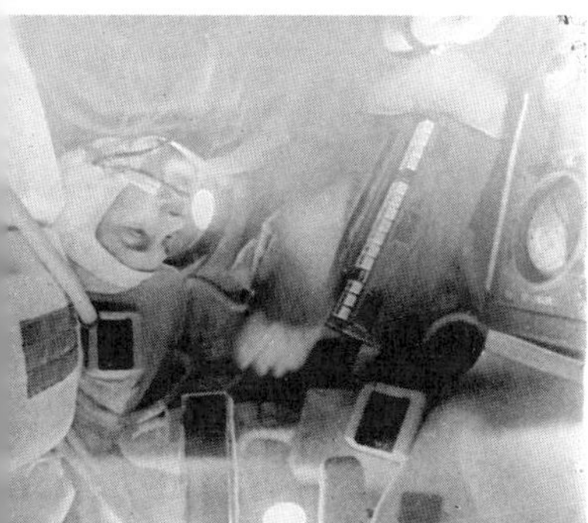

Russian cosmonauts: *V. Kubasov on left.*

34

The Modern Moon

BEFORE the start of the Space Age, observation of the Moon's surface was left mainly to amateurs. A few lunar photographs were taken with the Palomar reflector, and some magnificent pictures had come from Mount Wilson, but these had been produced more for public consumption than anything else. The Moon, after all, was a dead world.

Or – was it? Amateurs had their doubts, and persistently reported minor, localized obscurations and coloured patches, indicating that something in the form of gas was seeping out from beneath the surface layer. I was deeply involved in this kind of research (for that matter I still am), and one of the astronomers with whom I was in touch was Dr. Dinsmore Alter, at the Griffith Observatory in the United States. I suggested that one area where these localized obscurations had been seen was Alphonsus, a large walled plain between two others, Ptolemæus and Arzachel, near the centre of the Moon's Earth-turned face. In 1955 Alter took some photographs which seemed to provide confirmation, and on November 3, 1958 another of our correspondents, Nikolai Kozyrev, used the 80-centimetre reflector at the Crimean Astrophysical Observatory to photograph the spectrum of a bright red patch that had appeared briefly near Alphonsus' central mountain group.

Further observations came in during the next few years, both from professional astronomers (at the Lowell Observatory) and from amateurs. A network of observers was organized by the Lunar Section of the British Astronomical Association, of which I was then Director, and the results were very interesting indeed. My name for the events – Transient Lunar Phenomena, or TLP – has come into general use, and I think it is fair to say that the reality of the events themselves is no longer in question. Together with Barbara Middlehurst and her colleagues at the Lunar and Planetary Laboratory in Arizona, I published a list of several hundreds of recorded TLP; we also found that they occur in certain preferential areas, mainly round the edges of the circular *maria* and in regions riddled with rills. Obviously they are very minor, and there is no danger that any future lunar base will be shaken down by a violent ground tremor, but it does show that the Moon is not completely inert.

THE CRATERS ARZACHEL (UPPER) AND ALPHONSUS (LOWER).

Meanwhile, the rockets had begun to fly. The first attempted moon-shot was launched from Cape Canaveral in 1958. It failed, as did three more attempts in the same year, but in 1959 the Russians set up three Lunik (or Luna) vehicles, of which one went past the Moon at a distance of less than 6000 kilometres, one crash-landed on the surface on September 3, and the third – Lunik 3 – went on a round trip, sending back photographs of the far side of the Moon which had never before been accessible.

I have already noted that craters near the Moon's edge or limb will be foreshortened into ellipses if their true form is circular. It is difficult to map these limb regions, and it is only too easy to confuse a

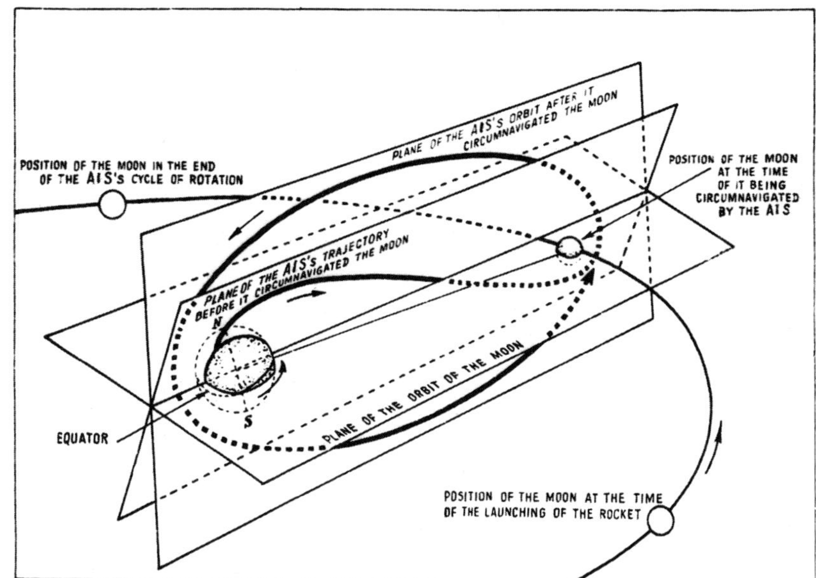

RUSSIAN DIAGRAM OF LUNIK 3's CIRCUMNAVIGATION OF THE MOON.

crater-wall with a ridge. For that matter, look at the conspicuous dark 'sea' of the Mare Crisium. It looks elongated north-south, but actually its north-south diameter is 450 kilometres and its east-west diameter as much as 563 kilometres. And the dark-floored 97-kilometre crater Plato is almost perfectly circular, though as seen from Earth it is an obvious oval.

The Moon keeps the same face turned towards us all the time, because its orbital period is the same as its axial rotation period: 27·3 days. There is no mystery about this behaviour; tidal friction over the ages has been responsible for it, but it was infuriating not to be able to see the 'other side'. Recall the poem allegedly written by a housemaid in the service of a well-known literary critic with astronomical leanings:

> O Moon, lovely Moon with the beautiful face,
> Careering throughout the bound'ries of space,
> Whenever I see you, I think in my mind
> Shall I ever, O ever, behold thy behind?

In fact we can examine a little over half the Moon's surface from Earth, because there are various effects known as *librations*. The Moon spins on its axis at a steady rate, but it does not move round the Earth (or, more accurately, round the barycentre, or centre of gravity of the Earth-Moon system) at a constant velocity. Obeying Kepler's Laws, it moves quickest when it is near *perigee*, its nearest point, and slowest at *apogee*, when it is furthest away. Consequently, the position in orbit and the amount of axial spin become periodically out of step, and we can see for some distance first round one mean limb and then round the other. This is called libration in longitude. There is also a libration in latitude, because the Moon's axis is tilted to the plane of its orbit, and a less important daily or diurnal libration. The overall result is that we can examine 59 per cent of the surface, though of course never more than 50 per cent at any one moment. It is the remaining 41 per cent which is permanently out of view. In the pre-rocket era, nothing could be done about it. The last-century Danish astronomer Hansen went so far as to suggest that the Moon was lop-sided in mass, so that all the air and

THE FAR SIDE OF THE MOON, *from Orbiter 2, showing the southern area. The far side contains no major 'seas'.*

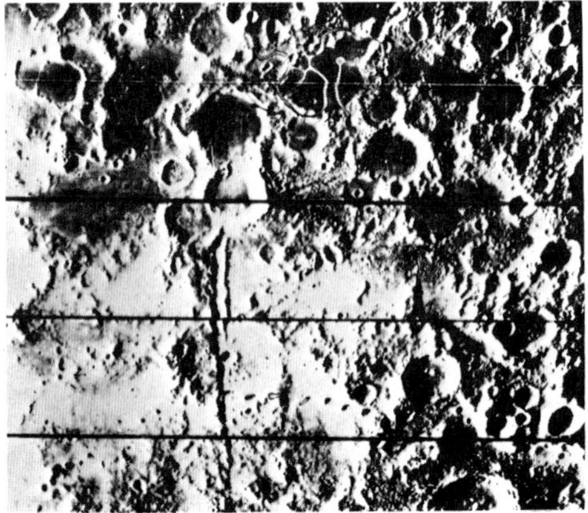

THE SCHRÖDINGER VALLEY, *a great valley on the Moon's far side, photographed from Orbiter.*

water had been drawn round to the far side, which might even be inhabited! Few people took this seriously, but direct exploration had to await the coming of the Luniks.

Lunik 3, sent up on October 4, 1959 (exactly two years after Sputnik 1) was a complex vehicle, launched by a step-arrangement in the usual way. All went well. By 4.30 G.M.T. on October 7 it had passed by the Moon, and lay well beyond it, at some 56,000 kilometres from the surface. From Earth, the Moon was then almost new, so that the hemisphere facing Lunik 3 was in full sunlight. The cameras were switched on, by remote control, and for the next forty minutes the pictures were taken, after which the films were automatically developed and fixed.

Lunik 3 was still moving away from the Earth. It reached apogee (470,000 kilometres) on October 10, and then started to swing back, reaching perigee (47,000 kilometres) on October 18. The pictures were transmitted by television techniques, and were successfully received by the Russian ground operators. Six days later they were released to the world, and I will always remember that I had the honour of being the first to show them to Britain, over the BBC that evening.

Very probably the Russians meant to keep in touch with the probe, and re-run the transmissions on the 'second time round', but contact was lost abruptly, and was never regained. It was suggested that the Lunik might have been hit by a meteoroid, but it is much more likely that the instruments developed some sort of fault. In any case, Lunik 3 had done its work well.

By modern standards the Lunik 3 pictures are very blurred and lacking in detail, but some features are clearly identifiable; in particular there is the dark plain which the Russians promptly christened the Mare Moscoviense or Moscow Sea, and the crater-plain which was appropriately named in honour of Tsiolkovskii, but all that could really be said was that the Moon's far side was just as barren and lifeless as the side we have always known. One feature, named by the Russians the Soviet Mountains, was taken for a long range of high peaks – but later proved to be nothing more than a bright ray; the Soviet Mountains do not exist.

The second Lunik had crash-landed on the Moon, though it had sent back no useful data. Other 'hard landers' followed, but were more ambitious. The American Ranger vehicles were intended to send back pictures just before destroying themselves on impact, and after six failures, between 1961 and 1964, success came. Ranger 7 landed in the Mare Nubium or Sea of Clouds, transmitting over four thousand photographs during the last few minutes of its career, and giving us our first really good close-up views of the Moon.

Crash-landing was one thing; soft-landing, by using rocket braking, was quite another, and some astronomers had questioned whether the lunar surface was firm enough to bear the weight of a space-probe. In 1955 Thomas Gold, then at the Royal Greenwich Observatory, had published a paper in which he claimed that the *maria* at least were covered with soft dust, so that a landing vehicle would simply 'sink into the dust with all its gear'. The theory did not seem to be very plausible, and practical observers had no faith in it, but there was only one way to settle the argument once and for all: send a probe to find out.

This is precisely what the Russians did, in February 1966. After several earlier failures they sent up Luna 9, which was slowed down by rocket braking while still above the Moon and then dropped gently on to the surface near the edge of the Oceanus Procellarum, not far from the dark-floored crater Grimaldi. Within a few minutes the first signals were being sent back direct from the lunar surface, and were picked up both in the Soviet Union and by Sir Bernard Lovell's team at Jodrell Bank. Luna 9 was standing on a hard layer, with no tendency to sink, so that Gold's dust theory was completely wrong. The scene was very much like that of a lava-field in Iceland or some similar place; there were rocks and boulders strewn around, and the whole moonscape was rough. Potential astronauts felt comforted. If a manned craft were to land, at least the Moon would refrain from swallowing it up.

By now the Apollo programme was under way, and President Kennedy had made his historic pronouncement that the United States would strive to put a man on the Moon before 1970. This meant drawing up really accurate maps of the whole of the lunar surface, and making quite sure that no landing would be attempted in an area which was potentially unsafe.

As so often happened, a Russian success was followed shortly afterwards by an American one. From 1966 to 1968 seven Surveyor space-craft were sent up to make controlled landings, and only two failed. The last three carried what may be called chemical sets, and were able to confirm the long-held suspicion that the Moon's surface is made up chiefly of the grey volcanic rock known to geologists as basalt. Surveyor 7 came down on the rim of Tycho, the great ray-crater in the southern uplands, and found that the rocks there were also volcanic. There was more aluminium in the highland rocks than in those of the *maria*, and less iron, but otherwise the composition was much the same.

Valuable though the Surveyors were, they were more than matched by the Orbiters, of which there were five between August 1966 and August 1967. All were successful. They were put into closed paths round the Moon, and sent back many thousands of high-quality photographs, including a superb view of the crater Copernicus which was widely acclaimed as 'the Picture of the Century'. The last Orbiter, No. 5, was deliberately crashed on to the Moon on January 31, 1968, its work done; and with it, we may say that Arago's 1840 dream had come true – though it had taken rather longer than he had expected!

The far side of the Moon differs from the near side in some respects. There are no large 'seas' comparable with the Mare Imbrium or the Mare Serenitatis, but there is one feature which proved to be of great importance: the Mare Orientale, or Eastern Sea.

I have a fatherly feeling for this vast structure, because, with H.P. Wilkins, I discovered it. In 1945 we were busy mapping the lunar limb, using Wilkins' 40-centimetre reflector, when we suddenly saw a feature which was highly foreshortened, but which looked like a small *mare*. We spent the rest of the night charting it as well as we could, and then sent in a report, with a chart and also a suggested name.* Of course we had no idea that it would prove to be of special interest, but the Orbiter photographs showed it to be multi-ringed and extremely complex.

*Since then the I.A.U. decree has reversed east and west, so that our *Eastern* Sea is now on the *western* limb of the Earth-turned hemisphere; but by the time the change in orientation was made, the name had been officially accepted.

RANGER 7

THE PICTURE OF THE CENTURY *(centre). The lunar crater, Copernicus, photographed from Orbiter 2.*
THE MOON FROM ORBITER 4 *(foot). Much of the area shown is permanently turned away from Earth.*

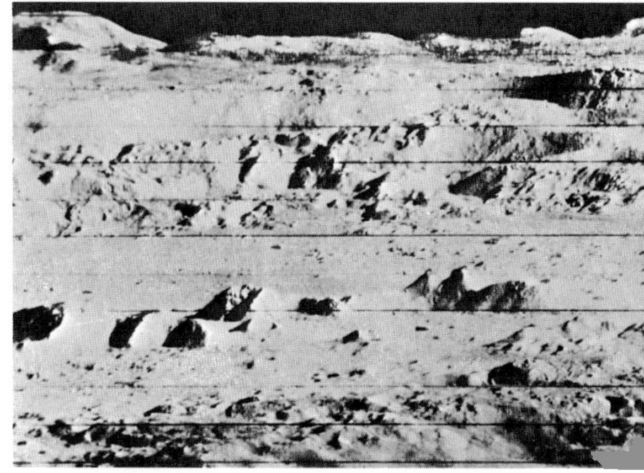

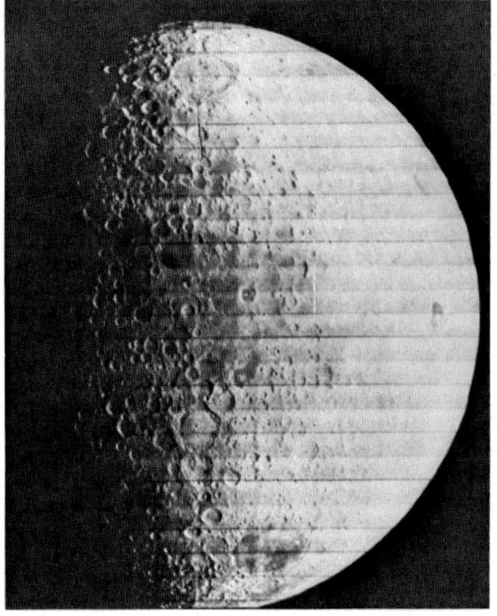

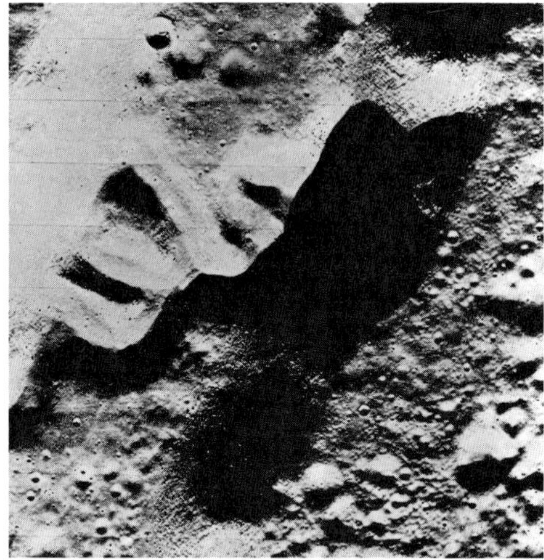

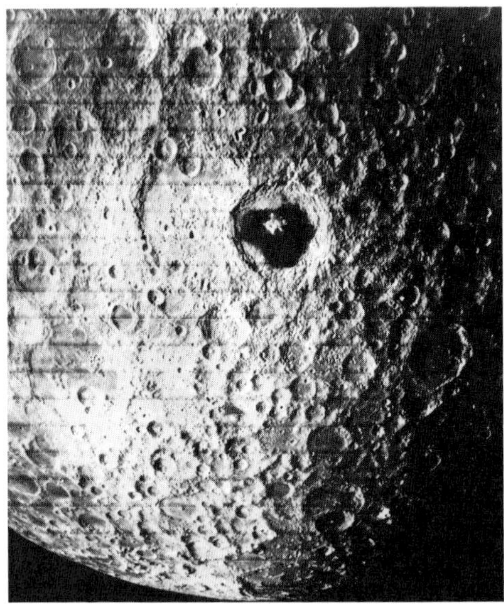

THE SCHRÖTER VALLEY *(top). High-resolution Orbiter 5 picture of part of the valley.*
TSIOLKOVSKII *(centre), the great crater on the Moon's far side, photographed from Orbiter. Note the blackness of the floor, which is due not to shadow, but probably to lava.*

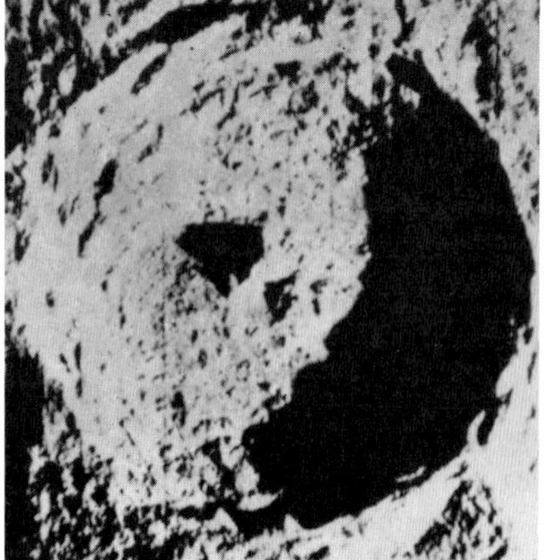

TYCHO, *photographed from Orbiter. The crater is 87 km in diameter.*

The far side of the Moon contains craters of all kinds, rather less regularly arranged than those on the near side; there are peaks, valleys and ray systems. The feature named after Tsiolkovskii is exceptional. It has a dark floor, certainly a dried lava-lake, from which rises a massive central mountain. It appears to be a sort of link between the craters and the seas.

By the time that the Apollo astronauts were ready to blast off we had learned a great deal about the Moon, but several problems remained unsolved. In particular, how were the craters produced? There were some bizarre theories (at times the craters were attributed to coral atolls, crystals, engineering structures and even bomb pits!) but there was only one real point at issue. Either the craters were of internal origin (volcanic), or else produced by meteoritic bombardment (impact).

The impact theory was first proposed more than a century and a half ago by Franz von Paula Gruithuisen. It was revived in the 1950s by the American astronomer Ralph Baldwin, and has remained popular, particularly in the United States. On the alternative theory, the craters are due to forces within the Moon. 'Volcanic' may be a misleading term, since a crater such as Tycho or Alphonsus is not in the least like our Vesuvius or Etna, but there is a distinct resemblance to the terrestrial volcanic structures known as calderas; the lunar craters are not randomly distributed, as they would be if due to impact; and the rocks brought home for analysis are of volcanic type. The arguments still go on. My own view is, and always has been, that internal action has played the main rôle in the moulding of the lunar surface, but both processes must have operated to some extent.

By 1968 the scene was set for the first manned round-the-Moon flight. Over the Christmas period Apollo 8, crewed by Astronauts Frank Borman, James Lovell and William Anders, was put into a circum-lunar path, and at one point the space-ship was a mere 113 kilometres above the surface. Apollo 9 was an Earth-orbiter, designed to test out the various components of the space-craft. Then, in May 1969, Apollo 10 carried Astronauts Stafford, Cernan and Young to within 16 kilometres of the Moon; all parts of the vehicle were given a thorough test, and final surveys were made of the intended landing-site on the Mare Tranquillitatis or Sea of Tranquillity.

The Apollo plan had been drawn up as the result of years of intensive study. The whole compound vehicle would be launched in the usual way, by a step-rocket arrangement, and put into lunar orbit. Next, two of the three crew members would transfer to a separate vehicle, the Lunar Module, which had been brought along together with the main Command Module and Service Module. The Lunar Module would be taken to the Moon by using its own descent engine. When the time for departure came, the Lunar Module would blast back to rejoin the main space-craft, after which it would be jettisoned. When the return journey had been almost completed, the Service Module would be jettisoned too, and the Command Module would splash down in the sea, using parachute braking.

Every possible precaution had been taken, but nobody under-estimated the dangers. If the landing on the Moon were faulty, or if the single ascent engine of the Lunar Module failed to work properly – and first time – the result would be disaster. Neither could there be any hope of rescue.

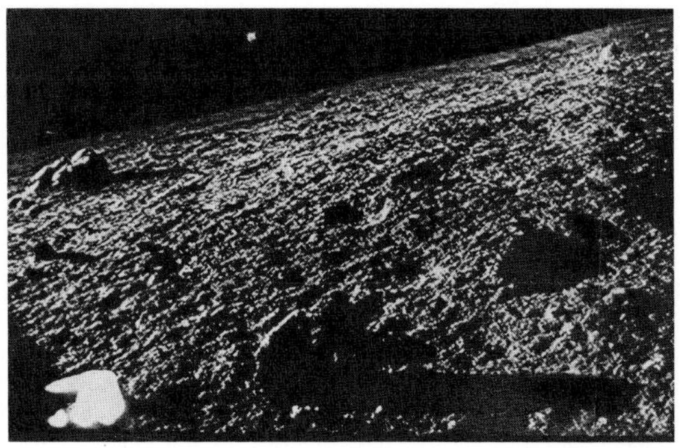

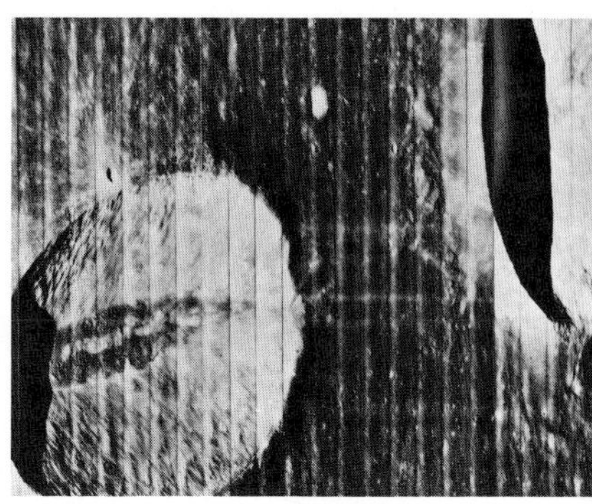

MESSIER AND MESSIER A. *(Right) The twin craters photographed from Orbiter 5, 1967. Each crater is less than 16 km in diameter.*
THE SCHRÖTER VALLEY *(left). Part of the great feature extending from the crater Herodotus (near Aristarchus), photographed from Orbiter 5, 1967.*
LUNA 13 LUNAR PANORAMA *(centre). Part of a picture transmitted by Luna 13 in December 1966.*

By July 1969 all was ready. Two of the Apollo 11 crew members were scheduled to land on the Moon: Neil Armstrong and Colonel Edwin Aldrin, leaving the third member, Michael Collins, in the circling Command Module. What they hoped to do was more dangerous than anything attempted before. Armstrong and Aldrin had to land their Lunar Module, *Eagle*, on the Sea of Tranquillity; go outside for a period of two to three hours, and finally blast back to rendezvous with Collins in the Command Module, *Columbia*. The eyes of the world were upon them.

There was something of a diversion on July 13, when, with their usual lack of warning, the Russians sent a new unmanned probe, Luna 15. Subsequently it entered a closed path round the Moon, and was switched about from one orbit to another, so that at one stage it and Apollo 11 were going round the Moon at the same time. On July 21 Jodrell Bank reported that the Luna was apparently descending, after which signals from it ceased; the Russians subsequently announced that it had crash-landed in the Mare Crisium. Meantime, the focus of attention was very much upon Armstrong, Aldrin and Collins.

Apollo 11 was launched on schedule, and swung out towards the Moon in the usual way. After the conventional manœuvres, it was put into a closed lunar path at a height of approximately 113 kilometres. Then, at last, the vital period began. Leaving Collins alone in the Command Module, Armstrong and Aldrin went into the Lunar Module, and *Eagle* began its descent. As the altitude dropped, the calm

SAFE LANDING AT THE END OF APOLLO FLIGHT – *splashdown in the ocean.*

EAGLE, FROM COLUMBIA *(top). The Lunar Module of Apollo 11 above the Moon, photographed by Michael Collins from the Command Module.*
MOON-ROCK *(centre). One of the first pieces brought back by Apollo 11. This was photographed by Patrick Moore at Houston, Texas, shortly after the return of the space-craft. The rock is enclosed in a glass dish, as it was still then in 'quarantine'.*

FOOTPRINT ON THE MOON, *Apollo 11, July, 1969. The print of the astronaut's boot is very plain.*

LAUNCHING OF APOLLO 15, *(opposite) in July, 1971, from Kennedy Space Center, Florida. Smoke billows as the huge Saturn 5 booster lifts off the pad.*

voices of Armstrong and Aldrin came through to Mission Control at Houston, Texas, to be relayed across the world. Finally, *Eagle* was hovering at only 46 metres above the surface. It could do so for only two minutes; there was insufficient fuel for a longer delay. Armstrong tilted the Module so that he could see the ground below; he was above a crater 'the size of a football pitch' with unpleasant-looking rocks, and he manœuvred *Eagle* down-range towards smoother ground. Then he brought the space-ship down. At last the wires projecting from the lower pads touched ground; a second later Armstrong cut the engines, and the *Eagle* came to rest. Man had reached the Moon.

'*Eagle* has landed,' came Armstrong's voice. It was an historic moment. At Mission Control both Oberth and von Braun were watching; their faith, and that of the other pioneers, had been justified.

At 1.53 Greenwich Mean Time on July 21, the cabin was depressurized. Half an hour later Armstrong and Aldrin were fully equipped; the hatch was opened, and Armstrong made his way down the ladder. By 2.55 he had reached the footpad, and paused briefly. His words came through, calmly and clearly:

'I'm at the foot of the ladder. The LM footpads are only depressed in the surface about one or two inches. Although the surface appears to be very finely grained as you get close to it, it's almost like a powder. Now and then it's very fine. I'm going to step off the LM now. That's one small step for a man, one giant leap for mankind.'

Seconds later, at 2.56 G.M.T., Armstrong was standing on the Moon.

No time was wasted. Armstrong could not stay outside for more than three hours – his oxygen supply was limited – and there was much to do. Aldrin joined him, and the two worked methodically, collecting samples of lunar 'soil' and setting up various pieces of equipment, including a seismometer to record any ground tremors or moonquakes. As they worked, they sent back a running commentary. This was Aldrin's description of walking: 'You have to be rather careful to keep track of where your centre of mass is. Sometimes it takes about two or three paces to make sure you've got your feet underneath you.' Armstrong reported that 'these boulders look like basalt, and they probably have two per cent white minerals in them'.

There was something eerie about it all, and yet the two astronauts were so very matter-of-fact and practical. 'Magnificent desolation' was Aldrin's description of the scene, and with good reason. But the two could not stay for long; the sample collection was completed, and they prepared to re-enter, Aldrin went back into the Module at 4.57, and Armstrong followed him at 5.09. The first Moon-walk was over; Armstrong had been outside for just over 2 hours 47 minutes.

The dangers were far from over. Following a somewhat fitful sleep, the astronauts made ready for blast-off. Again they depended solely upon *Eagle*'s ascent engine; again everything functioned perfectly. Using the bottom part of the Module as a launching pad, the top part was fired back into orbit, and made rendezvous with the waiting Collins. Docking was not so smooth as might have been wished, but it was accomplished safely – and the epic venture was complete. On the morning of July 22, Apollo 11 was put into an Earth-directed orbit, and splashed down in the Pacific, a few kilometres from the waiting recovery ship *Hornet*, at 4.49 p.m. on July 24. After a total of 195 hours in space, the final landing was a mere thirty seconds late.

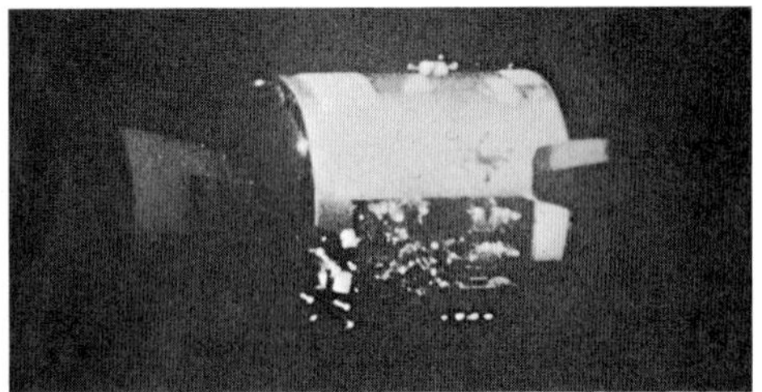

DAMAGED SERVICE MODULE OF APOLLO 13, *which was photographed after being finally jettisoned.*

I have described the Apollo 11 flight in some detail because it was an historical landmark, and I am, after all, writing an historical book. It would take up too many pages to discuss the later Apollos in the same way, but I can at least select some of the main highlights.

Apollo 12 (November 1969), crewed by Astronauts Conrad, Bean and Gordon, came down in the Oceanus Procellarum close to the old soft-lander Surveyor 3, which had been standing on the Moon for more than two years; parts of it were broken off and brought home for analysis. After Conrad and Bean had returned to the Command Module, their Lunar Module *Intrepid* was deliberately crashed on to the Moon, setting up an artificial moonquake. It was recorded by the seismometers which had been set up, and the vibrations continued for almost an hour; it was said that the Moon seemed to be 'ringing like a bell'.

Apollo 13 was a near-tragedy. On the outward journey there was an explosion in the Service Module, putting the main engines out of action. At 286,000 kilometres from the Earth, James Lovell, commander of the mission, was heard to say, 'Hey, we've got a problem here!' – perhaps the understatement of the century. It was later found that during the pre-flight checks too much current had been put through a heater circuit, so that a vital switch had been welded shut. During flight the temperature rose; the insulation system was destroyed, and fire broke out.

ALAN SHEPARD ON THE MOON, *Apollo 14, 1971. Note the intensely black shadow cast by the Lunar Module.*

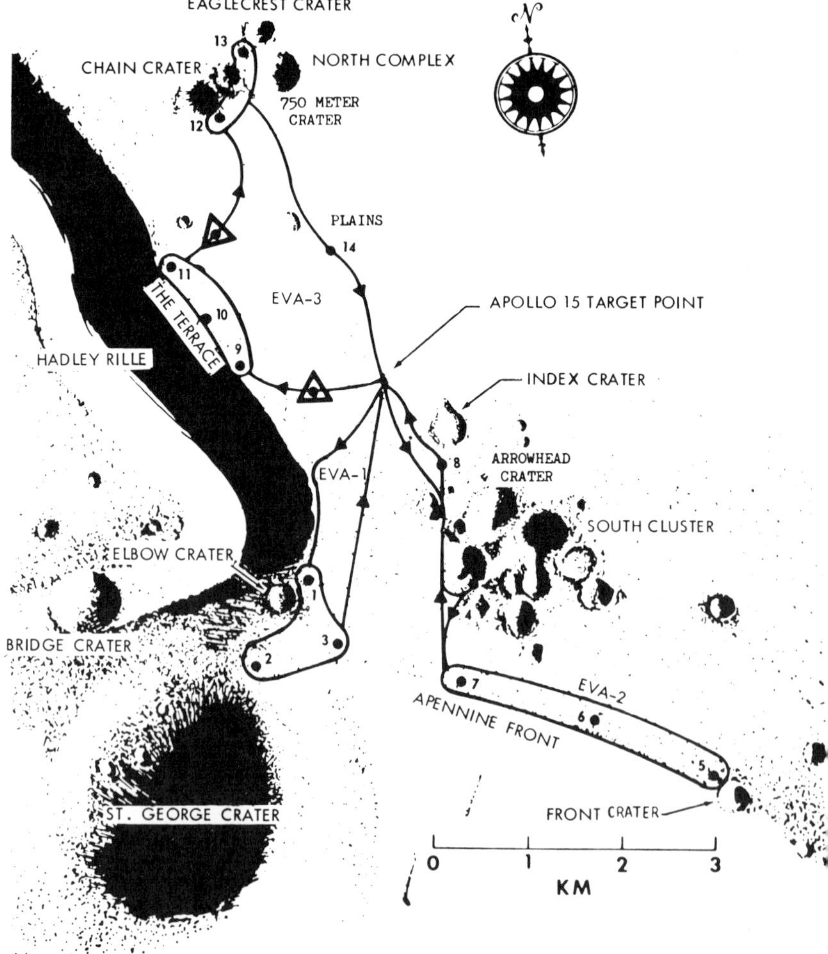

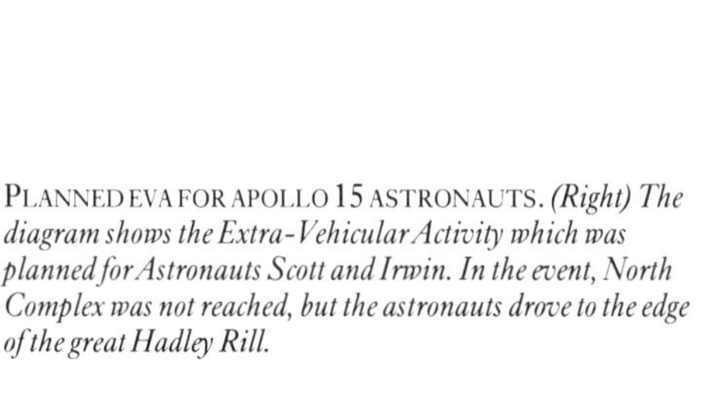

PLANNED EVA FOR APOLLO 15 ASTRONAUTS. *(Right) The diagram shows the Extra-Vehicular Activity which was planned for Astronauts Scott and Irwin. In the event, North Complex was not reached, but the astronauts drove to the edge of the great Hadley Rill.*

Inside the space-ship, the temperature dropped alarmingly; there was a potential shortage of water and – worse – oxygen. The Service Module, an essential part of the vehicle, had become a useless hulk. The only hope was to 'coast' round the Moon, and then use the descent engine of the Lunar Module *Aquarius* to accelerate the whole space-ship and bring it back to Earth. Thanks to the emergency procedures worked out by Mission Control technicians at breakneck speed, plus the coolness and courage of Lovell and his colleagues Fred Haise and Jack Swigert, the rescue plan was successful; but it was a salutary reminder that space-travel is always a risky business.

Apollo 14 (January 1971) was commanded by Alan Shepard, less than ten years after he had become America's first man in space; with him were Thomas Mitchell and Stuart Roosa. The landing was achieved near the walled plain Fra Mauro, and during their 'lunar walk' Shepard and Mitchell used a kind of cart, which enabled them to collect more samples than would have been possible otherwise. With Apollo 15, of the following summer (July 30 to August 7, 1971) Astronauts Scott, Irwin and Worden set up various new records. Scott and Irwin explored the foothills of the lunar Apennines, using an electrically-powered Lunar Roving Vehicle or moon-car to take them to the very rim of the 370-metre deep crack known as the Hadley Rill. The views of the Apennines were spectacular. Lunar mountains are not the jagged structures drawn by science-fiction enthusiasts not so long ago; they are comparatively smooth.

1972 saw the last of the Apollos. No.16, taking Astronauts Young and Duke to the Moon, was aimed at the highlands near the crater Descartes, a new type of terrain. Then, in December, Apollo 17 was dispatched to the Taurus-Littrow area at the edge of the Mare Serenitatis; of the two Moon-walkers, Eugene Cernan was a 'space veteran', while Harrison (Jack) Schmitt was a professional geologist who had been trained as an astronaut specifically for the mission.

APOLLO 15

APOLLO 15. *After the landing (right), James Irwin is seen saluting the U.S. flag. The mountain in the background is Hadley Delta. David Scott and James Irwin (above) are in the Lunar Rover and (below) David Scott drills into the lunar surface.*

Schmitt's most startling discovery was that of orange material round the rim of a small craterlet unofficially named Shorty. I was in Mission Control at the time, and I can well remember Dr. Schmitt's excited 'Gee! It's orange – crazy!' At first it was believed that the orange hue was due to recent volcanic activity, but later it was found that the colour was caused by very ancient glassy particles.

Throughout this period the Russians had been far from idle. In September 1970 they had managed to send an automatic probe to the Moon, soft-land it in the Mare Fœcunditatis or Sea of Fertility, and bring it home with 100 kilogrammes of samples. Even more important were the two Lunokhods, the first of which was taken to the Moon by Luna 17 in November 1970, and the second by Luna 21 in January 1973. The Lunokhods were 'crawlers' which moved around the Moon under the direction of the Soviet controllers. They obtained panoramic pictures, and carried out analyses of the surface material; Lunokhod 1 travelled a distance of over ten kilometres before its power failed. The Lunokhods looked frankly weird, and reminded me of a cross between a sardine tin and an antique taxicab, but they were brilliantly successful, and represented a real Russian triumph. They are still standing on the Moon; we know exactly where they are, and no doubt they will eventually be collected and taken to a lunar museum.

The manned missions have ended for the moment. It is not likely that the Americans will send any more men to the Moon until there is proper rescue provision, and the Russians have not announced any plans for lunar expeditions, though they may of course attempt something spectacular at any time, and may even set up a fully-fledged lunar base. Meanwhile, let me try to summarize what we have learned about the Moon itself.

Like the Earth, the Moon has proved to have a crust, a mantle and a core. Most of what we know about the interior has been derived from the Apollo experiments. The heat coming from below the surface was measured, showing that the Moon is not cold and rigid all the way through its globe, as some authorities had believed. There are frequent moonquakes, and, as I and others had suspected, there is a distinct link between the tremors and the observed Transient Lunar Phenomena.

Earthquakes set up 'waves' which pass through the Earth's globe, and give us information about what is happening below our feet. Some kinds of earthquake waves can pass through both solids and liquids, while others are stopped by liquid material – which is how we have been able to find out the size of the Earth's liquid, iron-rich core, which proves to be larger than the whole of the Moon.

The same procedure can be applied to the Moon. The Apollo astronauts set up seismometers, basically the same as those used on Earth, but much more sensitive. Moonquakes were found; some of them are centred not far below the crust, but others take place much lower down, about half-way between the Moon's surface and its core. From these results, we can make a reasonably good estimate of how large the Moon's liquid centre is. We have also made use of meteoritic impacts; one missile, which hit the Moon on July 17, 1972, is thought to have weighed about a ton. There are also the records obtained from the crash-landings of the jettisoned Apollo modules.

The outermost part of the Moon is called the *regolith*. It is a loose layer from 1 to 20 metres deep, shallower over the *maria* than over the highlands. Below the regolith is a kilometre-thick layer of broken-up rock, and below this again there is a layer of more solid rock, going down to about twenty-five kilometres. The thickness of the crust is generally between forty-five and sixty-five kilometres, though it may be more in some places, and it is definitely thicker on the far side than on the Earth-turned hemisphere. Below the crust comes the mantle, which goes down to around 1000 kilometres; then comes a region which is probably partially melted (the *asthenosphere*), and finally there is the heavy molten core, presumably rich in iron, at a temperature of more than 1000 degrees Centigrade.

Iron is a magnetic element, and it is therefore slightly surprising to find that the Moon has virtually no general magnetic field; Lunik 1 showed that, way back in 1959. There are regions of local magnetism, particularly on the far side, but it seems that the original overall magnetic field of the Moon has long since disappeared.

The seas or *maria* are lava-plains, in which the most common material is the familiar volcanic rock basalt. The highland rocks are less rich in iron than those of the seas, but contain more calcium and aluminium. No new materials have been found – and none were expected. There are, of course, some minerals not identical with anything on the Earth, and one of these has been named armalcolite, after the three Apollo 11 astronauts – *Arm*strong, *Al*drin and *Col*lins. But the elements making up the Moon are the same as those of the Earth, and the structural differences in the minerals are due simply to the fact that they have cooled and solidified under different conditions. The surface gravity is much lower; and if the Moon ever had an atmosphere, it certainly does not possess one now. To all intents and purposes the Moon is an airless world.

Slight irregularities in the movements of Orbiter 5 led to the discovery of regions of relatively dense material beneath some of the large craters and regular seas; these are called mascons, a convenient abbreviation of *mass con*centrations. Probably they are due to blocks of comparatively heavy volcanic rock. They are certainly not buried meteorites, as was once suggested.

DR. HARRISON SCHMITT BY A HUGE LUNAR BOULDER. *Apollo 17 photograph. Dr. Schmitt was the first qualified geologist to go to the Moon; this rock was one of the largest encountered by the two explorers (Schmitt and Eugene Cernan) during their EVAs, using the Lunar Roving Vehicle.*

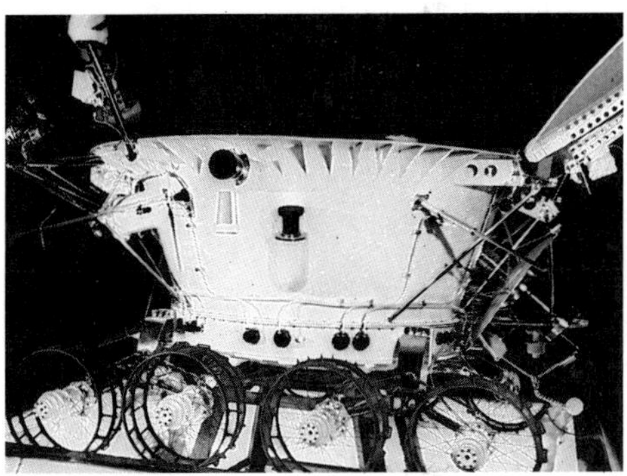

LUNOKHOD 1, *the Russian 'crawler' which was taken to the Moon.*

Armed with all this new information, we can speculate as to how the Moon evolved in the remote past. It never formed part of the Earth, as used to be thought; it has always been a separate world, and probably it was formed in the same way as the planets, by building up or accreting from a cloud of material surrounding the youthful Sun. There is no doubt that the Earth and the Moon are of about the same age; analyses of lunar rocks have left no doubt about this.

After its formation, over four and a half thousand million years ago, the Moon cooled down, and developed a comparatively firm crust. The great basins now filled by the *maria* were produced – either by vulcanism or by meteoritic impact, whichever theory you prefer; the oldest rocks date back at least four thousand million years. Next, molten magma poured out from the interior, making up the lava-sheets of today. This lava-flow stopped around 3·1 thousand million years ago, and then came the age of crater production. Yet we must remember that even the youngest large craters on the Moon date back for a thousand million years at least, so that by geological standards they are 'old'. The active history of the Moon came to a halt long before advanced forms of life developed on the Earth, and nothing much has happened since, though no doubt small impact craters are produced now and then.

One thing which has been proved without a shadow of doubt is that the Moon is lifeless, and has always been so. The lunar rocks contain no traces of living organisms, either past or present, and neither do they contain any traces of hydrated material (that is to say, material containing watery substances). Conditions were never suitable for life to begin, and so the Moon has remained sterile throughout its long history.

This does not make it any the less interesting. It has been described as a cosmical museum; it looks much the same now as it must have done when life on Earth was limited to tiny sea-creatures, so that when we study it we are delving back into the past. In the future it will no doubt be colonized, and a Lunar Base sounds much less fantastic today than an artificial satellite would have done a century ago. Meanwhile, the Moon remains the most spectacular object in the night sky. Anyone equipped with a telescope, or even a pair of binoculars, can follow the changes in appearance as the Sun rises over the lunar surface; a crater which looks prominent one night, when it is near the terminator and is filled with shadow, looks quite different a few hours later, while near full phase the brilliant rays from Tycho and Copernicus dominate the whole scene. There is still much that we do not know, despite the Lunas, the Rangers, the Surveyors, the Orbiters and the Apollo missions. The fascination of the Moon is as great as ever.

35

Travel Near the Sun

THE first attempt at interplanetary flight was made in February 1961, when the Russians launched an unmanned probe toward Venus. It was not a success. When it had receded to a distance of 7,500,000 kilometres from the Earth, signals from it stopped. Contact was never regained, and so nobody will ever know the final fate of Venera 1, though there is a good chance that it passed within 100,000 kilometres of Venus in May 1961, and is probably orbiting the Sun even now.

Over a year later the Americans made their first launching towards Venus. The initial probe, Mariner 1, was a complete failure, and within a few minutes it had come to an inglorious end in the sea; it was later found that in programming the instruments somebody had forgotten a minus sign, which shows that space technicians are human after all. Fortunately Mariner 2, launched on the following August 26, more than made amends for its luckless twin. It by-passed Venus at a mere 35,000 kilometres on December 14, and provided us with our first reliable information about that decidedly peculiar world.

Sending a space-craft to a planet is not so straightforward as with the Moon. The Moon, after all, is our companion, and keeps together with us as we travel round the Sun. A planet does not. Venus, the closest of them, never comes much within 39,000,000 kilometres of us, which is a hundred times as remote as the Moon, and it is quite impossible to wait until the Earth-Venus distance is least and then fire a rocket straight across the gap. Anything of the sort would use up far too much propellent. What has to be done is to use the Sun's gravitation to follow what is termed a transfer or Hohmann orbit. The probe is taken above the atmosphere by rocket power in the usual way, and is then slowed down relative to the Earth, after which the real journey can begin.

With Mariner 2, the launcher was an Atlas rocket, which rose vertically from Cape Canaveral and then moved off over South Africa. The second stage of the step-vehicle, an Agena rocket, then took over, and the Agena-Mariner combination was put into a 'parking orbit' at a height of 185 kilometres, moving at the regulation 29,000 kilometres per hour. As it reached a point over the African coast, the Agena motors fired again, and the total velocity rose to 41,034 kilometres per hour, about 1370 kilometres per hour more than the escape velocity at that altitude.

For a Venus probe, the direction of 'escape' has to be opposite to that in which the Earth is moving round the Sun. It was so on this occasion. As the Mariner moved away, it was slowed down by the Earth's pull until at a distance of 965,000 kilometres the velocity relative to the Earth had dropped to 11,060 kilometres per hour. In other words, Mariner was moving round the Sun at 11,060 kilometres per hour less than the Earth – and it started to swing inwards towards the Sun, picking up speed as it went. Meanwhile the Agena rocket had been separated from the Mariner, rotated through a wide angle and fired again, so that it was put into an entirely different orbit. Nobody knows

APPARENT SIZE OF VENUS, *from new to full, shown to the same relative scale.*

DRAWING OF VENERA 9 ON THE SURFACE OF VENUS. *This is an official Russian picture – for obvious reasons, no photographs of the vehicle on the surface of Venus are available!*

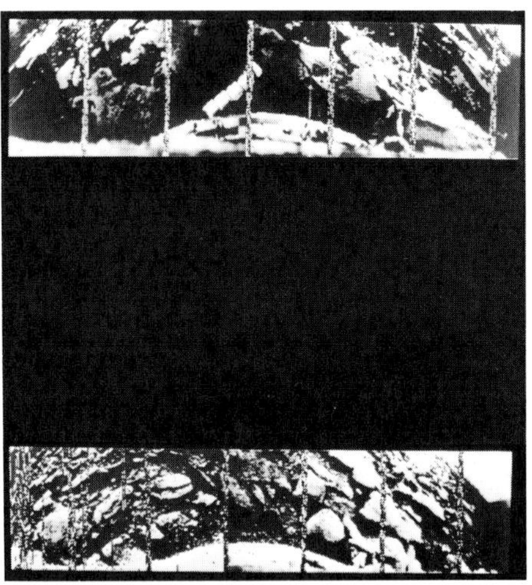

THE SURFACE OF VENUS, *as sent back by Venera 9 (upper) and Venera 10 (lower). Parts of the space-craft are shown.*

what happened to it, and nobody cares; it had completed its work, and on the journey to Venus it would have been simply a nuisance. Henceforth, Mariner 2 was on its own.

Following the transfer orbit, Mariner made its rendezvous with Venus less than four months later. Despite a mid-course correction, the minimum distance from Venus was rather greater than had been hoped, but all the instruments functioned well, and the results they sent back came as a distinct surprise.

Venus, remember, is permanently concealed by its dense atmosphere, and it had been nicknamed the Planet of Mystery. Working at Mount Wilson in 1932, W.S. Adams and T. Dunham had examined the spectrum, and had found that the top of the atmosphere, at least, contained a large quantity of the heavy gas carbon dioxide; but there was no reliable clue as to what conditions were like lower down, and no photographs or drawings showed more than very ill-defined patches which were due to nothing more substantial than clouds.

By 1962 there were two main theories about the surface of Venus. According to D.H. Menzel and F.L. Whipple, there were likely to be broad oceans, perhaps supporting primitive life-forms. This would mean that Venus might be a world in an early stage of evolution, just as the Earth used to be in the Coal Forest period; if so, then life could presumably develop there just as it has done here. Yet carbon dioxide tends to act in the manner of a greenhouse, shutting in the Sun's heat, and since Venus is forty million kilometres closer to the Sun than we are a high surface temperature was only to be expected. Moreover, the atmospheric carbon dioxide would have fouled the oceans, producing seas of soda-water (though not even the most confirmed optimist hoped to find any whisky to mix with it).

Alternatively, it was suggested that Venus could be a raging dust-desert, far too hot for water to exist. This was less inviting, but on the whole it appeared rather more probable. Another unknown factor was the length of Venus' rotation period, or 'day'. Photographic work by Gerard P. Kuiper, at Palomar, had led him to propose a period of a few weeks, but nobody was sure. In 1959 I wrote a monograph about Venus in which I listed all the rotation-period estimates made by observers using all kinds of methods; visual, photographic, spectroscopic and so on. There were more than a hundred of them, but there was a wide measure of disagreement; the estimates ranged between 22 hours and 224¾ days, the latter being, of course, the length of Venus' period of revolution round the Sun. We now know that every one of these estimates was wrong.

Mariner 2 provided the breakthrough. Instruments on board showed that the surface temperature was several hundreds of degrees, which at once disposed of the Whipple-Menzel marine theory. Carbon dioxide made up most of the atmosphere, and the rotation period was even longer than the 224¾-day 'year', which was totally unexpected.

Contact with Mariner 2 was finally lost on January 2, 1963, at a distance from the Earth of 87,000,000 kilometres. The next American fly-by, Mariner 5, was sent past the planet in 1967 at a distance of only 4000 kilometres, and confirmed all previous results. Then, in February 1974, Mariner 10 sent back excellent pictures of the cloud-tops. But meanwhile the Russians had been very active, and had achieved some truly magnificent successes.

Their aim was to soft-land their Venera probes, bringing them down through the dense atmosphere by parachute – a difficult manœuvre by any standards. It is not surprising that they had several failures, and it is now clear that some of the Veneras were literally crushed by the unexpectedly high pressures in the lower atmosphere, but in September 1970 Venera 7 came down safely, transmitting data for twenty minutes before being put permanently out of action by the intensely hostile conditions. Venera 8, of March 1972, lasted for fifty minutes after arrival, and in October 1975 Veneras 9 and 10 managed to send back one picture each. The scene was gloomy and rock-strewn; the level of illumination was said to be about the same as that at noon in Moscow on a cloudy winter day, and there were sluggish winds, which however had considerable force in so dense an atmosphere.

Then, in December 1978, several probes arrived at Venus. Two (Veneras 11 and 12) were Russian, and both soft-landed, transmitting for an hour after touch-down though without sending back any pictures. The others were American, known collectively as the Pioneers. The largest vehicle, nicknamed the 'Bus', dispatched four small probes on to the surface, while the Pioneer Orbiter was put into a closed path round the planet, and began a long spell of research which was still going on in 1983. The surface was mapped by radar techniques, and now, at last, we know what Venus is really like.

First, we have analyzed the atmosphere. Most of it is made up of carbon dioxide, and the clouds contain large quantities of sulphuric acid. The surface temperature is nearly 500 degrees Centigrade, and the rotation period is 243 days; this is longer than Venus' year, and to complete the picture Venus spins from east to west instead of from west to east. It has been termed 'the upside-down planet'; the reason for this remarkable state of affairs is completely unknown. To an observer on the surface the Sun would rise in the west, setting in the east 117 Earth-days later. Not, of course, that such an observer could see the Sun at all through those obscuring clouds; nothing would be visible apart, perhaps, from a dull glare.

The atmospheric structure is equally unusual, because the uppermost clouds, visible from Earth, have a rotation period of only four days. This had been suggested earlier by French astronomers at the Pic du Midi Observatory, but at the time of their announcement the result was greeted with some scepticism; I admit that I did not believe it – it seemed so curious! Below the clouds of Venus there is probably a deadly rain of sulphuric acid droplets, but the cloud deck ends some distance above the surface, and the lowest region of the atmosphere may be regarded as a super-heated smog. Lightning is continuous, and there is believed to be constant thunder as well. The conditions remind one of the conventional picture of hell.

Radar measures from the Pioneer Orbiter have provided a map of over ninety per cent of the total surface. There is a huge rolling plain stretching round the globe; there are two highland 'continents' which have been named Ishtar and Aphrodite; there are mountains, and there are valleys. The highest peaks, the Maxwell Mountains in the eastern part of Ishtar, rise to twelve kilometres above the main radius of the planet (we cannot use sea level as a reference, since Venus has no oceans). Aphrodite, the larger upland, has more than half the area of Africa, and at its eastern end is a vast rift valley nearly three kilometres deep, with a width of 280 kilometres and a length of over 2200 kilometres, so that it is much larger than anything on the Earth. There are many craters, most of them shallow, and several formations which are clearly volcanic. There are two objects, now named Rhea Mons and Theia Mons, which are almost certainly nothing more nor less than shield volcanoes – similar in type to the Earth's Mauna Kea and Mauna Loa, though much more massive. There is no reason to doubt that both Rhea and Theia are still active, and there is another probable active region – the so-called Scorpion's Tail – close to Aphrodite.

Venus certainly has a hot core, though it may be smaller than the Earth's, and there is no detectable magnetic field, presumably because of the slow rotation. If the crust is thicker than that of the Earth, as most investigators believe, we can explain the fact that active vulcanism is localized. On Earth, the study of what is termed plate tectonics has become all-important; broadly speaking, the continents move around like huge rafts or plates, and where two plates meet there is strong vulcanism. Venus may be a 'one-plate' planet, so that its evolution has been different.

Why is Venus so unlike the Earth? In size, mass and escape velocity they are almost twins, but they are indeed non-identical twins. The root cause must be Venus' lesser distance from the Sun. According to recent suggestions, the two worlds began to evolve along similar lines in the early days of the planetary system, when the Sun was less luminous than it is today by a factor of about thirty per cent. Gradually the Sun became more powerful. The Earth was sufficiently far out to escape serious damage; Venus was not. The young seas on the surface of Venus boiled away; the carbonates in the rocks were driven out to form a mainly carbon-dioxide atmosphere – and there was a kind of 'runaway greenhouse' effect, turning Venus into the furnace-like environment of today. (Carbon dioxide, of course, acts as a kind of blanket.) If the Earth had been a few million kilometres closer to the Sun than it actually is, things might have followed the same course, and you and I would not be here.

In some ways Venus has been a disappointment. It is utterly hostile, and there is no chance of its being visited, at least in the foreseeable future. When one looks at the lovely 'Evening Star', shining down like a small and friendly lamp, it is not easy to visualize the inferno-like conditions beneath the clouds.

I have mentioned Mariner 10, which by-passed Venus in February 1974. In fact Venus was merely incidental, since Mariner's main task was to investigate the other inner planet, Mercury. The gravitational pull of Venus was used to swing the probe in to a rendezvous with Mercury, following the plan which I have referred to as interplanetary billiards, and about which I will have more to say in Chapter 37. Mariner 10 made three active passes of Mercury, two in 1974 and the third in March 1975, before contact with it was lost; no doubt it is still orbiting the Sun and making periodical close approaches to Mercury, but we have no hope of finding it again.

Eugenios Antoniadi, using the Meudon 83-centimetre refractor, had published a map of Mercury in 1934; others were attempted later, but were no better than Antoniadi's, and we now know that none of them even approximated to the truth. Yet even before the flight of Mariner 10, one important discovery had been made. This was the length of the Mercurian rotation period.

MERCURY, FROM MARINER 10. *March 29 1974; the Mariner was 200,000 km from Mercury. Eighteen pictures, taken at 42-second intervals, were computer enchanced at JPL and fashioned into this photomosaic. About 2/3 of the area seen is in the southern hemisphere. The largest craters are about 200 km in diameter.*

MERCURIAN CRATER, *(above) 120 km in diameter; Photographed from Mariner 10 on March 29, 1974, from 33,760 km.*

Mercury takes eighty-eight days to go round the Sun. It had been assumed that the rotation period was the same, in which case Mercury would keep the same face turned sunward, just as the Moon does with respect to the Earth. There would be marked libration-type effects, but there would be an area of permanent day, a region of everlasting night, and a comparatively narrow 'twilight zone' in between. But if part of Mercury never received any sunlight, it would be bitterly cold; and in 1962 W.E. Howard, at Michigan, measured the long-wavelength radiations from the night side of the planet, proving it to be much warmer than it would be if it were permanently dark. Using Earth-based radar equipment in 1965, R. Dyce and G. Pettengill found the real rotation period to be 58·6 days. This is two-thirds of the orbital period, and, by a strange fact which may or may not be due to sheer coincidence, Mercury turns the same face towards the Earth every time it is best placed for observation. This is what had misled Antoniadi.

The 58·6-day rotation period leads to a very strange Mercurian calendar. There are two 'hot poles', one or the other of which will always receive the full fury of the Sun's heat when Mercury is at perihelion – and the orbit, remember, is decidely eccentric; the distance from the Sun ranges between over 67 million kilometres and less than 46 million kilometres. The Sun's journey across the sky will be decidedly erratic, because near perihelion the orbital angular velocity of Mercury is greater than the constant angular velocity of the globe due to rotation – and the Sun will move 'backwards' for a brief period before resuming its usual direction of travel. From some parts of the planet there will be two sunsets and two sunrises in quick succession.

One hates to think what a Mercurian would make of this, but in fact we can forget about the chances of any life there. There is virtually no atmosphere, and the surface temperature is a forbidding 350 degrees or so during the daytime when the Sun is high. Incidentally, Mercury is much denser than either Venus or the Moon, so that it seems to have a

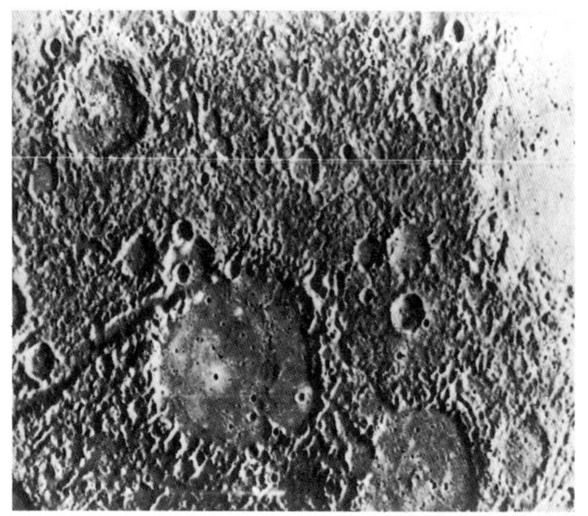

THE ROUGH MERCURIAN SURFACE. *Mariner 10 picture.*

relatively large, iron-rich core; the magnetic field is much weaker than the Earth's, but it is appreciable.

Superficially, at least, the surface of Mercury is very like that of the Moon, the most notable difference being the absence of dark regular 'seas' such as the lunar Mare Imbrium. There are craters in plenty, with peaks, ridges and intercrater plains; there are also ray-systems, and in fact the first crater to be identified on Mercury as Mariner 10 closed in is a ray-centre, named, as we have noted, in honour of Gerard P. Kuiper.

The most imposing feature on Mercury is the Caloris Basin, 1300 kilometres in diameter, which is bounded by a ring of smooth mountain blocks rising to an altitude of two kilometres. It has been compared with the Mare Orientale on the Moon, though the resemblance is not really close; it has been given its name – the 'Hot Basin' – because it lies at one of the hot poles. Unfortunately, only half of it has been mapped. Each time Mariner 10 by-passed Mercury the same regions were in sunlight, so that our charts are still incomplete.

It is fair to comment that almost all our detailed knowledge of Mercury comes from the results of the one probe, Mariner 10. No more missions there have been planned as yet; no doubt both orbiters and landers will be dispatched in the foreseeable future, if only to complete the surface mapping, but manned expeditions seem to be out of the question. The best we can say about Mercury is that hostile though it may be, conditions there are less unpleasant than they are on Venus.

36

The Craters of Mars

THE problems of Mars are quite different from those of Venus, because Mars is a completely different sort of world in size, mass and temperature. I have already said something about the canals, so let me take up the story again in more modern times, when the idea of intelligent Martians had been reluctantly discounted.

In 1956 I delivered a lecture about Mars to a scientific society in London. Among the statements I made were the following:

1. Mars has an atmosphere composed chiefly of nitrogen, with a ground pressure of about 90 millibars. This is equivalent to the pressure in the Earth's air at an altitude slightly less than twice that of the top of Mount Everest.
2. The reddish-ochre areas are deserts, coated with some coloured mineral.
3. The dark regions are old ocean basins, filled with primitive-type vegetation.
4. The canals have a basis of reality, even if they are not so straight or regular as Lowell had believed.
5. The polar caps are made up of a very thin layer of hoar-frost, only a centimetre or two deep at most.
6. In general Mars has a smooth surface, and there are no major mountain chains or deep valleys anywhere.
7. High, frost-covered plateaux are visible, such as Hellas, south of the V-shaped Syrtis Major.

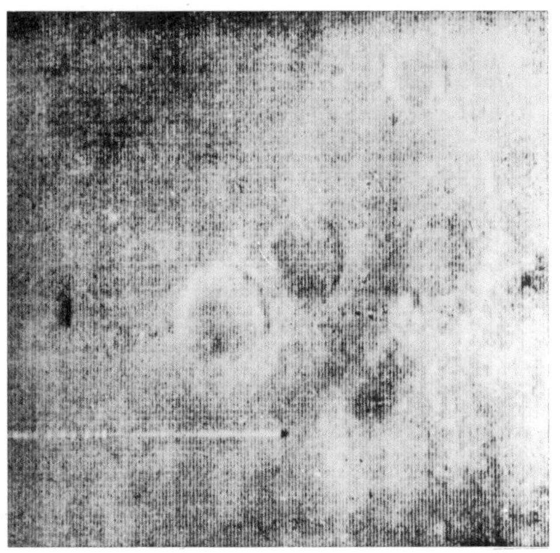

THE CRATERS OF MARS, *photographed from Mariner 4. This is Frame No. 8.*

All these statements were supported by the best available evidence – and all, apart from No. 2, have been found to be completely wrong. The space missions have led to a complete *volte-face* in our ideas about Mars.

Remember that Mars has only about twice the diameter of the Moon, and is always at least a hundred and fifty times as far away, so that no telescope will show it as clearly as the Moon is seen through ordinary binoculars; also, photography is of little help, because it is impossible to take pictures of Mars showing as much detail as I can see with the modest 39-centimetre reflector in my own observatory. Neither was the spectroscope really informative. A planet's spectrum is essentially a reflected solar spectrum. There are modifications caused by the fact that the light reaching us has passed through the Martian atmosphere twice, once going there from the Sun and again on its journey to the Earth, but nitrogen is very shy of showing itself spectroscopically, and the only gas definitely detected was carbon dioxide.

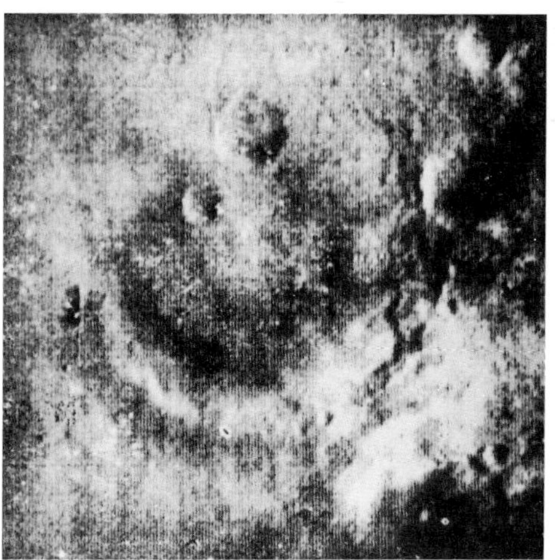

THE CRATERS OF MARS, *photographed from Mariner 4. This is the 11th Frame – the best obtained from Mariner.*

I must here refer to a rather strange episode, or, more accurately, pair of episodes. In 1892 Edward Emerson Barnard was observing Mars with the 91-centimetre Lick refractor when he recorded what he believed to be craters, and much later, in 1915, John Mellish also recorded craters, this time with the world's largest refractor, the

101·6-centimetre at Yerkes. Barnard suppressed his observations because he thought that 'people would laugh at them'; neither did Mellish publish any of his sketches, and they cannot now be located. In 1944 came a weird book by one Donald Lee Cyr, who believed the canals to be the fertility tracks of Martian animals(!) with crater-lakes along their courses; but on the whole the idea of a cratered Mars was not taken very seriously – until 1965, when Mariner 4 made its fly-by and sent back the first pictures of the planet from close range.

Once again a transfer orbit was followed, but this time the probe had to be speeded-up relative to the Earth, so that it would swing outwards to a rendezvous with Mars. Mariner 4 was launched on November 28, 1964 (its twin, Mariner 3, had been dispatched on November 5, but its guidance system had failed immediately, and it had sped off into space uncontrollable and uncontactable). By the following July 14 Mariner was only 10,000 kilometres from Mars; it had covered a total of over 670,000,000 kilometres, so that it had certainly not come by the shortest route. Altogether it transmitted twenty-one pictures from close range, several of which were of good quality.

The first and most surprising feature of all was that instead of being gently undulating, the Martian surface was cratered. Some of the craters even had central peaks, so that the overall impression was more lunar than terrestrial. The craters were found on both the bright and the dark areas; the dark regions were not depressions, and they were not covered with vegetation. The possibility of life on Mars began to look more remote.

But the most important results of all concerned the atmosphere. Mariner's path took it behind Mars relative to the Earth (though, of course, no telescope would show it). Just before the space-craft was occulted, and again just after it emerged, its radio signals came to us after having passed through the blanket of air above the Martian surface, and the way in which the signals were affected made it possible to estimate the electron density, i.e. the numbers of free electrons in the Martian atmosphere. From this, the ground density could be worked out.

The results were quite definite. Instead of having a pressure of over 80 millibars, the pressure of the atmosphere was less than 10 millibars; and instead of being composed of nitrogen, it was made up chiefly of carbon dioxide. This meant that the atmosphere of Mars was useless from our point of view. No Earth-type creature could possibly have breathed it even if it had been made up of pure oxygen, but there seemed to be almost no free oxygen at all.

Mariner 4 had done its work, and after its rendezvous it continued in a never-ending orbit round the Sun. The next missions were Mariners 6 and 7, launched from Cape Canaveral during the spring of 1969; both were successful, and both flew past Mars at distances of less than 3400 kilometres, only a few days after Neil Armstrong had stepped out on to the lunar Sea of Tranquillity. More craters were shown, and the extreme thinness of the atmosphere was fully confirmed. One formation, not far from the pole, was nicknamed the Giant's Footprint, though by now most astronomers were coming to the conclusion that Mars was just as sterile as the Moon. Hellas, once believed to be a plateau, was found to be a deep basin, but all in all, the new results tended to put Mars into the category of dead worlds of strictly limited interest. It was even thought that the polar caps might, after all, be made up of 'dry ice' (solid carbon dioxide) rather than water ice. There was certainly no trace of any canals.

What had actually happened was that by sheer bad luck, Mariners 4, 6 and 7 had passed over the least interesting regions of the whole planet. It was just as well that the next proposed missions, Mariners 8 and 9, were so far advanced in the planning stage that there was no thought of cancelling them.

The Russians had already launched two vehicles toward Mars; they dispatched another two in 1971, and four more in 1973. Yet for some reason or other they had practically no success, which was doubly surprising in view of what they had managed to find out about Venus, which should be a much more difficult target. The Soviet probes either went out of contact, missed the planet altogether, or stopped transmitting before sending back much information of value. No doubt the Soviet authorities will try again. Meantime, we depend entirely upon the results of the United States missions.

Mariner 8, launched in May 1971, was a prompt failure, because the second stage of the launching vehicle failed to ignite. Mariner 8 plummeted into the sea over five hundred kilometres north-west of Puerto Rico, and that was that. Mariner 9 was sent up later in the same month, on May 30, having been slightly modified to do the work of two vehicles instead of one, and by mid-November it had reached the neighbourhood of Mars. It was a new type of mission, because instead of merely by-passing Mars it was put into a closed orbit which brought it down to a minimum distance of 1370 kilometres above the surface. It worked splendidly; it returned over 7000 pictures, and went on functioning well after its expected lifetime. Contact with it was not finally lost until October 27, 1972.

When it first entered orbit, Mariner 9 could tell us little about the surface features. This was not the fault of the space-craft, but of Mars itself. There are periodical dust-storms, covering the surface completely, and one of these was in progress; I made some observations earlier in November, and could see nothing except an ochre, featureless disk. There was nothing to be done except wait for the dust to clear, which after some weeks it did. Then came the first of the spectacular pictures.

They were breathtaking indeed. Craters, mountains and valleys were shown, and there were also giant volcanoes, one of which – Olympus Mons, or Mount Olympus – towered to twenty-five kilometres above the mean level of the surface, and was topped by a huge volcanic crater or caldera sixty-four kilometres in diameter; the base of the whole structure is 600 kilometres broad. It is a shield volcano, similar in form to those of Hawaii, but very much larger. It was not unique. Along the ridge known as Tharsis there are other volcanoes: Ascræus Mons, Pavonis Mons, and Arsia Mons (seen as a speck by Schiaparelli, and named by him Nodus Gordii, or the Gordian Knot). Arsia Mons has the largest caldera of all, no less than 140 kilometres in diameter.

Extending away from these giant volcanoes are features which look like drainage systems; the great rift system Valles Marineris (Mariner Valleys) is 400 kilometres long, 200 kilometres wide at its broadest point, and 6 kilometres deep – equal to the rift valley on Venus (which was not then known, of course), and dwarfing our own much-vaunted Colorado Canyon. Elsewhere there were features which looked so like

MARINER 9, *the Mars probe launched in May 1971, and which reached the neighbourhood of Mars in the following November.*

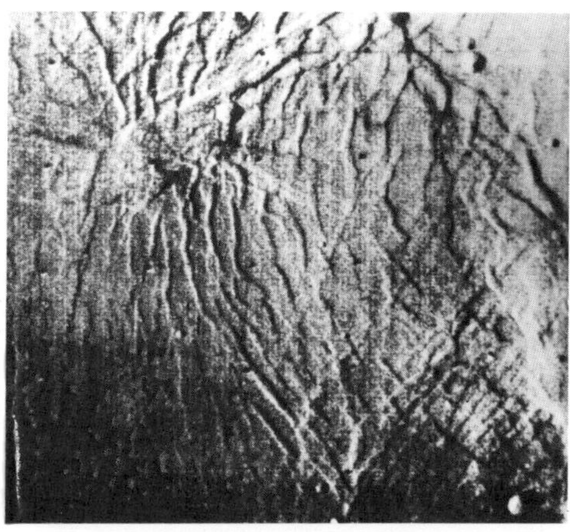

SURFACE FAULTS ON MARS. *The Phœnicis Lacus plateau area from Mariner 9 during its 67th orbit.*

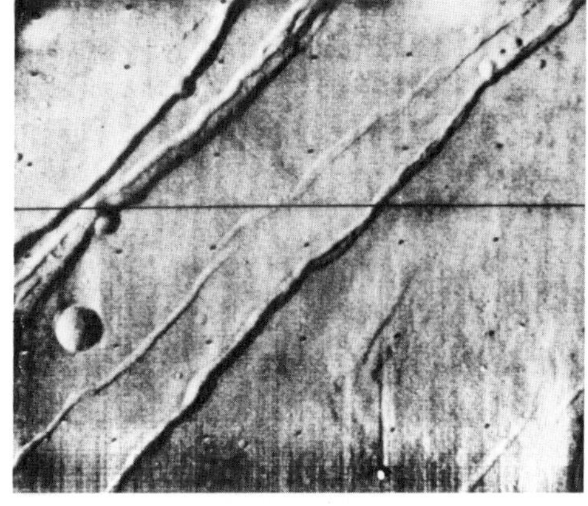

RILLS ON MARS, *in the Mare Sirenum, from Mariner 9.*

HIGH-RESOLUTION PHOTO OF THE MARTIAN SURFACE *near Viking 2 soon after landing at the beginning of September 1976. The vesicles, or small holes, common to most Martian rocks, are clearly to be seen.*

VALLEY IN RASENA AREA OF MARS, *photographed from Mariner 9 in 1972.*

LAUNCHING OF MARINER 9, *May, 1971. (Opposite) The probe begins its long flight to Mars.*

dry riverbeds that few people doubt that they really *are* dry riverbeds. If so, then water must have flowed on Mars at some time in the past.

The two hemispheres are not alike. The southern part of the planet contains the main volcanoes, and much of it is densely cratered; there are fewer craters in the north, but they are still fairly numerous, and isolated volcanoes are found too. Incidentally the Syrtis Major, the most famous of the dark areas, is a plateau sloping off to either side, so that the old picture of it as a vegetation-filled ocean bed could not possibly have been more wide of the mark. The Hellas basin is one of the few crater-free areas on the whole of Mars; another is the similar though smaller basin of Argyre.

(Note that in general, the names originally given by Schiaparelli, and adopted by Antoniadi, have been kept. The old maps were accurate enough for this, and it was not necessary to introduce a completely new nomenclature, as had to be done with Mercury.)

All this was fascinating enough, but there was one thing which Mariner 9 could not do: confirm or deny the existence of any life on Mars. The next step was to bring a probe down gently enough for it to send back direct information – which brings me on to the Vikings.

The Vikings were twins, virtually identical in every respect. Each was made up of two parts, an orbiter and a lander. The plan was to keep the two parts joined together until they had been put into an orbit round Mars; then, at a command from Earth, the lander would be separated, and would come down through the planet's atmosphere, checking its descent partly by parachute and partly by rocket power, finally touching down at less than ten kilometres per hour.

One problem was that the whole surface is rock-strewn, and there was no way of plotting the positions of rocks which might be small by Martian standards, but were quite big enough to topple or destroy any space-craft unfortunate enough to land on top of them. Luck was needed as well as skill, and on this occasion luck was on NASA's side. The first lander (Viking Lander 1, or VL1) came down in the ochre

FAULT ZONES *(opposite) break the Martian crust in this view obtained by Viking 1 in July 1976 of an area two degrees south of the equator and near a potential landing site for Viking 2. The fault valleys are widened by mass wasting and collapse. Mass wasting is the downslope movement of rocks due to gravity and possibly hastened by seismic shaking (Mars quakes).*

tract of Chryse, the 'Golden Plain', on July 20, 1976; the other (VL2) followed, arriving in the more northerly plain of Utopia on September 3. Actually, VL1 came down only eighty metres away from a boulder which would have caused fatal damage if it had been hit.

The very first picture from Chryse, sent back immediately after touchdown, showed a barren, rocky desert, with extensive sand-dunes as well as pebbles and boulders. The fine-grained material so much in evidence was rusty red, and the sky was pink, because of the very fine dust-particles suspended in the atmosphere. Meanwhile, other experiments were being started. Analyses of the ground material showed that in composition it was not very different from the lunar *maria*. Temperatures were low, ranging from −31°C near noon to −86°C before dawn; wind-speeds were light, little over twenty kilometres per hour. The atmosphere consisted of 95 per cent of carbon dioxide, with less than 3 per cent of nitrogen and smaller amounts of oxygen and other gases. Water vapour was detectable, though the atmospheric pressure was much too low for liquid water to exist at surface level.

Yet there was abundant evidence of past water activity, and this was confirmed by the orbiting section of Viking 1, which continued the work begun by Mariner 9. Moreover, it was established that the polar caps contain a large quantity of ice. The northern cap is mainly water ice, and may be several kilometres deep. (So much for the idea current less than two decades earlier, when it was believed that the frosty coating could not extend down for more than a centimetre or two!) The southern cap contains both water ice and carbon dioxide ice, but

OLYMPUS MONS, *the great Martian volcano, (below) was photographed by Viking 1 on July 31, 1976 from a distance of 8,000 km. The 24 km high mountain is seen in mid-morning, wreathed in clouds, and the multi-ringed caldera which is some 80 km across, appears cloud-free. Olympus Mons is about 600 km across at the base and would extend from San Francisco to Los Angeles, or London to Edinburgh.*

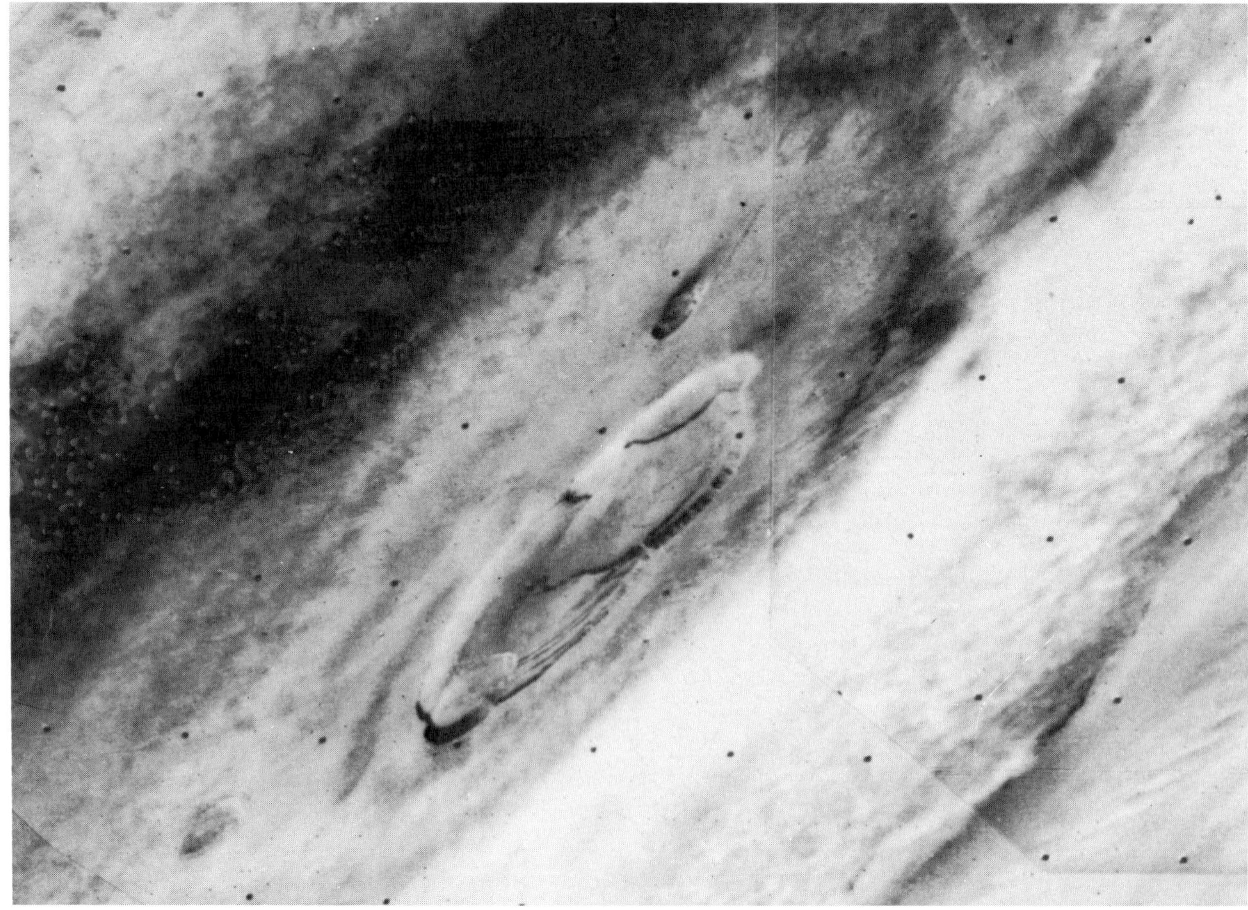

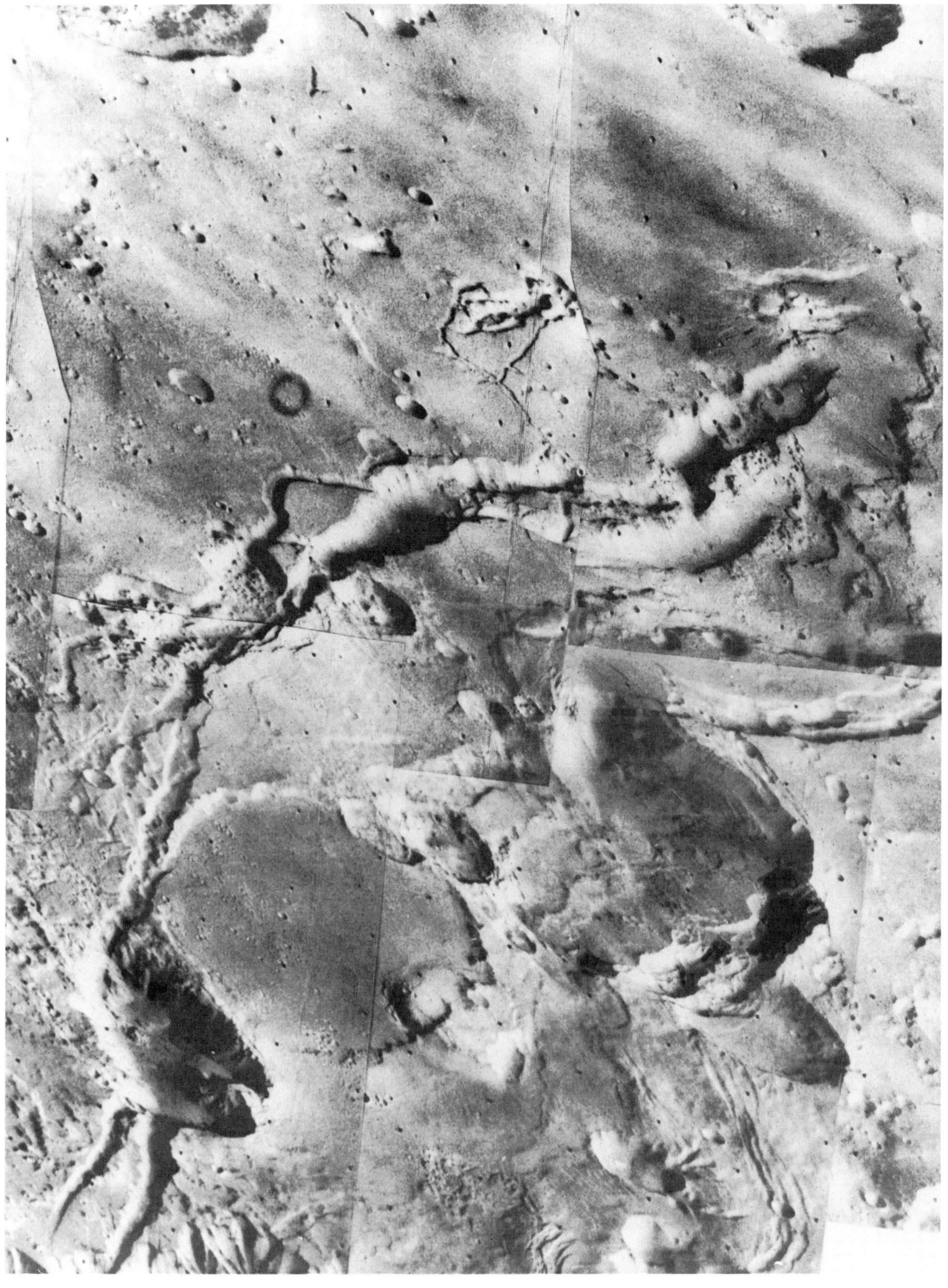

during Martian winter in either hemisphere some carbon dioxide is frozen out of the atmosphere and deposited at the polar cap, so that the atmospheric pressure as a whole falls slightly but perceptibly.

The method of searching for possible Martian life was to send out a scoop from the Lander, collect a sample of Martian 'soil', draw the sample back into the space-craft, and analyze it. There was an initial alarm with the grab of VL1, because a latchpin jammed and prevented the sample from being collected; after some manœuvring the pin was released and fell to the ground, where it was subsequently photographed. (It is ironical that a 7-centimetre pin almost wrecked one of the most complex and delicate experiments ever devised.) By July 28 a sample was secured, and testing began.

There were three experiments in all. What was done, in effect, was to provide the sample with nutrients. If any organisms were present, both the sample and the atmosphere in the chamber would be recognizably altered. Every possible precaution was taken, but the results showed no definite signs of any living thing on Mars. There

SUMMER DAY ON MARS, AUGUST 15, 1976. *Towards the left in this colour panorama of the Martian surface is a small dune of fine-grained material scarred by trenches dug by Viking 1's surface sampler on July 28 and August 3. The picture shows clearly the rocky terrain, and most particularly the bright reddish-orange colour of the surface of the planet, and the salmon-pink of the sky.*

VIKING 1 *(opposite) on the Martian surface, August 1976.*

PITS AND HOLLOWS ON MARS, *photographed from Mariner 9 in January 1971. The large closed basins to the top are each about 18 km across. It has been suggested, though not proved, that they are 'polar phenomena', perhaps due to the thawing of ice; the area is about 800 km from the Martian south pole.*

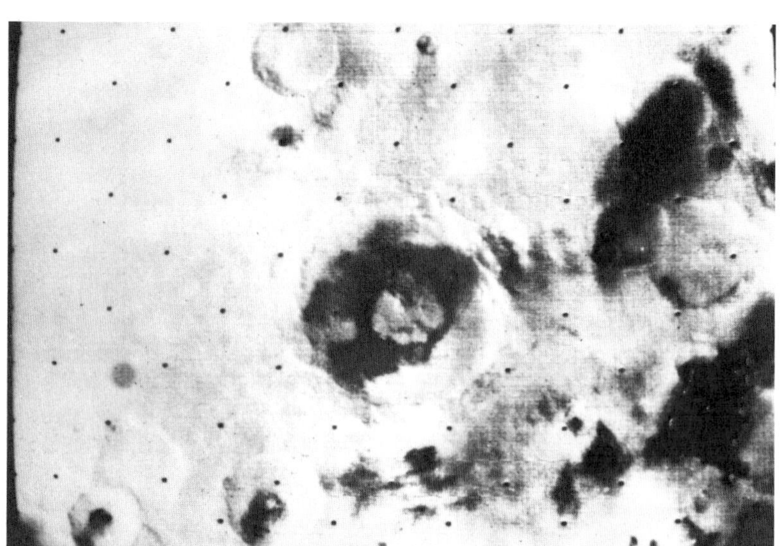

CRATERS ON MARS. *These dark-floored craters in the Phæthontis area of Mars were photographed from Mariner 9. The central crater has a diameter of over 110 km.*

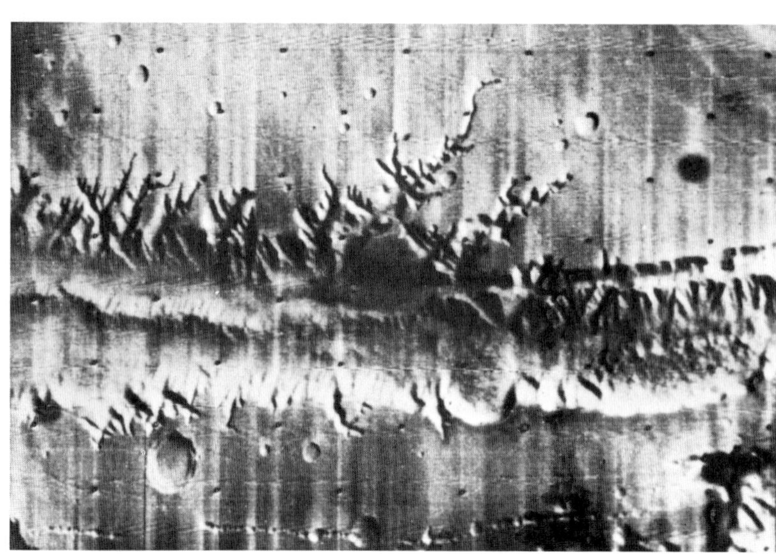

GREAT CHASM ON MARS. *This chasm in Tithonius Lacus, photographed from Mariner 9, is 480 km long and over 125 km wide. The 'tributaries' are in fact closed depressions*

were reactions indeed, but after making exhaustive tests the research team came to the conclusion that these reactions were chemical rather than biological.

It is always dangerous to be dogmatic, and we cannot say definitely that Mars is lifeless; but all the evidence seems to point that way, and if there is any Martian life it must be very lowly indeed.

What about past life? Here we cannot hope to give a final answer until we can obtain samples of the soil and analyze it in our laboratories, but this should be done within a few decades by using automatic probes, just as the Russians have done with the Moon. Meanwhile, it is a possibility which cannot be ruled out, because, as I have noted, there is strong evidence that Mars used to be less hostile than it is now. The dry riverbeds prove that.

Significantly, the riverbeds are not markedly eroded, which may indicate that by geological standards they are not very ancient; after all, the Martian atmosphere contains plenty of dust, and even fine dust is abrasive. According to one theory, Mars has a fluctuating climate because the angle of its axial tilt changes more noticeably than with the Earth. At present it is practically the same as ours (between 23 and 24 degrees), but over a cyclic period of about 50,000 years it ranges between 35 degrees and only 14 degrees. At times during this cycle both polar caps will be totally vaporized for part of the long Martian year, and Carl Sagan, who has more or less founded the new science of 'exobiology', has suggested that the atmosphere may be thickened sufficiently for rain to fall. Alternatively, it may be that Mars goes through periods of intense vulcanism, so that enough gas and vapour is sent out from the volcanoes to change the climate drastically.

If either of these ideas is correct, then Mars may again become less forbidding in the future, though subsequently things will revert to their present-day state. Personally, I doubt whether we will find Martian fossils, because life is notoriously slow to evolve, and it seems that the time-scale is wrong; Mars is never fertile enough for long enough to allow life to develop. But this again is something which we cannot hope to decide without handling samples in our laboratories.

As well as mapping the surface of Mars, the Viking orbiters also took close-range pictures of the two tiny satellites, Phobos and Deimos. From Earth they appear as nothing more than specks of light, but they were of special interest because of their movements. Phobos moves round Mars at only about 6000 kilometres above the surface, and its orbital period is only seven and a half hours; this is much shorter than a Martian day or 'sol', so that an observer on Mars would see Phobos rise in the west, cross the sky in an easterly direction, and set in the east only four-and-a-half hours later, while the interval between successive risings would be less than twelve hours. Deimos, moving at just over 20,000 kilometres above the surface of Mars, has a period of thirty hours; as Mars spins, therefore, Deimos almost keeps pace with it, and would remain above an observer's horizon for sixty hours consecutively, during which it would pass through its cycle of phases twice.

Mariner 9 obtained images of both, and the Vikings sent back close-range pictures which showed that both satellites are cratered. Phobos is an irregularly-shaped body, with a longest diameter of twenty-seven kilometres; Deimos, also irregular, has a longest diameter of no more than fifteen kilometres. Neither would be of much use as a source of illumination at night-time, and Deimos would look

AN OBLIQUE VIEW OF MARS, *photographed from Viking I on July 11, 1976, from a range of approximately 18,000 km. Argyre is the relatively smooth plain at top centre.*

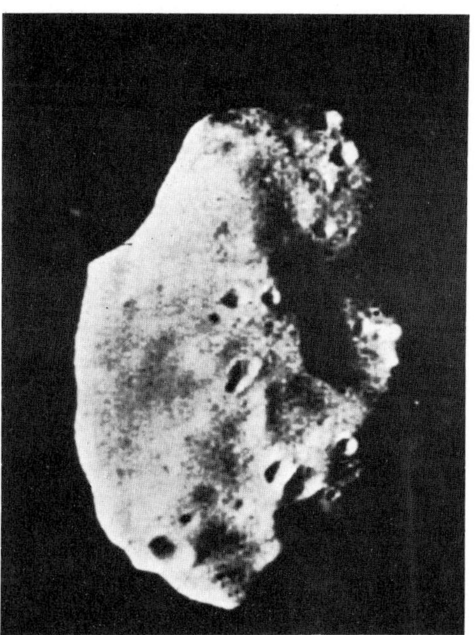

PHOBOS, *photographed from Mariner 9 in 1971.*

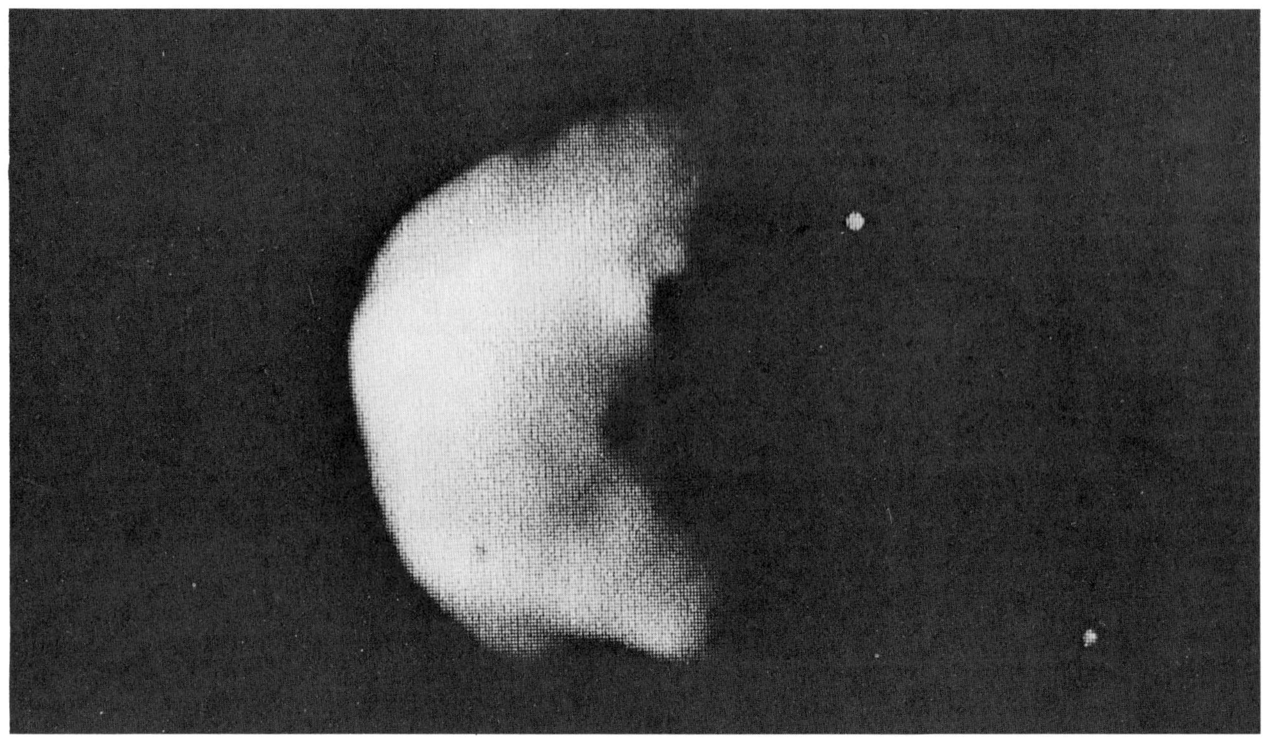

DEIMOS. *The outer satellite of Mars, photographed from Mariner 9. Its greatest diameter is only about 11 km.*

only slightly brighter than Venus does to us. They are quite unlike our massive Moon, and there is a good chance that they are former members of the asteroid belt which were captured by Mars in the remote past.

Way back in 1901, a Frenchwoman named Guzman offered a substantial prize to be given to the first person to establish contact with beings from another world – excluding Mars, which was regarded as too easy! We have learned a great deal since then, and we know that the Red Planet is not a world suited to life of our kind. Yet it is less unlike the Earth than any other planet in the Solar System, and if all goes well it will one day be reached. The new 'Martians' will have arrived.

37

Through the Asteroid Belt

A MISSION to any one of the inner planets takes a few months at most. When considering journeys to the outer members of the Sun's family, the time-scale is not only longer, but very much longer. To recapitulate: the giant planets are very widely spaced. Suppose that it is planned to go straight from the Earth to the orbit of Neptune, moving in a direct line (something which cannot be done in practice, of course); how much of the journey will have been completed when Jupiter's orbit is crossed? The answer is: only about one-sixth.

The asteroid belt is another hazard. Before the Pioneer missions of the early 1970s there was considerable apprehension about this swarm of dwarf worlds. The zone had to be crossed, and a collision between a space-ship and an asteroid could have only one result.

Fortunately the danger seems to have been overrated. Up to now four probes have been through the zone, two Pioneers and two Voyagers, and there has been no damage, so that the asteroids may be more widely separated than had been thought. The large ones were known, and could be avoided; the problem was with the small ones, many of which have not been discovered and which are probably too faint to be seen from Earth.

Actually, interest in the asteroids has recently been concentrated upon those with exceptional orbits, some of which can come close to the Earth. Eros was the first to be found (in 1898, by Witt of Berlin), but before long others were detected, and are known as the Apollo asteroids – the qualification being that an Apollo can cross the orbit of the Earth (which Eros, incidentally, cannot do, though it may approach us within 25 million kilometres). There are even a few asteroids, those of the Aten group, whose orbits lie wholly inside that of the Earth. Collisions may be expected occasionally, and could do a great deal of damage even though the average Apollo is only a few kilometres in diameter.

So far, the holder of the close-approach record is a midget asteroid, Hermes, which brushed past us in 1937 at a mere 780,000 kilometres. I still have a copy of the London *Daily Mirror* for the following January 10, when the news was released. The headline reads FIVE HOURS SAVED WORLD DISASTER, and the article continues: 'Five and a half hours saved the world from annihilation... Horror-stricken astronomers in different parts of the world saw the planet crashing straight toward the earth. They watched it until it was within 400,000 miles of the earth's crust. Then it veered off and rushed by.' In fact there was never any danger that Hermes would hit us, but can we always have been similarly immune? This brings me on to a revolutionary new theory which involves not only Apollo asteroids, but also comets.

The theory is due to two distinguished astronomers of the Royal Observatory Edinburgh; Victor Clube and Bill Napier. Unlike most astronomers, they believe that comets are not bona-fide original members of the Solar System, but are interstellar wanderers, many of which are captured by the Sun each time we pass through a spiral arm of the Galaxy. These captured comets are forced into orbits round the Sun at a distance of a light-year or so, and therefore there is a cloud or reservoir of comets at this distance. Such a reservoir had been suggested much earlier by the Dutch astronomer Jan Oort, and is always termed the Oort Cloud, though Oort himself believed that it had existed since the formation of the Solar System.

On the Clube-Napier theory, the Oort cloud is replenished at every passage through a spiral arm, but not all comets stay in the Cloud. If a comet is perturbed for any reason, it starts to fall inwards towards the Sun. It may even be destroyed – as happened to Comet 1979 XI, which was followed in towards perihelion but was not seen afterwards, so that it cannot have emerged; there have been several similar cases of cosmic hara-kiri. If the comet from the Oort Cloud escapes this fate, it may well swing back into a very elliptical orbit which will, after centuries or hundreds of centuries, take it back to the Cloud before it turns sunward again. But if the comet is violently perturbed by a planet, it may be forced into a smaller orbit, and becomes a periodical comet. (Jupiter, as the most massive planet, is usually the culprit.)

Every time a comet passes perihelion it loses mass, because of the evaporation of its ices. This is why most periodical comets are faint. Eventually the comet will lose all its gas, and nothing but the dead nucleus will remain. According to Clube and Napier, Apollo asteroids are nothing more nor less than ex-comets, in which case they are entirely different in nature from the larger asteroids in the zone between the orbits of Mars and Jupiter.

Once again following Clube and Napier: when the Sun emerges from a spiral arm, it is attended by large numbers of comets, and therefore, for the next few million years, comets are plentiful. So are Apollo asteroids. Now and then an Apollo will strike the Earth, and if it is large enough it may well cause violent climatic changes. It is suggested that some collision of this sort happened 65,000,000 years ago, at the time when there was a great change in life on Earth. The huge dinosaurs, which had been lords of the world for so long, died out – not gradually, but in a sudden 'extinction'. Dramatic cooling, caused by cloudiness as a result of an Apollo collision, could have been the basic cause. And it is true that the rocks on Earth laid down 65,000,000 years ago contain more than their fair share of the element iridium – which is a constituent of meteorites, and presumably of Apollo asteroids as well.

Clube and Napier believe that we can continue the story to near-historical times, and that men of thousands of years ago were periodically alarmed by a brilliant comet in an Earth-crossing orbit; they name this comet the 'Cosmic Serpent'. Each time it appeared, Apollo collisions would be likely, which could account for the age-old fear of comets – and the associated meteor stream might even cause long-term dimming of the light received from the Sun.

Centuries passed. The Cosmic Serpent faded; it lost its gas, and became a cosmic corpse, known to us today as the 10-kilometre Apollo asteroid Hephaistos, discoved in 1979. Calculations by F.L. Whipple in America had already indicated that there might have been a break-up of a single object in the third millennium B.C., so that today we see its remnants in Hephaistos, Encke's periodical comet, and the meteor shower known as the Beta Taurids. Moreover, it is suggested that a missile which landed in Siberia in 1908, blowing pine-trees flat over a wide area, was a fragment of Encke's Comet, which was mainly icy in nature and which therefore produced no impact crater.

The theory is as startling as it is spectacular. Whether or not it is correct remains to be seen, but if Clube and Napier are right there is every reason to assume that there will be more Serpents, more Apollo asteroids, and more disastrous collisions next time the Solar System passes through a spiral arm. Luckily this is not imminent, and will not happen for millions of years yet, so that eventually we may hope to be able to cope with the situation rather better than the dinosaurs did!

Returning now to the main asteroids, we must note the Trojans, which travel in the same path as Jupiter, though they keep prudently to points either 60 degrees ahead or 60 degrees behind the Giant Planet and are in no danger of being pulled into it. Points such as these are stable; they are known as Lagrangian Points after the famous 17th-century French mathematician Joseph Lagrange. Hector, the largest of the Trojans, seems to be very irregular in shape, with a maximum diameter of about 150 kilometres. It may even be made up of a pair of asteroids almost in contact.

Early in the 1970s, the planners at NASA worked out an ambitious scheme. It was known that for a few years around 1980 the giant planets would be so placed that it would be theoretically possible to send the same probe to each in turn. The original trajectory would take the space-craft from the Earth out of the neighbourhood of Jupiter. It would swing round Jupiter, picking up velocity in the process, and would use Jupiter's gravity as an extra propellent, 'throwing' it outwards toward a rendezvous with Saturn. Saturn would in turn send the probe on to Uranus, and Uranus to Neptune. This became known as the Grand Tour. Obviously it depended upon the giants being in the right place at the right time – something which happens only once in a couple of centuries, so that clearly the chance was not to be missed.

Then came one of the weirdest theories of modern times – the so-called Jupiter Effect, proposed by Dr. John Gribbin, of the New Scientist magazine, in 1972. With his co-author Stephen Plagemann, Gribbin described a close lining-up of planets which, he said, would occur in 1982. The combined gravitational pulls would affect the Sun, which would in turn affect the Earth and cause storms, cyclones, tornadoes and violent earthquakes. Gribbin's book was widely publicized, and caused so much alarm that official disclaimers had to be put out by various scientific bodies, including the Royal Greenwich Observatory and, in America, the Astronomical Society of the Pacific. In fact, any knowledge of rudimentary mathematics is enough to show that the effects of such an alignment would be absolutely undetectable – and in any case, there was no alignment in 1982; the planets were spread out through almost 90 degrees! Not to be daunted, Gribbin described another theory in October 1982. This time the planets were to be lined up on the same side of the Sun in what he termed a 'synod', so that each time the Earth came round to this side of the Sun it would be slightly speeded-up; the effect would make for a cold spell which would in time develop into a full-scale Ice Age.

Frankly, this idea must be put into the same category as the first, and it is not easy to see how either could be taken seriously, but any theory, however peculiar, will have its supporters. Let us remember Dr.

JUPITER FROM PIONEER 10, *December 1973. The Great Red Spot is well shown, and also a satellite shadow.*

Immanuel Velikovsky, a Russian-born psychiatrist who emigrated to America, and in the 1950s revised the whole of cosmology on the grounds that the planet Venus was originally ejected from Jupiter and became a comet, periodically approaching the Earth and causing such events as the Biblical Flood before having its tail chopped off during an encounter with Mars and turning into a planet. For a while, indeed, Velikovsky became something of a cult figure, and his ignorance of astronomical science was so profound that it was impossible to argue with him.

With these asides, let us return to the space-probes.

Pioneer 10 was launched from Cape Canaveral on March 2, 1972, with Jupiter as its target. There was some apprehension at the idea of passing through the main asteroid belt, which was impossible to avoid, but no disasters resulted, and since three more probes have since made safe crossings it may well be that small asteroids are less numerous than had been feared. On December 3, 1973 Pioneer 10 by-passed Jupiter at a distance of 131,400 kilometres, and sent back more than three hundred pictures, as well as carrying out detailed investigations of all kinds. Various long-standing problems were cleared up at once, and our knowledge of Jupiter was increased beyond all recognition.

The idea of Jupiter as a dwarf sun had been abandoned, following a brilliant series of papers by Sir Harold Jeffreys in the 1920s, but the internal make-up of the planet was still not known with any certainty. Spectroscopic observations carried out in 1932 by Rupert Wildt, in America, showed that the outer gases contained large amounts of ammonia and methane, which was not surprising; both these are hydrogen compounds, and it was reasonable to assume that Jupiter consisted largely of hydrogen. On Wildt's model there was a rocky core 60,000 kilometres in diameter, overlaid by an icy shell 27,000 kilometre thick, above which lay the hydrogen-rich atmosphere. Then, in 1951, W. Ramsey in England proposed a different model. This time the 120,000-kilometre core was made up of hydrogen, so compressed that it acquired the characteristics of a metal. The core was surrounded by an 8,000-kilometre layer of ordinary solid hydrogen, above which came the atmosphere.

Pioneer confirmed that neither of these models is correct. Jupiter seems to have a relatively small solid core, with a mass of from ten to twenty times that of the Earth. Above this is a thick shell of liquid metallic hydrogen. At about 46,000 kilometres from the centre of the planet there is a sudden transition from liquid metallic hydrogen to ordinary liquid molecular hydrogen, and then comes the gaseous atmosphere, about 1,000 kilometres deep, made up of 82 per cent hydrogen, 17 per cent helium, and only about 1 per cent of other elements. It contains water droplets, ice crystals, and crystals of ammonia and other compounds. Therefore, most of Jupiter is liquid; nothing could be more different from the small, solid inner planets.

The dark belts were found to be regions where gases are sinking. The bright zones mark areas where gases, warmed by Jupiter's internal heat, are rising into the upper atmosphere, where they cool down to form clouds of ammonia crystals floating in the gaseous hydrogen. The Great Red Spot, once thought to be a kind of 'island', has proved to be a whirling storm – a phenomenon of Jovian meteorology, spinning round anticlockwise and dominating the whole of that latitude of the planet. The colour may be due to nothing more nor less than red

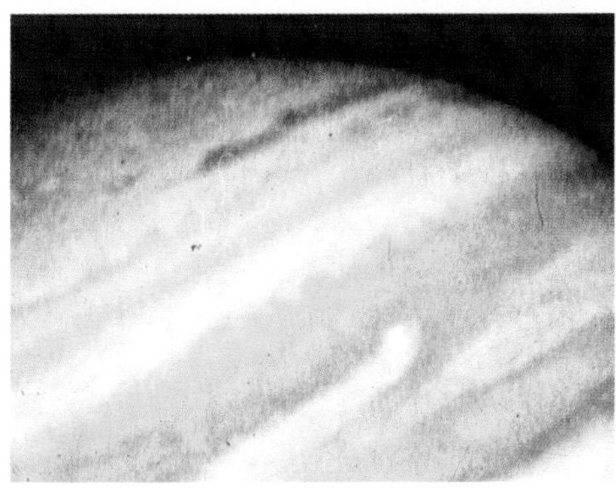

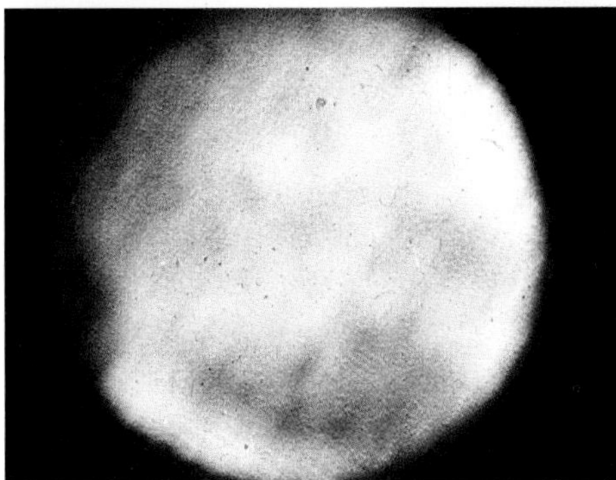

JUPITER FROM PIONEER 10. *(Top) Crescent Jupiter – a view never seen from Earth (2) Surface, showing one of the white 'plumes' (3) Another surface view (4) Ganymede, the third satellite of Jupiter, showing some distinct dark and brighter areas.*

phosphorus. The Pioneer 10 pictures also showed several of the long-lasting white ovals which had been traced by visual observers ever since the 1930s.

It was also confirmed that Jupiter sends out more energy than it would do if it depended solely upon what it receives from the Sun. The central temperature may be as much as 30,000 degrees, which is much too low to trigger off nuclear reactions and turn Jupiter into a star, but is still very appreciable. It had been suggested that this might be due to a slow contraction of the globe, which would release energy gravitationally, but if Jupiter is mainly liquid some other cause must be found. One cannot compress a liquid; if you doubt me, try squashing a glass of water! It now seems likely that the inner heat has been 'left over', so to speak, from the time when Jupiter was being formed from the solar nebula.

Jupiter has a very strong magnetic field. The magnetic axis is displaced from the rotational axis by 11 degrees, and is reversed relative to ours, so that a compass needle would point south instead of north. In addition there are zones of dangerous radiation, not unlike our own Van Allen belts, but much more powerful. Pioneer 10 passed within about 129,000 kilometres of the cloud-tops; had it gone any closer, the instruments would have been put out of action, and the main mission would have failed. Orbits of later space-craft were altered, so that the time spent over the most dangerous zone (that of Jupiter's equator) was as short as possible.

Jupiter, then, was a world of ceaseless activity, and when Pioneer 11 made its pass of the planet almost exactly a year later, on December 2, 1974, marked changes in the surface features had occurred. In general Pioneer 11 confirmed all the findings of its predecessor, but its work was not over. Using the 'interplanetary billiards' technique, it was put into a path which swung it back across the Solar System and then out to a rendezvous with Saturn in 1979. But by then, two even more ambitious space-craft were on their way.

The Voyagers were sent up in 1977. No 2 went off first, on August 20; Voyager 1 was launched on September 5, but was following a more economical orbit, so that it reached Jupiter on March 5, 1979, while Voyager 2 did not arrive until the following July 9. Once again the vividly-coloured, restless face of Jupiter was studied, together with the magnetic field, the radio emission and much else; but the most spectacular findings came not from the planet itself, but from its surroundings.

Jupiter proved to have a ring. It is not the same as Saturn's; it is thin and obscure, made up of darkish particles, and is too dim to be seen from Earth. It may not even be permanent. But it certainly exists, and has caused the theorists to do some very hard thinking.

As the Voyagers passed round the night side of Jupiter they recorded brilliant auroræ. There were also lightning flashes, so that evidently thunderstorms are going on all the time; Jupiter is a noisy world.

Twelve satellites were known by this time. Of course there were the four Galileans (Io, Europa, Ganymede and Callisto), together with Amalthea, which had been discovered by Barnard at the Lick Observatory in 1892. The other satellites, all very faint, had been found photographically. They fell into two groups. Leda, Himalia, Lysithea and Elara moved at between 11 and 12 million kilometres from Jupiter; Ananke, Carme, Pasiphaë and Sinope at between 20 and

24 million kilometres. The outer four had retrograde motion, so that they were often regarded as captured asteroids rather than bona-fide satellites. Only Himalia was as much as 100 kilometres in diameter, and the orbits of all these midget attendants were so strongly perturbed by the Sun that they cannot be regarded as even approximately circular.

The Voyager pictures showed up several more small inner satellites. The present total is sixteen, and no doubt more remain to be discovered. Amalthea was irregularly-shaped, reddish and cratered. But the main attention was focused upon the Galileans.

All these, you will recall, are of planetary size. Europa is only a little smaller than our Moon, Io rather larger, and Ganymede and Callisto much larger; Ganymede exceeds Mercury in size, though admittedly not in mass. A certain amount of surface detail had been recorded on all four, mainly by Audouin Dollfus and his colleagues using the 61-centimetre refractor at the Pic du Midi, but it was very vague. Most astronomers tacitly assumed that the Galileans must be very like each other – probably icy and cratered, though there were nagging doubts in the case of Io, which was known to have a definite effect upon the long-wavelength radio radiations coming from Jupiter.

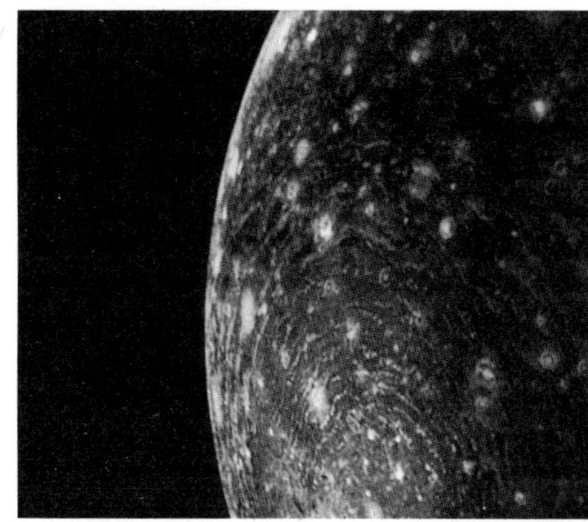

CALLISTO. *Voyager picture showing part of the huge ringed structure, Valhalla.*

Nothing could have been further from the truth. The Galileans are not alike, and in some ways they are the most astonishing bodies in the whole of the Solar System.

Callisto, the outermost of the four, proved to have an icy surface, so crowded with craters that there seems no room for any more; there is a huge ringed structure, now named Valhalla, and a rather smaller one, Asgard. It is thought that there is a thick ice and rock crust going down to 300 kilometres, below which there is a water or ice mantle surrounding a solid, rocky core. Undoubtedly Callisto is absolutely inert, and has been so for thousands of million of years, so that it is truly a cosmical museum.

Ganymede, rather larger and slightly denser, also has an icy surface, but as well as craters there are darker patches, the largest of which has been fittingly named Galileo Regio. There are also many grooves between five and ten kilometres wide, and though no activity takes place now it seems that there have been crustal movements in the past, so that the surface is less ancient than Callisto's. The icy crust has a probable depth of 100 kilometres or so, and the rocky core may have a diameter of at least 4000 kilometres.

Next, Europa – and a major shock. It was not well seen from Voyager 1, but Voyager 2 passed within 205,000 kilometres of it, and revealed a terrain which was completely unexpected. There are almost no craters; it has been commented that the surface is as smooth as a billiard-ball, and though there are immense numbers of bright and dark lines, criss-crossing each other, there is a startling lack of vertical relief. Again the surface is icy, but the crust may give way at a depth of around seventy kilometres to a region of slushy water-ice surrounding the rocky core. We do not pretend to know why Europa is so smooth. If there have ever been craters, they must have been destroyed – but how? Soft ice or water welling up from below the crust is one suggestion, but in its own way Europa is as puzzling a body as we have ever found.

Yet even Europa pales before Io. This time the surface is brilliant red, and has been compared with an Italian pizza! Sulphur is everywhere; there are calderas, lava-lakes and active volcanoes, hurling

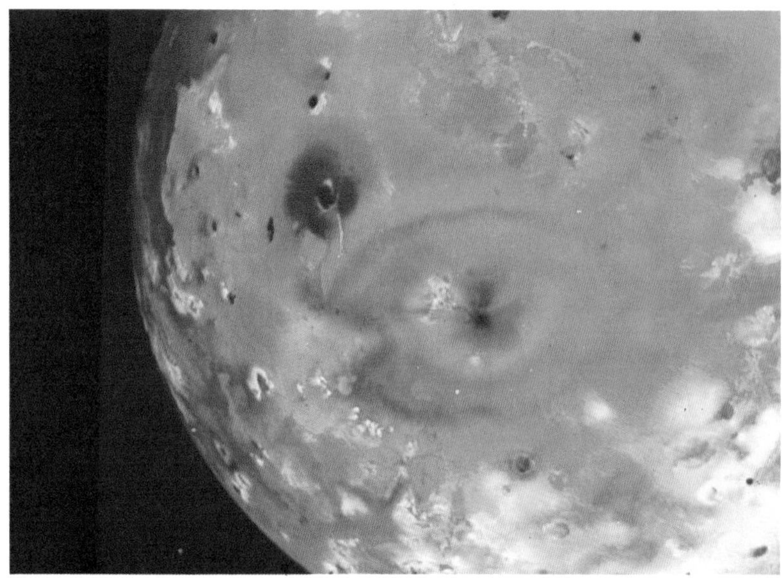

Io. *A Voyager view showing the brilliant red sulphur surface and the active volcanoes.*

material high above the surface. The volcanoes, subsequently given picturesque names such as Pele, Loki and Prometheus, altered both in form and in level of activity between the two Voyager passes, but it is abundantly clear that Io is a world in constant turmoil. According to one model there is an upper crust of frozen sulphur about a kilometre deep; below this is a three-kilometre deep liquid sulphur ocean, below which again is a molten silicate interior overlying a rocky core.

It is not easy to explain why Io is so highly volcanic – always bearing in mind that Ionian volcanoes are not of the same type as those of the Earth (or, for that matter, Mars or Venus). The only reasonable suggestion made so far is that the interior is being constantly 'churned' by the changing gravitational pull of Jupiter, but there is no sign of anything comparable with Europa, which is less than 300,000 kilometres further out. At any rate, Io is a world to be avoided. It moves in the thick of Jupiter's radiation zones, and contributes to them by its volcanic activity; it may even play a major rôle in the production of Jovian auroræ, and there is a very powerful electric current flowing between it and the planet.

If it were possible to step out on to the surface, an astronaut would be greeted with a spectacle which defies description, Sulphur dioxide produces an incredibly thin, evil-smelling atmosphere; the surface is pelted with pastel yellow, orange and blue 'snow' as blobs of sulphur rain down, and nearer the equator there may be patches of sulphur dioxide frost, together with crusted-over black lakes produced by floods of black, molten sulphur flowing out from fissures and then freezing. All in all, Io may lay claim to being the most lethal world in the entire Solar System.

Such were the findings of the two Voyagers. They had been magnificently successful, and had more than fulfilled the hopes of their makers, but their main work was not over. Jupiter, with its belts, its ring, its Red Spot and its incredible satellite family, was left behind. Saturn was next on the list.

38

Saturn

On August 3, 1933, Saturn was being studied by a most unusual astronomer: William Thompson Hay, whose observatory had been built at Norbury in Outer London and equipped with a fine 15-centimetre refractor. Hay was a skilful amateur astronomer, but he was better known as Will Hay, the stage and screen comedian. Many people remember him today for his rôles in films such as *Oh, Mr. Porter!*

On this occasion Will Hay was particularly interested in what he saw. Near Saturn's equator there was a bright white spot, unlike anything that had been previously recorded. During the next few weeks the spot gradually lengthened, and the portion of the disk following it darkened, as though material were being thrown up from beneath the clouds. After a month or two the spot had lost its identity, but it had shown once again that Saturn, like Jupiter, is a restless world.

As with Jupiter, so with Saturn; Jeffreys' papers of the 1920s showed that it could not be a miniature sun, and Wildt proposed the same kind of model as with Jupiter, involving a rocky core, a thick layer of ice, and then the hydrogen-rich atmosphere. The rings were believed to be composed of ice particles, and this was supported by some experiments in 1972, when the rings were first contacted by radar. Meanwhile, Gerard Kuiper had looked at Saturn with the Hale reflector, on one of the rare occasions when it was used visually instead of photographically, and had concluded that the famous Cassini Division was the only genuine gap; the others, including Encke's Division, were dismissed either as mere ripples or else as illusions.

The belts are much less prominent than those of Jupiter, and there are few well-marked spots; Will Hay's was very much an exception, and even that did not last for long. Nine satellites were known. Phœbe had been added to the list in 1898 by W.H. Pickering (incidentally, the first satellite discovery to be made photographically), and Pickering had also reported another satellite; it was named Themis, but was never seen again, and probably does not exist. In 1966, when the ring system was edgewise-on to the Earth and the satellites were particularly well placed for observation, Audoin Dollfus at the Pic du Midi announced the detection of yet another satellite, moving within the orbit of Mimas, the closest satellite previously known. It was given the name of Janus, and included in most official lists, though some astronomers remained dubious about it.

Extra rings, too, had been reported. In 1909 the French astronomer Fournier announced the discovery of a ring outside the main system, and there was also the controversial Ring D, between the Crêpe Ring and the cloud-tops. The first account of it was given by another French astronomer, P. Guèrin, and was regarded as definite, but I admit that I was sceptical. I failed to find Ring D even when using the Lowell and Meudon refractors, and I felt that I would have seen it had it really been there.

The first space-craft to fly past Saturn was Pioneer 11, in September 1979. It passed 20,200 kilometres above the outer clouds, and sent back some tantalizing pictures. So far as Pioneer 11 was concerned, Saturn was very much of an afterthought; it had been put into an encounter orbit because following the success with Jupiter, there was sufficient power left, and the opportunity was really too good to be missed. The outer F ring was confirmed, though not Guèrin's D ring. The two inner satellites were found, but neither corresponded to Dollfus' 'Janus'.

Pioneer showed that although Saturn is in some ways similar to Jupiter, it is not simply a smaller version of it with rings added. There is a solid core, surrounded by a shell of liquid metallic hydrogen which is smaller than Jupiter's both absolutely and relatively; outside this comes a deep layer of liquid molecular hydrogen, and then the outer atmosphere. The core is hot, though the temperature is lower than that of Jupiter. Also, the basic cause of the excess of energy sent out seems to be different. There has been ample time for Saturn to have cooled down since its formation, and it is possible that the extra energy is produced by helium in the outer layers sinking down through the lighter hydrogen, releasing gravitational energy as it does so.

As Pioneer 11 moved away from Saturn to begin its never-ending journey among the stars, the Voyagers were closing in. No. 1 was well in the lead, and by the beginning of November 1980 it was coming within range. I was at the main control centre, the Jet Propulsion Laboratory at Pasadena, California, and neither I nor anyone else who was present will ever forget the excitement as one incredible discovery followed another.

The face of Saturn was very much as had been expected; it was relatively bland, because of the greater amount of 'haze' above the cloud-tops as compared with Jupiter, but various belts were seen, as well as spots and ovals. One distinctly reddish spot was 12,000 kilometres long, much the same size as some of Jupiter's white ovals. The wind-speeds in the equatorial zone were very high indeed, reaching over 1500 kilometres per hour. But it was the rings which attracted most attention.

There was not simply one division, as Kuiper had believed, but many hundreds! It was said that the rings had more 'grooves' than a gramophone record, and there were even ringlets in the Cassini Division, which had been expected to be more or less empty. Two of the minor ringlets, one inside the main Crêpe Ring and the other in a dark gap at the outer edge of the Cassini Division, were slightly eccentric instead of being perfectly circular, and the outer F ring was 'clumpy'; it seemed almost to be braided, as though two or three separate rings had become intertwined. Another dim ring, lettered E, lay outside the F ring. Guèrin's D ring did not exist as such (which made me glad that I had not tricked myself into 'seeing' it, as some observers had done). There is ring material between the Crêpe Ring and the cloud-tops, and this is sometimes known as the D ring, but it is quite invisible from Earth.

The discovery of the 'grooves' effectively disproved a theory which had been almost universally accepted. It had been thought that the Cassini and Encke Divisions* were due to the gravitational pulls of the inner satellites, particularly Mimas, so that every time a particle moved into the 'forbidden zone' it was systematically perturbed until it had

*Some astronomers now call Encke's Division the 'Keeler Gap', after the American astronomer J.E. Keeler, who made careful studies of it, but there seems no reason to alter the long-established name.

moved out again. But this cannot account for the hundreds of minor divisions, and even though Mimas and the other satellites may be involved in some way they cannot be solely responsible. We must think again.

Curious radial spokes were seen crossing the brightest ring, B. (I am here referring to what we normally call the B ring, now known to be made up of many individual ringlets.) They persisted for hours; they came into view upon emerging from Saturn's shadow, and when they disappeared they were replaced by new ones. Logically they should not exist, because, following Kepler's Laws, the inner ring particles move round Saturn faster than the outer ones, so that any radial feature should be broken up quickly. At the time I suggested that they might be due to small particles 'lifted' away from the main plane of the rings by magnetic or electrostatic forces, but they are certainly very odd indeed. Voyager 1 was unable to show individual particles in any of the rings, but it seems that the average diameter for a particle in the main system is of the order of 8 to 10 metres, though in the Crêpe Ring it is rather less.

Several new inner satellites of Saturn were found. Two of them, known provisionally as S.13 and S.14, move close to opposite boundaries of the curious F ring, and 'box in' the particles, so to speak, forcing them back into the ring area if they try to escape. Each satellite is around 200 kilometres in diameter, and, like the ring particles, they are probably made up chiefly of ice. Satellite S.15 moves at only 800 kilometres outside the edge of the main bright ring, A. Further out (149,400 kilometres from the centre of Saturn) are the two satellites Calypso and Telesto, which are in almost identical orbits. Periodically they approach each other, come within a few kilometres or tens of kilometres of each other, and then actually exchange orbits. Both are very irregular in form, with longest diameters of not more than 100 kilometres. It seems highly probable that they are the two halves of a former larger satellite which met with disaster. One of them may have been Dollfus' original 'Janus'.

Another satellite, S.12, was found to move in the same orbit as Dione, one of the familiar satellites – discovered by Cassini as long ago as 1686. S.12 is at one of Dione's Lagrangian points, so that it keeps at a safe distance. Subsequently, examination of the Voyager pictures revealed two more small satellites, S.16 and S.17, moving in the orbit of Tethys; S.17 is 60 degrees ahead of Tethys, and S.16 is 60 degrees behind it.

Saturn's satellite system is completely different from that of Jupiter. Much the largest member of the family is Titan, which has a diameter of 5140 kilometres – slightly smaller than was thought before the Voyager mission, and not quite the equal of Ganymede in Jupiter's system, but still comparable with the planet Mercury. The overall density of the globe is only about twice that of water, so that there may be equal amounts of ice and rock. It was already known that Titan has an atmosphere; this had been discovered by Kuiper spectroscopically, as early as 1944, and it was tacitly assumed to be made up of methane (marshgas). As Voyager 1 closed in, we at the Jet Propulsion Laboratory hoped to be able to see details, particularly since the space-craft was to pass at only 5000 kilometres from Titan's surface.

We were disappointed. Nothing could be seen except an orange blob. Several distinct, detached haze layers could be made out above

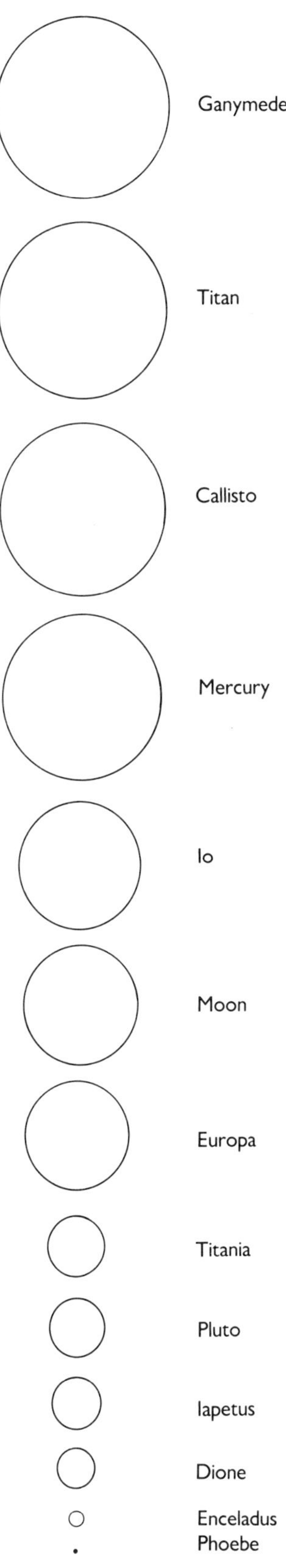

RELATIVE SIZES OF SOME SATELLITES *together with the planets Mercury and Pluto.*

the opaque layer, and these merged into a dark hood over Titan's north pole, but that was all, though it looked as though the southern hemisphere was slightly brighter than the northern.

Instead of being composed of methane, the Titanian atmosphere proved to be made up chiefly of nitrogen, with methane present in smaller quantities. The atmosphere is denser than ours; the ground pressure is about 1·6 times that of the Earth's air at sea-level. The surface temperature of about –180 degrees Centigrade is close to the 'triple point' of methane, i.e. the temperature at which methane can exist either as a solid, liquid or gas. On Titan, therefore, methane may play the same rôle as water does on Earth, as rain, snow, ice and vapour; there may be a methane rain from the clouds, with rivers of liquid methane and methane-ice cliffs. Organic molecules based upon methane no doubt exist, but the low temperature is an effective barrier to the development of life as we know it. Finally, Titan itself has no detectable magnetic field. It lies at the edge of Saturn's somewhat variable magnetosphere, so that it is sometimes inside and sometimes outside.

MIMAS, *from Voyager 1. Note the large crater.*

The other satellites studied from close range from Voyager 1 were Rhea, Dione and Mimas; less distinct views were obtained of Iapetus, Tethys and Enceladus. Rhea and Dione proved to be cratered, but Dione also showed curious branching valleys and some smooth plains, so that internal activity had presumably occurred in the past; moreover, Dione is considerably denser than most of the other satellites, so that it seems to contain more rock. Rhea, too, showed wispy streaks on its bright surface. Mimas was something of a surprise. Most of its icy surface is cratered in the usual way, but there is one huge crater with a central peak, about one-third the diameter of the satellite itself.* If this crater were formed by an impacting body, it is hard to see how Mimas itself escaped being shattered into fragments, particularly as its overall density is only about 1·2 times that of water. Tantalizing pictures of Tethys, at lower resolution, showed a tremendous valley over 700 kilometres long and at least 60 kilometres wide, while Enceladus appeared to be smooth and almost craterless, as with Europa in Jupiter's system. Iapetus was confirmed as having one bright and one dark hemisphere, as had long been suspected from its regular variations in brightness.

Voyager 1 had completed its task. It had approached Saturn within 126,400 kilometres of the cloud-tops, and had performed almost perfectly throughout. It would encounter no more planets, but would simply move out of the Solar System, never to return. We should continue to keep in touch with it until the early 1990s, but eventually we will lose it. Millions of years hence, it may still be wandering in the depths of interstellar space, unseen, unheard, and untrackable. Meanwhile, its twin, Voyager 2, was well on its way to a rendezvous with Saturn, and there was ample time to re-programme it so as to take full advantage of what Voyager 1 had found.

The Voyager 2 pass took place on August 24, 1981. The rings were seen at much higher resolution, and the numbers of individual ringlets were increased; there were thousands of them, not only hundreds.

*I was in the Press Room at J.P.L. when the picture of Mimas was first shown. It did look amazingly like the *Death Star* in the famous film *Star Wars*, and one reporter called out: 'Gee, that crater must be the engine!' The crater has now been named 'Herschel'.

The edges of the extraordinary spokes in the ring system were found to be very sharp, and the idea of particles elevated by magnetic or electrostatic forces gained in favour. (Incidentally, Saturn's magnetic axis is within one degree of the axis of rotation, so that the situation is markedly different from that on Jupiter.) There had been perceptible changes in the surface features since the previous encounter, but some of the spots were identifiable, particularly one large, bright, 2500-kilometre oval. Another huge brown spot appears to be in the nature of an anticyclonic storm, moving eastward at thirty metres per second inside a generally westward flow.

This time Iapetus was seen from close range. As had been expected, one hemisphere was as bright as ice and the other blacker than a blackboard. Since the overall density of Iapetus is so low, it seems certain that the true surface is bright and that the dark patch is caused by a deposit – but from where does this deposit come? One suggestion was that it came from the remoter satellite Phœbe, and that the dark material had been simply 'floated in' to Iapetus; but this does not seem very probable, partly because Phœbe is so small, partly because it is a long way out and partly because it is hard to see how so much material could have been removed from it. Phœbe, like the outer small satellites of Jupiter, has retrograde motion, so that it may well be a captured asteroid.

If no external force has been responsible, then the dark material must have welled up from inside Iapetus, but we may not know the answer until we can measure the thickness of the deposit, which Voyager 2 was unable to do.

The next surprise came with Hyperion, closer-in than Iapetus, which proved to be irregular in shape; as I have noted earlier, it was even likened to a hamburger! It measured about 120 kilometres by 60 kilometres, and there were suggestions that it might be the half of a larger body; but, as I at once asked, if this is true, then where is the other half? Enceladus was seen to be smooth, with only a few craters; if Mimas and the other satellites had been so bombarded with meteorites it is hard to see how Enceladus can have escaped, and once more we may have to do some drastic re-thinking. Tethys produced a huge crater 400 kilometres in diameter, which is larger than the whole of Mimas, and is the largest crater so far found anywhere in the Solar System.

During the Saturn pass, the controllers at J.P.L. had an unpleasant shock. The main scan platform of the cameras on Voyager 2 jammed, and some vital information was lost. Ironically, the trouble was put right because of a faulty command. The platform should have been ordered to turn 1½ degrees in one direction, in an attempt to release it; the command ordered a ten-degree turn in the other direction – and the scan platform freed itself! This was just as well, because after moving away from the Saturnian system Voyager 2 still had two important missions ahead of it.

Many problems have been solved, but on the whole it is fair to say that the Voyagers have made us realize that we know far less about Saturn than we had fondly believed. With its rings and its diverse family of satellites, it is one of the most puzzling members of the Sun's family as well as being unquestionably the loveliest.

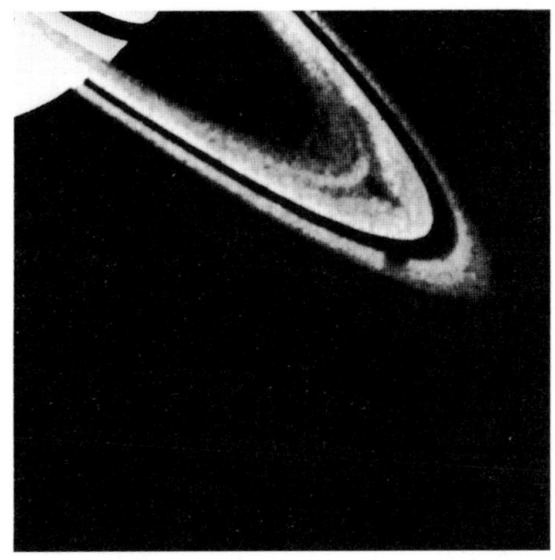

RADIAL 'SPOKES' IN SATURN'S RINGS *seen from Voyager.*

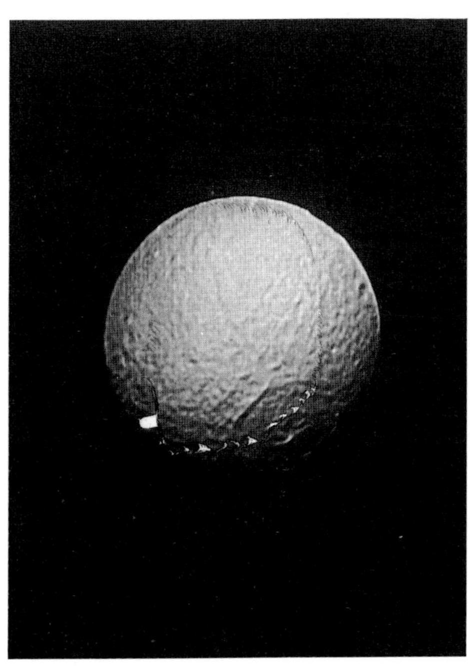

TETHYS. *Note the huge crater, now named Odysseus.*

39

The Depths of the Solar System

IN planning space missions, it is always important to be flexible. It was so with the Pioneers and Voyagers. As we have seen, Pioneer 11 was diverted on to a rendezvous with Saturn only after it had become clear that such a thing was practicable, and in view of the success of Pioneer 10. With the Voyagers, two alternative plans were worked out before launching. Titan was regarded as an important objective – almost as important, indeed, as Saturn itself – and Voyager 1 was scheduled to pass close to it. If it had failed to do so, then Voyager 2 would have had to rendezvous with Titan as well as with Saturn, and that would have been the end of its main task. It could not 'take in' Titan and then go on to the outer planets. If, on the other hand, Voyager 1 was successful with Titan, it would leave Voyager 2 free to ignore Titan and move on to, first, Uranus (January 1986) and then Neptune (August 1989).

Fortunately Voyager 1 did succeed, and Voyager 2 was able to follow the second alternative. At the moment it is on its way to Uranus. But first let me backtrack to say what we have already learned about the two outer giants.

They are not merely small editions of Jupiter or Saturn. Their diameters are less; Uranus is slightly the larger of the two, with a diameter of 51,800 kilometres as against 49,500 kilometres for Neptune, but Neptune is the more massive. According to the latest models, each has a rocky core surrounded by a liquid mantle of water, methane and ammonia, above which comes a low-density gas layer made up chiefly of hydrogen and helium. The central temperatures are of the order of 7000 degrees Centigrade, but they are not the same, because from what we can gather there is an important difference between the two: Neptune emits more energy than it ought to do if it depended only upon the Sun, while Uranus does not. This means that Uranus lacks an internal heat-source. The reason for this difference is unknown. Possibly Uranus also has an inner source of heat which is somehow blanketed in, but at any rate the two are not identical twins even though they are so alike in size. Also, the atmosphere of Uranus is very clear down to great depths, while that of Neptune contains a variable haze made up of particles of unknown composition.

Telescopes will not tell us much. Vague indications of a bright equatorial zone have been seen on Uranus, but virtually nothing on Neptune; the main difference is in colour, since Uranus is sea-green, while Neptune is bluish. The rotation periods are longer than was previously believed, and probably lie between seventeen and twenty-two hours, but no exact values can be given as yet.

Perhaps the most obvious difference between the two is that while Neptune's axis of rotation is inclined to the perpendicular by an angle of 29 degrees, only slightly more than ours, that of Uranus has a tilt of 98 degrees, more than a right angle. This means that the rotation is technically retrograde, though it is not generally regarded as such. At times, as in 1985, one of the poles faces the Sun, and we have a

MODEL OF VOYAGER 1 AT J.P.L. *Photograph by Patrick Moore.*

bird's-eye view of it; the orbits of the five satellites then appear circular as seen from the Earth. By 2007 the equator will be presented, so that the satellites will seem to move 'up and down'.

Because of this curious tilt, the seasons on Uranus are most peculiar. During the Uranian year, 84 times as long as ours, each pole will have a 'day' lasting for twenty-one Earth-years and a 'night' of equal length, but from Uranus the Sun appears less than twice as large as Jupiter does to us.

Uranus has a ring-system. It was discovered more or less by chance. Occasionally the planet passes in front of a star, and hides or occults it. This provides a good opportunity for measuring the apparent diameter, which of course determines the duration of the occultation – a method pioneered by Gordon Taylor of the Royal Greenwich Observatory. An occultation of a ninth-magnitude star was predicted for March 10, 1977, and was duly observed. Both before and after occultation the star 'winked' regularly, so that it was obviously being hidden by rings of material circling the planet. Since then the rings have been fully confirmed, and it has been found that there are at least nine of them. All are narrow (only one, the outer-most, spans as much as a hundred kilometres); six of them are slightly eccentric, and all are made up of darkish material, quite unlike the ice-particles in the Saturnian system. Normal telescopes will not show them, but they have been detected by using infra-red equipment on the Hale reflector.

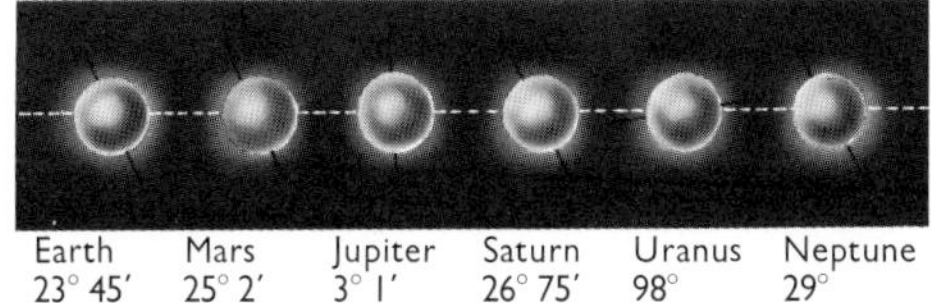

INCLINATION OF THE ROTATION AXES OF THE PLANETS. *The inclinations of Mercury, Venus and Pluto are not known with any certainty. The unusual tilt of Uranus is very evident.*

The five satellites – Miranda, Ariel, Umbriel, Titania and Oberon – are unremarkable; all are below 2000 kilometres in diameter, and are presumably icy and cratered. Their orbits are almost perfectly circular, and they move in the plane of the Uranian equator.

Neptune has two satellites, probably three. One of them, Triton, is of planetary size, perhaps as large as any satellite in the Solar System, and it may well have an atmosphere, though until Voyager 2 makes its pass in 1989 we have little hope of finding out for certain one way or the other. Triton moves round Neptune in a period of 5·9 days at a distance of 355,000 kilometres, and its orbit is circular; what makes it unusual is that it moves in a retrograde direction. All the other retrograde satellites in the Solar System are small and probably asteroidal, so that Triton is in a class of its own. The other proven Neptunian satellite, Nereid, is less than 1000 kilometres in diameter, and has an orbit more like that of a comet than a moon; the distance from Neptune ranges between 1,390,500 kilometres out to as much as 9,733,500 kilometres, and the revolution period is almost a year.

Since Jupiter, Saturn and Uranus all have rings, though admittedly of different types, it is generally thought that Neptune must have a ring also. I am not so sure. The presence of a large satellite, orbiting in a direction opposite to that in which Neptune rotates, may make conditions unstable, and I am inclined to believe that Neptune, alone of the giants, may be ringless.

Lastly, in this necessarily somewhat breathless tour of the Solar system, we come to Pluto. Again I must backtrack, because the story of this strange little body goes back for over seventy years, to the time of Percival Lowell.

As you will recall, Neptune was discovered in 1846 because of its effects upon Uranus. The Sun's family was again regarded as complete. Yet over the years very minor irregularities again began to show

CLYDE TOMBAUGH *at his Las Cruces home, 1982, with Patrick Moore.*

up; could there be yet another planet, moving at a greater distance from the Sun and making itself felt?

Periodical suggestions were made. The famous French astronomer Camille Flammarion discussed the matter as early as 1879, but the first serious investigations were made by Lowell in the early part of our own century. Lowell's calculations were based upon the movements of Uranus rather than those of Neptune, because Uranus had already completed more than a full circuit of the Sun since its discovery, and its orbit was the better-known of the two. Lowell made various predictions, and even started to search photographically from his observatory at Flagstaff, but nothing was found, and when Lowell died in 1916 his 'Planet X' remained unknown. Similar searches made by Milton Humason at Mount Wilson in 1919, based upon mathematical calculations by W.H. Pickering, were similarly fruitless.

After a lapse of nearly ten years Vesto Melvin Slipher, Director of the Lowell Observatory, decided to return to the problem. For the purpose he acquired a fine 33-centimetre refractor, specially designed. It was then that Clyde William Tombaugh came into the story.

Tombaugh was born on February 4, 1906, on a farm in Illinois. He was interested in astronomy from boyhood; he obtained a small refractor, made some sketches of Mars, and sent them to the Lowell Observatory. Slipher saw them, and was impressed. It was true that Tombaugh was untrained; equally obviously he had unusual talent, and Slipher invited him to join the Lowell staff. Tombaugh accepted. Only when he arrived did he learn that his principal task was to search for Planet X.

The method involved taking two photographs of the same region of the sky, at an interval of a few nights, and then comparing them – just as had been done in asteroid-hunting. The two plates were put side by side into an instrument known as a blink-comparator, and were examined in quick succession by flicking a switch to and fro. A star would remain still, but any object which had moved during the interval would jump to and fro. Tombaugh had to take the photographs and then 'blink' them – a truly Herculean task. To complicate matters still further, he had no real idea of the position of Planet X, even if it existed at all. So he started working his way round the Zodiac, developing new techniques as he went, and by January 30 he had come to the rich Milky Way region in Gemini.

On February 18 he started examining plates taken previously on January 21 and 29. Suddenly a moving speck near the third-magnitude star Delta Geminorum came into view. It leaped to and fro by exactly the right amount, and it was very close to the position which Lowell had given. Tombaugh checked and re-checked; only when he was absolutely confident did he go and break the news to Slipher.

The excitement at the Lowell Observatory was intense. During the next clear night the new planet was identified visually; it showed no disk, but its movement proved that it really was orbiting the Sun at a distance greater than that of Neptune. On March 12 Slipher sent the following historic telegram to Harvard Observatory:

SYSTEMATIC SEARCH BEGUN YEARS AGO
SUPPLEMENTING LOWELL'S INVESTIGATIONS FOR
TRANS-NEPTUNIAN PLANET HAS REVEALED OBJECT

WHICH SINCE SEVEN WEEKS HAS IN RATE OF MOTION AND PATH CONSISTENTLY CONFORMED TO TRANS-NEPTUNIAN BODY AT APPROXIMATE DISTANCE HE ASSIGNED. POSITION MARCH 12d 3h GMT WAS SEVEN SECONDS OF TIME WEST FROM DELTA GEMINORUM, AGREEING WITH LOWELL'S PREDICTED LONGITUDE.

The search was over at last. At the suggestion of an eleven-year-old Oxford girl, Venetia Burney (now Mrs. Maxwell Phair), the planet was named Pluto, after the God of the Underworld. The name seemed appropriate; Pluto is a dark, gloomy world, and the first two letters, PL, are the initials of Percival Lowell.

Yet before long, doubts began to arise. Certainly Pluto was trans-Neptunian, but was it really Lowell's Planet X? Measurements showed that it was unexpectedly small and faint, which is why Lowell himself had not found it; it was later detected on two of the plates taken at Flagstaff during Lowell's lifetime, and also on two plates exposed by Humason during his brief search in 1919. A small planet could not pull measurably upon giants such as Uranus or Neptune unless it were improbably dense. Also, Pluto's orbit was unique. The period was 247¾ years, and the path was so eccentric that when near perihelion Pluto came within the orbit of Neptune – though because Pluto's orbit was also tilted at the comparatively sharp angle of 17 degrees, there was no danger of collision. Between 1979 and 1999 Pluto is closer to the Sun than Neptune, so that it has temporarily forfeited its title of 'the outermost planet'. Perihelion is due in 1989.

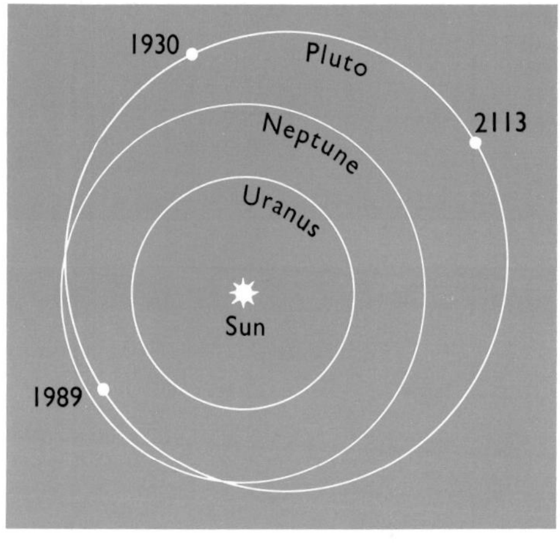

ORBITS OF THE OUTERMOST PLANETS. *Pluto's minimum distance from the Sun is less than that of Neptune, and the planet will not next reach perihelion until 1989. However, there is virtually no danger of a collision, since Pluto's orbit is inclined to the ecliptic at the relatively large angle of 17 degrees.*

No telescope will show Pluto as anything more than a dot of light. Gerard Kuiper's efforts to measure its apparent diameter were unsuccessful, even though he used the Hale reflector at Palomar. The only possibility seemed to be to use Gordon Taylor's method, and wait until Pluto passed in front of a star. The time taken for the planet to pass over the star would then give a real clue to Pluto's diameter.

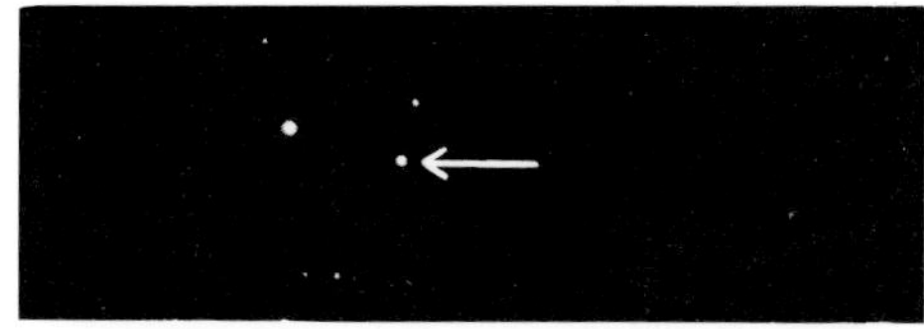

PLUTO, SHOWING APPARENT MOTION OVER 24 HOURS. *Photographed with the 508-cm Hale reflector at Palomar.*

Obviously it was essential to know just when Pluto would be obliging enough to do this, and in 1978 James W. Christy, of the United States Naval Observatory at Flagstaff, began to examine plates which had been taken over the years with the 155-centimetre reflector there. To his surprise, he found that Pluto's image was irregular. There seemed to be a 'bump', and Christy came to the conclusion that this bump was nothing more nor less than a comparatively large satellite, so close to Pluto that the two images merged into one. Christy contacted the Cerro Tololo Observatory in Chile, and, using the 158-centimetre reflector there, John Graham confirmed the irregular shape. Finally, in April 1981, D. Bonneau and R. Foy, using the 3·6-metre reflector at Hawaii, obtained a picture showing Pluto and its satellite separately. So the satellite really does exist. It has been named Charon, after the gloomy ferryman who used to take departing souls across the river Styx on their way to Pluto's kingdom.

I may add that there was a near-occultation of a star by Pluto on April 6, 1981. From my observatory at Selsey, I used my 39-centimetre reflector to watch Pluto as it tracked past the star, missing it by a very small angle. In South Africa, Alistair Walker, at Sutherland, saw the star disappear for a few seconds. Pluto was not in the right position to cause an actual occultation, but Charon was.

Pluto and Charon make up a pair which is unique in many respects. First, Charon's orbital period is 6.3 days – and it had already been found, from slight fluctuations in its brightness, that Pluto has an axial rotation period of the same length, so that as seen from one hemisphere of Pluto Charon would seem to stand still in the sky; from the other hemisphere it would never be seen at all. The distance between the two is only about 19,000 kilometres. Moreover, Charon has a diameter about one-third that of Pluto itself, and this means that the two make up something more like a binary planet than a planet-and-satellite combination. Finally, the small size and low mass of Pluto was confirmed. It is probably smaller than the Moon; and if it is composed largely of ice, as is very likely, it is far too lightweight to cause detectable perturbations in the motions of Uranus or Neptune. There can be no doubt that whatever Pluto may be, it is not Lowell's Planet X.

In March 1980, exactly half a century after the discovery of Pluto had been announced, there was an historic symposium at the University of New Mexico in Las Cruces, where Clyde Tombaugh is now Emeritus Professor. It was a joyous occasion. Technical papers were given, summing up most of what has been discovered about Pluto; for instance the surface is apparently covered with methane frost, with a very tenuous atmosphere made up partly of methane and partly of some heavier gas such as neon. The symposium was followed by a banquet, at which Clyde Tombaugh was guest of honour. He was presented with the Regent's Medal, the University's highest honour, and an asteroid – No 1604 – was named after him, leading him to comment that at least he now owned a piece of real estate that nobody could touch! The citation on the medal is worth quoting:

DISCOVERY OF PLUTO, 1930. *Pluto is indicated by arrows. (Upper) March 2. (Lower) March 5. The planet has moved appreciably in the interval. The bright, over-exposed star to the left is Delta Geminorum. Lowell Observatory photograph.*

'The Board of Regents of the New Mexico State University take great pride in presenting the Regent's Medal to Clyde William Tombaugh, discoverer of the planet Pluto, the ninth major planet in the Solar System. Awarded on the fiftieth Anniversary of his discovery, in recognition of that universal event and in appreciation of his efforts in establishing an internationally recognized planetary research program at the New Mexico State University.'

Yet if Pluto is not Planet X, as now seems certain, can Lowell's fairly correct prediction have been due to sheer luck? This seems to be the accepted view, but frankly I doubt it. Coincidences can and do occur, but this would be a truly amazing one. It is more likely, I believe, that the real Planet X remains to be discovered. Unfortunately we have no idea of where it may be, and so its detection will be largely a matter of chance.

If I am right, then Pluto may be only one of a whole swarm of similar-type bodies in the far reaches of the Solar System, and this idea has been supported by the discovery, in 1977, of a small world moving mainly between the orbits of Saturn and Uranus. The discovery was made by Charles Kowal, using the Schmidt telescope at Palomar. The object was given an asteroidal number (2060) and also a name – Chiron, after the kindly centaur who was the teacher of Jason, leader of the mythological Argonauts. (Do not confuse Ch*i*ron with Ch*a*ron, the satellite of Pluto; it is a pity that the two names are so alike.) Chiron has an orbital period of 50 years. Its diameter may be as much as 600 or 700 kilometres; its nature is unknown, but it and Pluto may be of the same basic type. Other Chirons, and other Plutos, may well exist.

And this, so far, takes us to the known limit of the Solar System – apart from the comets, many of which have elliptical orbits taking them out much further. Neither can we estimate the extent of the Solar System. If the Oort Cloud really exists, then the Sun's dominance spreads out for a radius of at least a light-year, but that is as much as we can say at present. Yet one thing is certain: we have learned more about the Solar System during the past twenty years than we had been able to do over the previous twenty centuries.

40

Into the Future

BEFORE the end of the century we may confidently hope that apart from Pluto, all the planets of the Solar System will have been contacted by unmanned space-craft. Financial constraints always have to be taken into account; but even if the United States does not continue with planetary research, the Soviet Union will unquestionably do so.

Manned research is different, because for the moment at least it is confined to the neighbourhood of the Earth-Moon system. The main drawback is the lack of full international co-operation. I have never had much faith in the idea of a space-race, but it is certainly true that so far as rocketry is concerned both the great powers are highly secretive, because the ability to send a probe to the Moon goes along with the ability to launch a nuclear missile at an enemy country. Until Mankind comes to its senses, we cannot hope to realize our full potential in space. There have been a few encouraging signs. One full-scale USA-USSR mission has been carried out, when, on July 17, 1975, an American Apollo-type vehicle linked up with a Soyuz craft from the Soviet Union, and the two crews joined forces. This was much more than a mere propaganda exercise, as was unkindly suggested at the time. But as yet, the two programmes remain separate.

The Russians have had major successes with their space-stations, and have shown that a period of several months under zero gravity produces no permanent ill-effects. From Cape Canaveral in Florida has come the Space-Shuttle; a recoverable vehicle which takes off like a rocket, flies like a space-ship and lands like a glider. The first Shuttle flight was triumphant, and the sight of the vast, now unpowered craft gliding in, piloted with the utmost skill by John Young (late of Apollo 16) was impressive by any standards. The Shuttle will make orbital flight not only cheaper, but also safer. Plans are now well advanced for setting up a large space telescope in 1985, and within a couple of decades after that there is every chance that there will be a permanent Lunar Base.

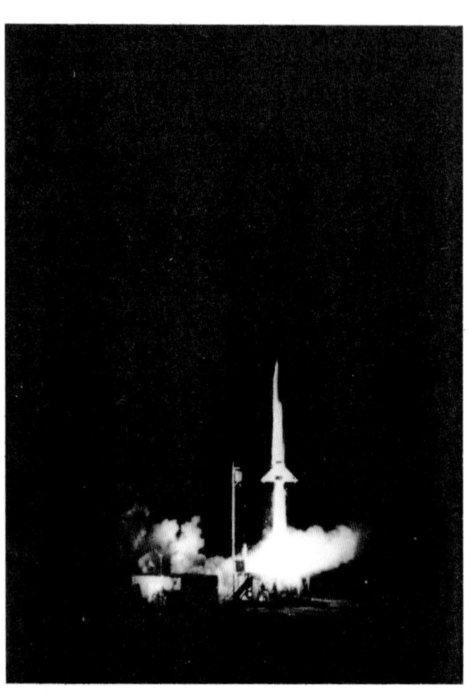

NIGHT LAUNCHING OF VIKING 13, *in the United States.*

Such a base will have unique advantages. Experiments can be carried out there, under reduced gravity and high vacuum, which cannot be attempted on Earth. The Moon's far side will be particularly suitable for radio astronomy, because it is shielded from artificial transmissions from home, and is 'radio quiet'. If Lunar Bases are established, they must be on an international basis, subsequently if not at first. It may be idealistic to suggest that by colonizing the Moon we may go a long way toward uniting the Earth, but I have an instinctive feeling that it may happen.

The Space Telescope, will be almost as large as the Hooker reflector which Hale master-minded less than eighty years ago. Obviously it will have tremendous potential, because it will be operating from beyond the unsteady, obscuring atmosphere. One thing it may possibly do is to detect planets of other stars – and this brings me on to the all-important question of life beyond the Earth.

I must narrow the problem down at once, by confining it to the possibility of life of the kind we can understand. This is reasonable enough. All living matter is made up of familiar atoms, and by now we are sure that no atom-types remain undiscovered. In nature they range from hydrogen, with a nucleus and one circling electron, up to uranium, with 92 electrons; and as I have pointed out, the 'extra' elements tacked on to the heavy end of the sequence are unstable, so that in this context we can forget about them.

Only one atom is suitable for building up the large, complex molecules that are needed for life. This is carbon. Silicon can make some sort of attempt, but we have no evidence of silicon-based life, and if we rule it out we are left with carbon. It follows that life in the universe, wherever it may be found, is carbon-based.

I agree that there may be a fundamental flaw in this argument; but if so, then most of our modern science is wrong. When presented with a set of facts which is incomplete, all one can really do is collect what facts there are, and then interpret them in the most reasonable possible way. When we do this, all the evidence is against alien life of the sort beloved by science-fiction writers. Personally, I have no faith in the idea of

ARTIST'S IMPRESSION OF THE SPACE TELESCOPE

WOOMERA. *This is the main Commonwealth rocket range, and is situated in Australia. The photograph gives a general view.*

SKYLARK ROCKET. *One of the British rockets being launched from the Commonwealth testing ground at Woomera, Australia. These vehicles were not designed to launch satellites, but to carry out upper-atmosphere research.*

silicon-based organisms swimming around inside Jupiter, or civilizations so unlike ourselves that they would come into the category of what are termed BEMs or bug-eyed monsters. Unless any evidence to the contrary turns up, I will continue to be sceptical. In my view, bug-eyed monsters may be classed together with astrology, UFOlogy, and the believers in a flat Earth.

On the other hand, this is not to suggest that life must be unique to our world. Every time Man has tried to set himself up on a pedestal, he has been unceremoniously knocked off it, and it would surely be conceited to suggest that we alone populate the universe.

The Solar System is unpromising; the space-craft have shown us that no terrestrial-type life can exist away from the Earth itself (the only loophole being very primitive organisms on Mars, and even this now seems unlikely). So we must look further afield, and the first step is to decide how many suitable planets there may be. The essential requirements are: a tolerable temperature, suitable surface gravity, an oxygen-bearing atmosphere, and an adequate supply of water.

Much depends upon how planetary systems came into being. A theory championed by Sir James Jeans, well-remembered today both for his pioneer work in astrophysics and for his popular books and broadcasts, assumed that the planets were pulled off the Sun by the action of a passing star. The encounter led to a tongue of solar material being dragged off, and after the intruder moved away again a whirling, cigar-shaped tongue was left, which subsequently broke up into planets –the largest planets, Jupiter and Saturn, being in the middle, where the thickest part of the cigar would have been. This sounded plausible enough, but the mathematical objections to it seem to be insuperable. Sir Fred Hoyle proposed that the Sun used to have a binary companion, which exploded as a supernova and scattered débris from which the planets were built up, but there are serious problems here too. Modern theories go back, in very modified form, to the famous Nebular Theory of Laplace, put forward as long ago as 1796; here, a shrinking gas-cloud threw off rings, each of which condensed into a planet and left the remnant of the cloud to make up the modern Sun.

Gaseous rings could not, in fact, be produced in this manner, but nowadays it seems certain that the planets built up by gravitational accretion from a solar nebula. The fact that the inner planets are small and solid, while the outer planets are not, is a logical consequence of their having been formed in different regions. Also, during its contraction toward the Main Sequence, the Sun went through a T Tauri stage when the solar wind was extremely powerful, and this also had marked effects upon the newly-formed planets, particularly the nearer ones. Many details remain to be worked out, and the Swedish astronomer Hannes Alfvén has rightly stressed that magnetic forces must have played an important rôle, but – and this is the vital point – on this sort of theory, planetary systems are likely to be common. What can happen to the Sun can presumably happen with any other solar-type star.

Very massive stars are not suitable candidates for planetary centres. They are too violent and too short-lived, and moreover life is slow to evolve, so that it would have insufficient time to develop before changes in the parent star became unpleasantly obtrusive. So we must cast around for milder stars.

Unfortunately a planet is so small and faint compared with a normal star that no telescope yet built has any chance of seeing even a relatively large planet associated with a lightweight star, but there are other methods. At the Sproule Observatory in America, Peter van de Kamp and his colleagues have studied the proper motion of Barnard's Star, a dim red dwarf in Ophiuchus, which is a mere six light-years away (closer than any other star beyond the Sun, apart from the members of the Alpha Centauri group), and have concluded that it is 'weaving' its way along, so that it is being perturbed by at least one body – perhaps two – too low in mass to be stellar. Doubts have been expressed recently, but Barnard's Star is not the only candidate, and there is nothing at all unreasonable in the idea.

If Barnard's Star is attended by a Jupiter-sized planet or planets, the Space Telescope may be able to show it. This would be a highly significant breakthrough.

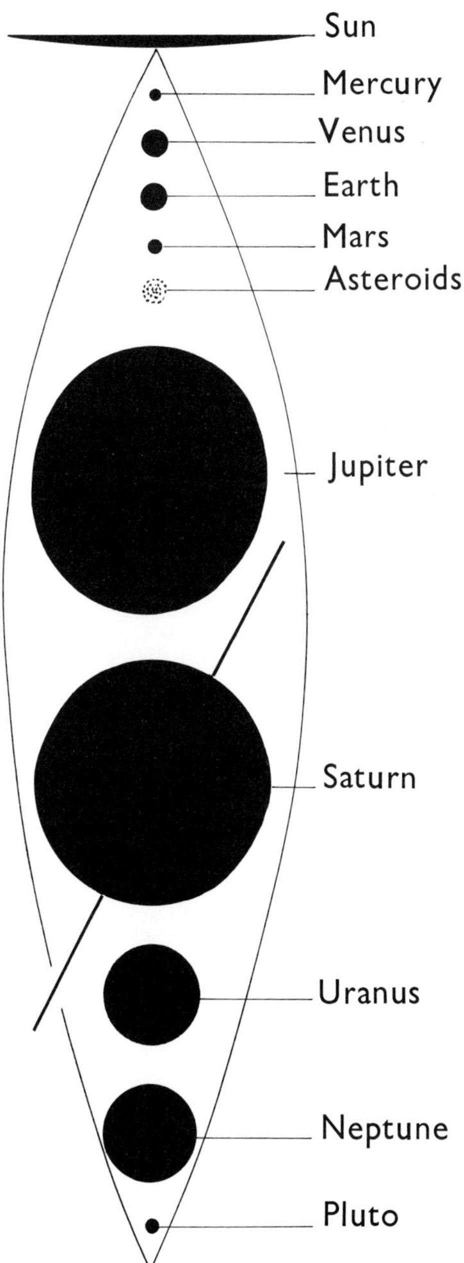

JEANS' TIDAL THEORY OF THE ORIGIN OF THE SOLAR SYSTEM. *The theory was widely accepted for some years, but mathematical investigations have revealed so many weaknesses in it that it has now been rejected.*

APOLLO 15. *(Left) The astronauts and their LRV are on the plain at the foothills of the Apennines, with Hadley Delta, the 'uniform mountain', in the background.*

SEISMOMETER *which was taken to the Moon in Apollo 11 in 1969. It was made on the same principle as an Earth seismometer, but was more sensitive.*

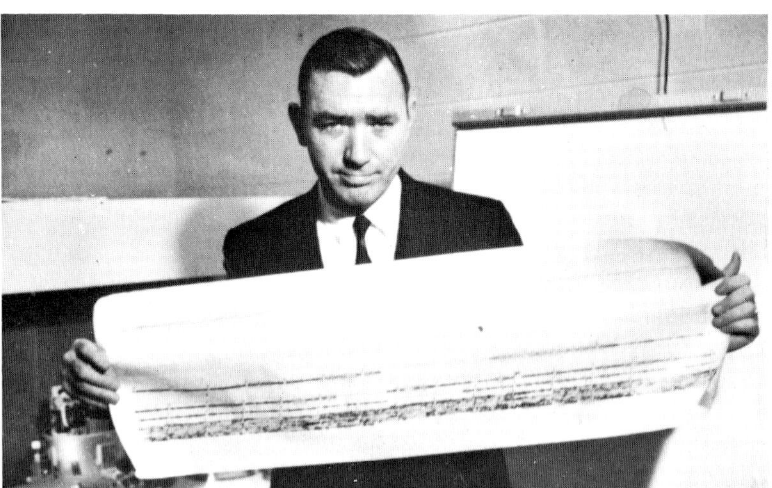

MOONQUAKE TRACE-SIGNALS. *Dr. Gary Lathom, Principal Investigator for the lunar seismic experiment, with the record of the first trace-signals of 'moonquakes' sent back by the Apollo 11 seismometer from the Mare Tranquillitatis. Photograph by Patrick Moore, 1969.*

Next, what about the origin of life? We have no definite knowledge of how life began on Earth, though there are various useful pointers. Alternative theories, according to which life was brought here from space, have been proposed from time to time; Svante Arrhenius, earlier in our century, believed that life was carried Earthwards by a meteorite, but this 'panspermia' theory never met with much support, because it seemed to raise more difficulties than it solved. Much more recently, in 1978, Sir Fred Hoyle and his Indian-born colleague Chandra Wickramasinghe have revived the theory in modified form. They point out that organic molecules exist in interstellar space, so that it is probably in space that life has originated, subsequently being brought to the Earth not by a meteorite, but by a comet. They even consider that comets in general (and Halley's in particular) have brought the water which now fills our oceans, and they go on to suggest that even today comets may deposit viruses in the upper atmosphere, leading to epidemics of diseases such as influenza.

The idea has been coldly received by medical researchers, and indeed by astronomers as well, but it has led to a good deal of interesting discussion. In any case, one thing is definite: life on Earth exists, and a planet-wide civilization has developed. In this case, the same process may well have operated elsewhere.

Interstellar travel lies far in the future as yet. To send a Voyager-type rocket to any other stellar system would take many thousands of years, and I for one have no faith in the concepts found in so many novels – interstellar arks, deep-freezing of astronauts and so on. I believe that if we are ever to go to other systems we must await some fundamental breakthrough, about which we cannot even speculate as yet. I will merely add that though teleportation, thought-travel and the like are sheer fiction today, they sound no more improbable than television would have done to King Canute.

This being so, the only chance of contact seems to be to use radio waves, which travel at the same velocity as light. The two nearest stars which are reasonably like the Sun are Tau Ceti and Epsilon Eridani, both of which are smaller and cooler than the Sun but are not unreasonable candidates. In 1960 Frank Drake and his colleagues used the radio telescope at Green Bank, in West Virginia, to 'listen out' at a wavelength of 21·1 centimetres, in the hope of picking up signals which were too rhythmical to be natural, and the concentration was upon these two selected stars. The 21·1-centimetre wavelength was selected because, as we have noted, it is the wavelength emitted by the clouds of cold hydrogen in the Galaxy, and radio astronomers anywhere would presumably know about it. Not surprisingly, nothing positive emerged, but it is a measure of our changed attitude that it was thought worth trying at all, and other similar experiments have been conducted since, including some with the great 1000-metre natural bowl radio telescope at Arecibo.

Obviously the method is limited. If an astronomer on, say, Tau Ceti picked up a signal transmitted from Earth in 1993, it would have been sent out in 1982. If he replied immediately, we would have the answer by 2004 – a total delay of 22 years, which would make quick-fire repartee rather difficult. But we have to admit that the experiment is the longest of long shots in any case.

Of course, if other civilizations exist they may be far more advanced than ourselves, in which case they might be able to contact us. The idea of visiting space-craft is not out of the question, even though there is not the slightest evidence that it has happened yet. It is interesting to speculate as to what would be the general reaction to an alien craft

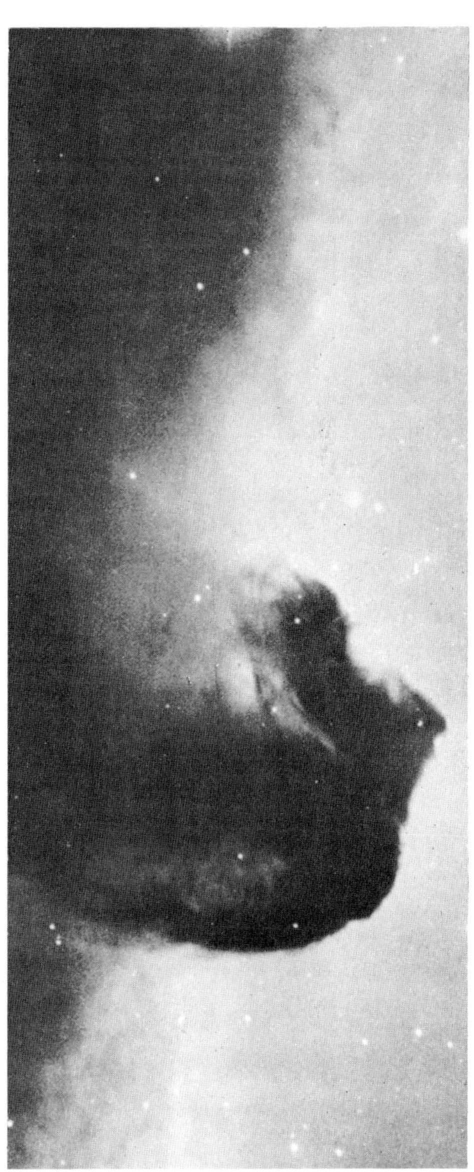

THE HORSE'S-HEAD NEBULA IN ORION, *near the bright star Zeta Orionis. This nebula, known officially as I.C. 434 or Barnard 33, is made up of dark material, and it is easy to see why it has been nicknamed the 'Horse's-Head'. The photograph was taken in red light with the 508-cm Hale reflector at Palomar.*

INSIDE THE DEEP SPACE NETWORK AT J.P.L. *Photograph by Patrick Moore, 1982.*

IRAS (INFRA-RED ASTRONOMICAL SATELLITE) *(right) being tested at the Fokker BV plant, Holland.*

THE 12-M ANTENNA AT THE IRAS GROUND STATION *at Chilton, Oxfordshire.*

landing on Earth. I hope that our reception of the visitors would be friendly, though I admit that I am not confident.

A few astronomers have protested that it would be unwise to make our presence known by transmitting obviously artificial signals. I cannot see the sense in this argument. If another civilization shows itself capable of crossing interstellar space to visit us, it is bound to be more advanced, and therefore more sensible, than we are. We would have nothing to fear, and probably much to learn. If no contacts are ever made either way, it will show either that there are no other beings within range, that they are unable to get in touch with us, or that they know so much about us that they have come to the wise decision to leave us well alone!

The idea of life distributed widely through the universe is popular among astronomers, but it is not universal. In particular, Sir Bernard Lovell has pointed out that the initial build-up of a living organism from inorganic matter involves a whole series of events, each of which is most improbable. He feels, therefore, that life must be very rare, and that the existence of a planet which *could* support life is no guarantee that it *does* support life. This, too, is an eminently reasonable attitude, and at present the whole question remains completely open.

Another possibility is that as a general rule, life on a suitable planet evolves to the stage in which it has the technical knowledge to destroy itself – and inevitably does so. This would explain the lack of contact from afar, and our own experience shows, unfortunately, that there is considerable support for it. Wars on Earth used to be localized. Thus Rome destroyed Carthage, and drove a plough over its site; if Carthage had destroyed Rome instead, you would still be reading this book. But since the first atom-bomb fell on Hiroshima in 1945 the whole situation has changed. Weapons now exist, stock-piled in the armouries of nuclear powers, which could wipe out all civilization here, and would probably render the Earth permanently uninhabitable. It is a terrifying prospect, but the next two or three centuries should be decisive. If we avoid a major war during that period,

PHOTOGRAPH OF THE EARTH, *as seen by the Apollo 8 astronauts, during their orbital flight around the Moon.*

THE FIRST LAUNCH OF THE SHUTTLE – *the recoverable space-craft.*

Mankind will have become too advanced, and too civilized, to destroy itself. In that case there is no limit to what we may achieve. And elsewhere in the universe, intelligent beings may exist who are far in advance of ourselves. One of the world's greatest scientists – Francis Crick, Nobel prize winner and a brilliant biologist – has gone so far as to suggest that life on Earth might have been deliberately started by an unpiloted, spore-carrying probe sent deliberately to our world from some far-away system.

If we use our common-sense, Man has a long future. Eventually, of course, conditions will change. As the Sun leaves the Main Sequence and enters the red giant stage, our oceans will boil, the atmosphere will be stripped away, and the Earth will be scorched – probably destroyed. Yet who knows? – by then, we may well have learned enough to save ourselves.

Meanwhile, we are making steady progress in our exploration of space. Less than 600 years ago it was still believed that the Earth lay at rest in the exact centre of the universe. By now we have found out that our world is an unimportant planet circling an unimportant star in an unimportant galaxy; we have reached the Moon, and we have been able to study objects so remote that the light we now receive from them started on its journey long before the Earth existed.

The story of astronomy is not over. It may be only just beginning.

Some Landmarks in the History of Astronomy

B.C. *(All dates approximate only.)*

585 Thales correctly predicts an eclipse.
550 Pythagoras speculates about the movements of the planets.
350 Aristotle gives observational proofs that the Earth is not flat.
280 Aristarchus suggests that the Earth is in orbit round the Sun.
240 Eratosthenes measures the size of the Earth
140 Hipparchus draws up his star-catalogue.

A.D

c.150 Ptolemy writes the *Almagest.*
c.150 Lucian's *True History* (first Moon-voyage story).
813 Al-Mamun founds the Baghdad school of astronomy, and has the *Almagest* translated into Arabic.
903 Al-Sûfi draws up his star catalogue
1006 Appearance of the brilliant supernova in Lupus.
1054 Appearance of the supernova in Taurus (the Crab supernova).
1232 Rockets used at the Battle of Kai-fung-fu, between the Chinese and the Mongols.
1270 Alphonsine Tables published, by order of Alphonso X of Castile.
1433 Ulugh Beigh sets up an observatory at Samarkand.
1440 Nikolaus Krebs (Nicholas of Cusa) speculates about the motion of the Earth.
1474 Regiomontanus suggests the 'lunar distances' method of determining longtitude.
1543 Publication of Copernicus' *De Revolutionibus.*
1572 Tycho Brahe's supernova in Cassiopeia.
1576 Tycho founds his observatory at Uraniborg, and begins work on his star catalogue.
1595 Mira Ceti observed by Fabricius.
1596 Tycho leaves Hven; Uraniborg abandoned.
1600 Giordano Bruno burned at the stake in Rome, partly for his Copernican views.
1603 Publication of Johann Bayer's star catalogue, *Uranometria.*
1604 Kepler's supernova in Ophiuchus.
1608 Lippershey invents the telescope. (Telescopes may have been invented earlier, but we have no definite proof.)
1609 Thomas Harriot draws the first telescopic map of the Moon.
1609 Kepler publishes his first two Laws of Planetary Motion.
1610 Galileo begins his telescopic observations, and makes a series of major discoveries.
1611 Sunspots observed by Galileo, Scheiner and J. Fabricius.
1612 Orion Nebula discovered by N. Peiresc.
1618 Kepler's Third Law published.
1627 Kepler publishes the Rudolphine Tables of planetary motion.
1631 Gassendi observes the transit of Mercury, predicted by Kepler.
1632 Publication of Galileo's *Dialogue.*
1633 Galileo summoned to Rome, and forced to recant.
1634 Publication (posthumously) of Kepler's novel *Somnium.*
1637 Founding of the first national observatory (Copenhagen, Denmark).
1638 Phocylides Holwarda discovers the variability of Mira Ceti.
1639 Horrocks and Crabtree observe the transit of Venus.
1647 Publication of Hevelius' map of the Moon.
1651 Publication of Riccioli's lunar map, introducing the modern system of nomenclature.
1655 Discovery of Titan, by Huygens; Huygens also gave the correct explanation of Saturn's ring system.
1656 Founding of the second Copenhagen Observatory.
1659 First markings on Mars seen (the Syrtis Major, by Huygens).
1663 Principle of the reflecting telescope proposed, by James Gregory.
1665 Newton's pioneer experiments at Woolsthorpe.
1666 Polar caps on Mars observed by G.D. Cassini.
1667 Founding of the Paris Observatory (completed in 1671). Cassini appointed Director.
1668 Newton builds the first reflector. (This is the probable date. The telescope was presented to the Royal Society in 1671.)
1669 Variability of Algol discovered, by Montanari.
1671 Iapetus discovered by Cassini.
1672 Rhea discovered by Cassini.
1675 Royal Greenwich Observatory founded.
Cassini discovers the main division in Saturn's rings.
Velocity of light measured, by O. Rømer.
1676 Halley goes to St. Helena to catalogue the southern stars.
1682 Bright comet observed by Halley (= Halley's Comet).
1685 First astronomical observations made from South Africa (Father Guy Tachard, at the Cape).
1687 Publication of Newton's *Principia.*
1704 Publication of Newton's *Opticks.*
1705 Halley predicts the return of his comet, for 1758.
1723 Construction of the first really good reflecting telescope (a 15-centimetre, by Hadley).
1725 Publication of the final version of Flamsteed's star catalogue.
1728 James Bradley discovers the aberration of light.
1729 Chester More Hall discovers the principle of the achromatic object-glass.
1744 Appearance of de Chéseaux' multi-tailed comet.
1750-2 Lacaille, at the Cape, draws up his star catalogue.
1750 T. Wright speculates about the origin of the Solar System.
1758 Palitzsch recovers Halley's Comet. (Perihelion, 1759.)
1761 Transit of Venus; the atmosphere of the planet discovered by M.V. Lomonosov.
1762 Completion of Bradley's measurements of the positions of 60,000 stars.
1767 Nevil Maskelyne founds the *Nautical Almanac.*
1769 Transit of Venus observed by Captain Cook (and others!).
1772 Bode's Law, discovered by Titius, popularized by J.E. Bode.
1774 First recorded observation by William Herschel.
1776 Small but good lunar map published by Tobias Mayer.
1779 Johann Schröter founds his observatory at Lilienthal.
1781 Herschel discovers Uranus.
1783 Goodricke correctly explains the variability of Algol.
1784 Goodricke discovers the variability of Delta Cephei.
1786 William Herschel gives the first reasonably good explanation of the shape of the Galaxy.
1789 Completion of Herschel's

greatest reflector. With it, Herschel discovers Mimas and Enceladus.

1796 Publication of Laplace's Nebular Hypothesis (origin of the Solar System).

1798 Brandes and Benzenberg use the triangulation method to measure the altitudes of meteors.

1799 Great Leonid meteor shower.

1800 William Herschel detects infra-red radiation from the Sun.

1801 G. Piazzi discovers the first asteroid, Ceres. Publication of Lalande's catalogue of 47,380 stars.

1802 Second asteroid (Pallas) discovered, by Olbers.
Herschel announces his discovery of binary systems.
Wollaston observes dark lines in the solar spectrum.

1803 Fall of meteorites at L'Aigle. Nature of meteorites established.

1804 Harding discovers Juno.

1807 Olbers discovers Vesta.

1813 Destruction of Schröter's observatory at Lilienthal.

1815 Fraunhofer's first detailed map of the solar spectrum, showing 324 absorption lines.

1818 Encke predicts the return of a comet discovered by Pons.

1819 Bessel completes the reduction of Bradley's star positions.

1820 Foundation of the Royal Astronomical Society (originally the Astronomical Society of London).

1821 Fearon Fallows arrives in South Africa to direct the new observatory at the Cape.

1821 Sir Thomas Brisbane, in Australia, founds an observatory at Paramatta.

1822 First predicted return of Encke's Comet (recovered by Rümker, at Paramatta).

1824 First clock-driven, equatorial telescope set up (Fraunhofer's telescope made for the Dorpat Observatory).

1826 Biela's Comet discovered, independently by von Biela and Gambart.

1827 First calculation of the orbit of a binary star (Xi Ursæ Majoris, by Savary).

1829 Cape Observatory completed.

1833 Brilliant Leonid meteor shower.

1834-8 John Herschel, at the Cape, carries out the first really detailed survey of the far southern stars.

1834 Bessel discovers irregularities in the motion of Sirius, and attributes them to an unseen companion.

1835 Second predicted return of Halley's Comet.

1837 Publication of the lunar map and book by Beer and Mädler.

1837 Publication of the first good catalogue of double stars (F.G.W. Struve's *Mensuræ Micrometricæ*) from Dorpat.

1838 Bessel measures the distance of a star (61 Cygni).

1839 Pulkovo Observatory in Russia completed.

1840 J.W. Draper takes the first photograph of the Moon.

1842 First attempts to photograph a solar eclipse (by Majocci).
C. Doppler announces Doppler's Principle.

1843 Draper obtains the first daguerreotype of the solar spectrum.

1843 H. Schwabe announces the discovery of the 11-year solar cycle.

1844 Harvard College Observatory founded (first official observatory in the United States). The first large telescope was installed there in 1847.

1845 Completion of Lord Rosse's great reflector at Birr Castle. Spiral forms of galaxies discovered.
Hencke discovers the fifth asteroid, Astræa.
Solar photographs taken by Fizeau and Foucault.

1846 Discovery of Neptune, by Galle and D'Arrest at Berlin, from calculations provided by Le Verrier.
Triton discovered by W. Lassell.

1848 Roche proves that Saturn's rings cannot be solid (Roche's Limit).

1850 Discovery of Saturn's Crêpe Ring, by Bond at Harvard.
First stars photographed (Vega and Castor).

1851 First photograph of a total solar eclipse (by Berkowski).

1852 Last recorded appearance of Biela's Comet.

1857 Clerk Maxwell proved theoretically that Saturn's rings must be made up of discrete particles.

1857 Founding of the Sydney Observatory.

1857 First good photograph of a double star obtained (the Mizar-Alcor pair, by Bond, Whipple and Black).

1858 Appearance of Donati's Comet.

1859 Kirchhoff correctly interprets the dark lines in the spectra of the Sun and stars.

1861 Spörer announces his law of sunspot distribution.

1862 Discovery of the Companion of Sirius, by Clark, in the position predicted by Bessel.
Construction of the first great refractors, including the Newall 63·5-centimetre, formerly at Cambridge and now in Athens.
Completion of Argelander's *Bonner Dürchmusterung.*

1863 Secchi first classifies the stars into different spectral types.

1864 Huggins proves spectroscopically that some nebulæ are gaseous.
First spectroscopic examination of a comet (Tempel's, by Donati).
Founding of the Melbourne Observatory. (The large reflector was completed in 1869).

1865 Publication of Jules Verne's *From the Earth to the Moon.*

1866 Brilliant Leonid meteor shower.
Association between comets and meteors established, by Schiaparelli.

1867 Wolf-Rayet stars first described in detail (G. Wolf and G.A. Rayet).

1868 Janssen and Lockyer independently show how to observe solar prominences without waiting for an eclipse.
A. Ångström publishes his detailed map of the solar spectrum.

1870 C.A. Young takes the first photograph of a solar prominence.

1872 H. Draper photographs the first stellar spectrum (of Vega).

1874 Transit of Venus; new measures of the astronomical unit.
Observatories founded at Meudon (France) and Adelaide (Australia).

1877 Schiaparelli first describes the Martian 'canals'.
Asaph Hall discovers Phobos and Deimos.

1878 Julius Schmidt publishes his elaborate map of the Moon.
Jupiter's Great Red Spot becomes very conspicuous.

1879 Founding of the Brisbane Observatory.

1880 Draper takes the first good photograph of a gaseous nebula (M.42).
Nice Observatory 76·2-cm refractor completed.

1881 Early 'rocket aircraft' design by

Kibaltchitch.

1882 Gill's classic photograph of a bright comet, showing so many stars that the idea of star-cataloguing by photography was born.
Last transit of Venus until the year 2004.

1885 Founding of the Tokyo Observatory.
Appearance of the supernova S Andromedæ, in M.31.

1887 Completion of the Lick 91·4-centimetre refractor.

1888 Publication of Dreyer's *New General Catalogue* of clusters and nebulæ.

1889 First spectroscopic binaries discovered (Mizar and Beta Aurigæ).

1890 British Astronomical Association founded.
Lockyer announces his theory of stellar evolution.
E.C. Pickering gives a more detailed classification of stellar spectra.
H. Vogel finally establishes the existence of spectroscopic binaries.

1891 Completion of the Arequipa southern station of the Harvard College Observatory.
The spectroheliograph invented by G.E. Hale (and independently by H. Deslandres in France).
First discovery of an asteroid by photography (by Max Wolf).
Hermann Ganswindt delivers a public lecture about space-travel.

1892 First photographic discovery of a comet, and last visual discovery of a satellite (Amalthea), both by Barnard at Lick.

1893 Completion of the 71-centimetre Greenwich refractor.

1894 Lowell founds the Lowell Observatory at Flagstaff, Arizona.

1895 Tsiolkovskii submits his first space-flight article.

1896 Completion of the Meudon 83-centimetre refractor.
Founding of the Perth Observatory (Australia).
Publication of the first lunar photographic atlas.
Completion of the Royal Observatory at Blackford Hill, Edinburgh.
Discovery of the white dwarf companion of Procyon (by Schaeberle).

1897 Completion of the Yerkes Observatory.

1898 Witt discovers the first asteroid to come within the orbit of Mars (No. 433, Eros).

1900 Burnham publishes an important catalogue of double stars.

1901 Brilliant nova seen in Perseus.

1903 Publication of Tsiolkovskii's first important article rocketry.

1905 Mount Wilson Observatory founded.

1908 Hertzsprung describes the giant and dwarf divisions of stellar types.
Completion of the Mount Wilson 152-centimetre reflector.
Fall of the Siberian 'meteorite' (possibly the nucleus of a small comet).

1910 Return of Halley's Comet.

1911 First HR Diagram drawn.

1912 Miss Leavitt's studies of short-period variables in the Small Magellanic Cloud lead to the discovery of the Cepheid period-luminosity law.

1913 H.N. Russell puts forward his theory of stellar revolution.
Founding of the Dominion Astrophysical Observatory at Victoria (British Columbia).

1914 R.H. Goddard's first practical rocket experiments.

1915 W.S. Adams studies the spectrum of Sirius B, leading on to the discovery of White Dwarfs.

1917 Completion of the 254-centimetre Hooker reflector at Mount Wilson.

1918 H. Shapley gives the first accurate estimate of the size of the Galaxy.
Brilliant nova in Aquila.

1919 Important catalogue of dark nebulæ published by Barnard.
Goddard suggests that a vehicle could be sent to the Moon.

1920 V. Slipher announces his results concerning the Red Shifts in the spectra of galaxies.
'Great Debate' between Shapley and Curtis.

1923 E. Hubble proves that the galaxies are external systems.
Hale invents the spectrohelioscope.

1924 H. Oberth publishes his classic book on interplanetary flight.

1926 Goddard fires the first liquid-propellent rocket.

1927 Formation of the VfR (Society for Space-Travel).
Studies by J.H. Oort show that the centre of the Galaxy lies in the direction of the Sagittarius star-clouds.
Completion of the Boyden Observatory, Bloemfontein.

1930 Discovery of Pluto, by Clyde Tombaugh.
Bernhard Schmidt invents the Schmidt telescope.

1931 First experiments by Jansky, leading to the detection of radio waves from space (published in 1932).
Close approach of Eros; astronomical unit re-measured.
H. Winkler, in Germany, fires the first European liquid-propellent rocket.

1932 T. Dunham discovers carbon dioxide in the atmosphere of Venus.

1934 Will Hay discovers a white spot on Saturn.
Bright nova in Hercules, discovered by J.P.M. Prentice.

1937 Grote Reber builds the first intentional radio telescope.
First rocket tests at Peenemünde.

1938 New theory of stellar energy announced independently by H. Bethe and G. Gamow.

1942 M.H. Hey and his colleagues detect radio waves from the Sun.

1944 Atmosphere of Titan discovered, by Kuiper.
H.C. van de Hulst suggests that interstellar hydrogen should emit radio waves at 21·1 centimetres.
First V.2 rockets sent against England.

1945-6 Radar echoes received from the Moon.
White Sands Proving Ground established.
Arthur Clark's paper about communications satellites published in *Wireless World*.
Preliminary work at Jodrell Bank begun.

1946 Hey, Parsons and Phillips identify the radio source Cygnus A.

1947 Woomera rocket range established in Australia.

1948 Completion of the Hale 508-centimetre reflector at Palomar.
Identification of the radio source Cassiopeia A (M. Ryle, F. Graham Smith).
H. Bond and T. Gold propose the steady-state cosmological theory.

1949 Identification of further radio sources, including the Crab Nebula, M.87, and Centaurus A.
Rocket testing ground established at Cape Canaveral.

1950 Ryle, Smith and B. Elsmore detect M.31 at radio wavelengths.

Funds for building the Jodrell Bank radio telescope obtained by Sir Bernard Lovell.

1951 H. Ewen and E. Purcell detect the 21·1-centimetre radio emission predicted by van de Hulst.

1952 W. Baade announces revision of distance-scale for the galaxies.

Tycho's supernova identified at radio wavelengths (Hanbury Brown and Hazard).

1955 Completion of the great 'dish' at Jodrell Bank.

First detection of radio waves from Jupiter, by Burke and Franklin.

M. Ryle constructs a radio interferometer.

Project Vanguard announced by the United States.

1957 Russia launches Sputnik 1; beginning of the Space Age.

1958 First American satellite (Explorer 1); discovery of the Van Allen zones.

N. Kozyrev observes a red TLP in the lunar crater Alphonsus.

1959 Three Russian probes sent to the Moon; first pictures of the Moon's far side.

1960 First weather satellite (Tiros 1).

1961 Completion of the Parkes radio telescope, west of Sydney.

Venera 1 (first Venus probe) launched by the Russians, but contact soon lost.

First manned space-flight; Yuri Gagarin (USSR).

First American in space (A. Shephard).

1962 Thermal radio emission detected from Mercury (Howard, Barrett and Haddock, from Michigan).

First radar contact with Mercury (Kotelnikov, USSR.)

First American orbital flight (John Glenn).

First attempted Mars probe (USSR); it was unsuccessful.

First successful planetary probe; Mariner 2 to Venus.

Transatlantic television programmes relayed by way of the Telstar satellite.

First X-ray source discovered (Scorpio X-1).

1963 P. van de Kamp announces the detection of a planet orbiting Barnard's Star.

Identification of quasars, by M. Schmidt at Palomar.

1964 First good close-range photographs of the Moon sent back by Ranger 7.

1965 First 'space-walk', by A. Leonov (USSR).

Close-range pictures of Mars sent back from Mariner 4.

Identification of the 3°K microwave radiation, as a result of theoretical work by Dicke and experiments by Penzias and Wilson.

Rotation period of Mercury determined by Dyce and Pettingill.

1966 First controlled landing of an automatic probe on the Moon (Luna 9, USSR).

First landing of a probe on Venus (Venera 3), though contact with it was lost.

First American controlled landing on the Moon (Surveyor 1).

First circumlunar probe (Luna 10, USSR).

First really detailed close-range pictures from Orbiter 1.

1967 Discovery of pulsars, by Jocelyn Bell-Burnell at Cambridge.

Completion of the 249-centimetre Isaac Newton Reflector at Herstmonceux.

First gamma-ray satellite launched (OSO-III).

1968 Orbiter programme completed.

Identification of the Vela pulsar as a radio source.

First manned flight round the Moon; Lovell, Borman and Anders, in Apollo 8.

1969 First men on the Moon; Aldrin and Armstrong, Apollo 11.

First optical identification of a pulsar (the Crab pulsar, by Cocke, Taylor and Disney at Steward Observatory, USA).

First gamma-ray source identified.

1970 Completion of the 100-metre radio 'dish' at Bonn, Germany.

Completion of the large reflectors at Kitt Peak (Arizona) and Cerro Tololo (Chile), each 401-centimetres aperture and the first large reflector (a 224-centimetre) erected on Mauna Kea in Hawaii.

First successful controlled landing on Venus (Venera 7, USSR).

1971 Uhuru X-ray satellite launched.

1971-2 First detailed close-range pictures of Mars (Mariner 9).

1971 End of the Apollo programme, with Apollo 17.

Completion of the 381-cm telescope at Mount Stromlo, Australia.

1973 Opening of the Sutherland station of the South African Observatories.

Launch of Skylab (manned by three successive crews, 1973-4).

First fly-by of Jupiter (Pioneer 10).

1974 Completion of the 389-cm reflector at Siding Spring, Australia.

First detailed views of the cloud-tops of Venus (Mariner 10).

First close-range pictures of Mercury (also Mariner 10).

Second fly-by of Jupiter (Pioneer 11).

1975 Completion of the 382-cm reflector at La Silla, Chile.

First USA-USSR space mission (Apollo-Soyuz).

First pictures obtained from the surface of Venus (Veneras 9 and 10, USSR).

1976 First successful controlled landings on Mars (Vikings 1 and 2).

Completion of the 600-cm reflector at Zelenchukskaya (USSR).

1977 Optical identification of the Vela pulsar, from Siding Spring.

Discovery of Chiron, by C. Kowal.

Discovery of the rings of Uranus.

1978 Einstein observatory launched (X-ray studies).

Completion of the Russian underground neutrino telescope.

Discovery of Pluto's satellite, Charon, by J. Christy.

First reliable maps of Venus (Pioneer orbiter).

1979 Two fly-by missions to Jupiter (Voyagers 1 and 2).

First fly-by of Saturn (Pioneer 11).

Opening of the observatory at La Palma.

Commissioning of the UKIRT telescope on Mauna Kea.

1980 First detailed pictures of Saturn and its system (Voyager 1).

1981 Second fly-by of Saturn (Voyager 2).

Successful initial test of the Space Shuttle (U.S.A.).

1982 Discovery of the most remote known quasar, 2000-330 (D. Jauncey and A. Wright, Parkes, NSW).

First infra-red emissions from a quasar detected (P. Harvey).

Recovery of Halley's Comet (G.E. Danielson, D. Jewitt).

List of Illustrations

Acknowledgements

Illustrations are keyed by page number followed by positioning symbol T for top, C for centre, B for bottom, L for left, R for right. For abbreviations, see list below.

Cover: Robert P. Kirschner/University of Michigan, Ann Arbor.
Frontispiece: CIT.
8, 9T: MWP; 9B, 10, 11: M; 12T: Allen Lanham; 12B: M; 13: MWP; 14, 15: M; 16T: *Daily Express*; 16B: M; 17T: F.J. Ackfield, Forest Hall Observatory, Northumberland; 17B: H.E. Wood, Union Observatory, Johannesburg; 19T, 19B, 21, 22: M; 23T: BBC Hulton; 23B: Deutsches Museum, Munich; 24, 25: M; 27L, 27R: SM; 28, 29T, 29B: M; 30 (both): W.T. O'Dea; 31, 32, 33: M; 34T, 34B: BBC Hulton; 35: SM; 36T: M; 36B, 37: SM; 38T: Gösta Persson; 38B: M; 39T: SM; 39B, 40T, 40B, 41, 42T, 42B, 43T, 43C, 43B, 44, 45T, 45B, 46T, 46B: M; 47T: MWP; 47C, 47B: M; 49T: MWP; 49B: BBC Hulton; 50T: Dominic Fidler; 50B: BBC Hulton; 51T: M; 51B: PM; 52T: Lick Observatory; 52B, 53L, 53R: M; 53B: Lowell Observatory; 54: M; 55: W.M. Baxter; 56T; 56B: SM; 57T: M; 57B: W.M. Baxter; 58: D.J.K. O'Connell, Vatican Observatory; 59T: PM; 59B: D.J.K. O'Connell, Vatican Observatory; 60, 61T, 61B: MWP; 62L: *Eastern Daily Press;* 62R: W.M. Baxter; 63T: M; 63B: Lick Observatory; 64: M; 65T: SM; 65L, 65R: M; 66: Lick Observatory; 67: M; 68T: SM; 68BL: Lowell Observatory; 68BR: M; 69T: M; 69C: MWP; 69B: M; 70L, 70R, 70C: K.S.G. Stocker; 70B: M; 71T: PM; 71B: M; 72T: Lick Observatory; 72B: PM; 73T: Courtesy of The Royal Society; 73B: British Library; 74L: PM; 74R: MWP; 74B, 75, 76L, 76R: M; 77: British Library; 78: PM; 79T, 79B: SM; 80T: SM; 80B: PM; 81: British Library; 82T: RAS; 82B: M; 83: Lowell Observatory; 84-85T: MWP; 84L(3), 84R(2): M; 85C: PM; 85B: National Maritime Museum; 86, 87T, 87B: M; 88-89T: PM; 88C: M; 88B: PM; 90T: M; 90B: PM; 91: CIT; 92T, 92C: PM; 92B: MWP; 93T: SM; 93B: B. Warner & T. Saemundsson, University of London Observatory; 94L, 94R: M; 94B, 95T: PM; 95B: H.R. Hatfield; 96T: M; 96B: SM; 97: MWP; 98L, 98R: M; 99: MWP; 100T, 100B: SM; 101L, 101C, 101R: PM; 102: M; 103: MWP; 104, 105T: M; 105C: H.P. Wilkins; 105B, 106T, 106B, 107T, 107C: M; 107B: Paul Doherty; 108: SM; 109, 110: M; 111: Deutsches Museum, Munich; 112: Royal Greenwich Observatory; 113T: E.A. Whitaker; 113B: M; 114: J. Bennett; 115T, 115C: G.E.D. Alcock; 115B, 116T, 116B, 117: M; 118T: H.B. Ridley; 118C: D.R. McLean; 118B: K. Kennedy; 119T: Paul Doherty; 119B: G.J.H. McCall; 120T: H.B. Ridley; 120C: M; 120B: SM; 121T: PM; 121B: D.E. Blackwell & M.F. Ingham; 122L: PM; 122R, 123: Birr Castle; 124: MWP; 125T, 125B: Birr Castle; 127: J. McBain; 129, 130, 131L, 131R, 131B, 132T, 132B: M; 133: NASA; 134T: PM; 134B: Lick Observatory; 135T: Brian Gulley; 135B: PM; 136: M; 138T: Lick Observatory; 138C: Pic du Midi Observatory; 138B, 139T, 139B: M; 140T: NASA; 140B: M; 141L: Pic du Midi Observatory; 141R: NASA; 141B: Ramon Lane; 142T: NASA; 142B: Yerkes Observatory; 143T: SM; 143C: Lowell Observatory; 143B: SM; 144T: M; 144B: NASA; 145T: MWP; 145C: NASA; 145CB: JPL; 145B: NASA; 146T, 146B, 147T: M; 147L: MWP; 147R: M; 148L: RAS; 148C: G.V. Schiaparelli; 148R, 148B: M; 149T: W.T. Hay; 149C: Lowell Observatory; 149B: H.R. Hatfield; 150T: M; 150R: MWP; 150L: H.R. Hatfield; 151: PM; 152T, 152B, 153: MWP; 154T: Boyden Observatory, Bloemfontein; 154B: PM; 155: U.S. Naval Observatory, Washington; 156T: MWP; 156B, 157T, 157C: PM; 157B: ANB; 158: PM; 159: ANB; 160L, 160R: PM; 160B: ANB; 161T, 161B, 162, 163: PM; 165: MWP; 167: PM; 168T, 168B: W.M. Baxter; 169T, 169B: MWP; 170: P.A. Reuter; 171: Ake Wallenquist; 172, 173T: MWP; 173B: Royal Greenwich Observatory; 175T, 175L: M; 175R, 176: MWP; 177L: H.B. Ridley; 177R, 178, 179T: M; 179B, MWP; 180T, 180B, 181T, 181B, 182: M; 183: MWP; 184, 185, 186, 187: M; 188T: H.R. Hatfield; 188B: K. Kennedy; 189T, 189B, 190T: M; 190C: MWP; 190B: M; 191L: MWP; 191R: M; 192L: W.S. Franks; 192R: M; 194, 195: MWP; 196, 197L, 197R: M; 198: Elliott & Fry Ltd.; 200-201: M; 203: MWP; 205: M; 206: NASA; 207L, 207R: M; 208L: Lick Observatory; 208R, 209, 210: MWP; 211T: M; 211C: Royal Observatory, Cape; 211B, 212, 213T, 213B: MWP; 214T, 214B, 215T, 215B: CIT; 216, 217T, 217B: MWP; 218: M; 219: CIT; 220T, 220C, 220B: RAS; 220R, 221L, 221R, 221B: MWP; 222T: CIT; 222B: MWP; 223T: CIT; 223B, 224L, 224R: MWP; 225, 226L: M; 226R: Karl Jansky; 227: COI; 228T: ANB; 228B: PM; 229T: National Radio Astronomy Observatory, Green Bank, WV; 229B: Pieter Morpurgo; 230: COI; 231T, 231L, 231R, 232T: PM; 232B, 233T: ANB; 233B: M; 234L: Colin Ronan; 234R: McDonnell Aircraft Corporation; 234B: PM; 235T, 235B: M; 236T: ANB; 236B: MWP; 237T, 237B, 238: PM; 239: CIT; 240T, 240B, 241T, 241B: MWP; 242: PM; 247: NASA; 248: British Library; 249L, 249R: NASA; 250, 251T, 251B: M; 252T: NASA; 252B: *Soviet Weekly;* 253T: Morris Allan/Photocraft; 253B, 254T: NASA; 254B: *Soviet Weekly;* 255T: Novosti; 255L, 255R: USSR Academy of Sciences; 256: NASA; 257: USSR Academy of Sciences; 258T, 258B, 259T, 259C, 259B, 260T, 260C, 260B, 261TL: NASA; 261TC: Novosti; 261TR, 261BL, 261BR, 262T: NASA; 262C: PM; 262B, 263, 264T, 264L, 264R, 265, 266L, 266R, 266B, 267T: NASA; 267B: PM; 269: M; 270T: Novosti; 270B: NASA; 273L: JPL; 273R, 274: NASA; 275T, 275B, 277T, 277C, 277B, 278T, 278B, 279, 280, 281, 282, 283, 284T, 284C, 284B, 285T, 285B, 286, 290, 291T, 291C(2), 291B, 293, 294: NASA; 297: M; 298, 299T, 299B: NASA; 300: PM; 301: M; 302 PM; 303T: M; 303B: MWP; 304: Lowell Observatory; 306: U.S. Air Force; 307: NASA; 308T: ANB; 308B: COI; 309T: M; 309B, 310T: NASA; 310B: PM; 311T: MWP; 311B, 312L, 312R: PM; 313L, 313R: NASA.

Abbreviations:
ANB: Australian News and Information Bureau
CIT: California Institute of Technology & Carnegie Institute of Washington
COI: Central Office of Information
JPL: Jet Propulsion Laboratory
M: Macdonald & Co. (Publishers) Ltd.
MWP: Mount Wilson and Palomar Observatory
PM: Patrick Moore
RAS: Royal Astronomical Society
SM: Science Museum

PRINTED IN BELGIUM BY
proost
INTERNATIONAL BOOK PRODUCTION